Aqueous Phase Adsorption

Theory, Simulations and Experiments

Aqueous Phase Adsorption

Theory, Simulations and Experiments

Edited by

Jayant K. Singh
Nishith Verma

CRC Press
Taylor & Francis Group
Boca Raton London New York

CRC Press is an imprint of the
Taylor & Francis Group, an **informa** business

CRC Press
Taylor & Francis Group
6000 Broken Sound Parkway NW, Suite 300
Boca Raton, FL 33487-2742

First issued in paperback 2020

ISBN 13: 978-0-367-57093-4 (pbk)
ISBN 13: 978-1-138-57521-9 (hbk)

Library of Congress Cataloging-in-Publication Data

Names: Singh, Jayant K., editor. | Verma, Nishith, editor.
Title: Aqueous phase adsorption : theory, simulations, and experiments / edited by Jayant K Singh, Nishith Verma.
Description: Boca Raton : Taylor & Francis, CRC Press, 2019. | Includes bibliographical references and index.
Identifiers: LCCN 2018018459| ISBN 9781138575219 (hardback : alk. paper) | ISBN 9781351272520 (ebook)
Subjects: LCSH: Adsorption.
Classification: LCC TP156.A35 A77 2019 | DDC 660/.284235--dc23
LC record available at https://lccn.loc.gov/2018018459

Visit the Taylor & Francis Web site at
http://www.taylorandfrancis.com

and the CRC Press Web site at
http://www.crcpress.com

Contents

Preface

Adsorption is a commonly used technique for treating wastewater. It is energy efficient and considered to be effective for removing a wide range of pollutants from water, including arsenic, fluoride, volatile organic compounds, and toxic metals. With the increasing production of a number of pharmaceutical, healthcare, and agricultural products, and simultaneously continuous decrease in the global water table, several aqueous species have been identified as "emerging pollutants," for which new adsorbents are yet to be produced.

Through various chapters contributed by accomplished researchers working in the broad area of adsorption, this book provides necessary fundamental backgrounds required for an academician, industrial scientist, or engineer to initiate studies in this area. In particular, this book covers theoretical aspects of adsorption, including an introduction to molecular simulations and other numerical techniques that have become extremely useful as an engineering tool in recent times to understand the interplay of different mechanistic steps of adsorption. Further, the book provides brief experimental methodologies to use, test, and evaluate different types of adsorbents for water pollutants.

The editors of this book want to make it clear that this book is not to be considered as a textbook for undergraduate students who want to get firsthand experience of adsorption experiments or/and simulations, nor it is there to provide ready-made solutions to the buyers of wastewater treatment plants. Rather, the book is useful to those who want to make an advancement in developing novel remediation technologies, as its chapters systematically span the mechanistic steps of adsorption in water under batch and flow conditions and a wide range of applications in the critical areas of aqueous phase adsorption via the illustration of various case studies.

Chapter 1 provides a solid background of the fundamentals of adsorption, including thermodynamic aspects of adsorption and classical adsorption isotherms used to explain the adsorption data. It moves on to describe the efficacy of computer modeling currently used to understand complex adsorption systems at a fundamental (molecular) level. It also shows how computer simulations can be used as a means to predict the adsorption behavior and other relevant materials properties, from the microscopic detail viz. atomic masses and atomic charges, etc., via the Monte Carlo and molecular dynamics approaches. The second part of Chapter 1 provides an overview of the mesoscopic lattice Boltzmann approach, including computational steps to predict flow field and associated concentration and temperature profiles in a packed bed adsorber. The chapter ends with a brief description of the experimental methodologies in practice to measure the aqueous phase adsorption capacity of a solid under batch or static conditions, breakthrough experiments, and challenges in conducting the adsorption tests.

Chapter 2 presents an overview of graphene and graphene oxide-based materials for membrane separation process. In particular, this chapter summarizes key molecular simulations studies illustrating the high selectively and mass transport of graphene-based membranes. Chapter 3 illustrates the use of computational chemistry and molecular dynamic simulations in understanding the ligand-aided metal ion transfer process aiming towards separation of metal ions from the aqueous phase using liquid-liquid separation. The chapter explains how interaction strength of ligand molecules with radionuclide can be calculated from computational chemistry calculations that can aid in selecting suitable extractants. Chapter 4 presents recent experimental and simulation studies using metal organic frameworks. In particular, molecular simulation studies are presented demonstrating the use of MOFs in the removal of organic and inorganic molecules from aqueous phase. Chapter 5 presents the use of liquid adsorption to modulate the properties of coating on nanoparticles for various applications. The chapter provides an insight into the coating of polymer chains on inorganic nanoparticles and its effect on the aggregation and dispersion properties of nanoparticles. In addition, solvation behavior of coated nanoparticles are also discussed.

Chapter 6 describes the computational steps of implementing the lattice Boltzmann methods–based models for predicting the adsorption breakthrough profiles in a tubular column packed with porous adsorbent particles. The description is sufficient in that readers can easily grasp the development of mesoscopic or lattice Boltzmann methods–based model equations analogous to the macroscopic conservation equations for the tubular packed bed adsorber. The chapter goes on to provide detailed numerical scheme, model validation, and the characteristic simulation results for concentration distributions in the macro voids of the adsorber and within pores of the particles. The chapter ends with a brief description of the numerical approach to simulate non-isothermal adsorption. Chapter 7 presents an experimental study of the synthesis and practical application of nanozeolites-based adsorbents in removing the commonly found toxic phenolic and metal contaminants in many industrial wastewater effluents. This study assumes significance from the perspective of determining the effects of physico-chemical properties of the materials, as well as operating conditions on the removal efficiency of the adsorbents. Chapter 8 emphasizes the treatment of the pharmaceutical compound containing wastewater using a hybrid approach combining adsorption and photocatalytic reaction, rather than a single technique. Several new antibiotic compounds have been developed in the last decade, which can potentially enter the surface water through various channels endangering the aquatic lives, present even at trace levels. Therefore, the remediation of such water invariably requires a carefully designed and tested approach. The most useful aspect of the chapter is the description of a case study of treating the oxytetracycline contaminated water and toxicity tests on the treated water. Chapter 9 takes up the most interesting but challenging aspect of wastewater pollution and control: Can we not only treat wastewater but also recover the waste as a useful product? Well, it can be done. The study described in this chapter shows that although agricultural activities and food industry over the last decade have tremendously increased the generation of a huge volume of residues such as husks, seeds, and peel, these residues can be reused or revalorized to produce bioadsorbents for organic and inorganic pollutants in wastewater. The last chapter, Chapter 10, provides an industrial perspective of adsorption applied as an efficient technique for providing clean drinking water. The study describes the development of a water filter based on activated carbon which is the most commonly used adsorbent material worldwide. Besides emphasizing the roles of surface functional groups present in the carbon substrate in removing different pollutants in water, the study recommends the surface modifications of carbonoceous material with a view to specifically removing toxic metals and organic contaminants from water. The chapter illustrates the findings with the laboratory and field data collected for water filters fabricated from the catalytic carbon.

Editors

Prof. Jayant K. Singh is a professor in the Department of Chemical Engineering at Indian Institute of Technology Kanpur, Kanpur, India. Dr. Singh's current research interest is in material modeling, self assembly, energy storage, wetting transition, nanotribology and selective adsorption. Dr. Singh has co-authored more than 100 peer-reviewed articles in international journals of repute. He is a recipient of prestigious awards such as JSPS Invitation Fellowship, Alexander von Humboldt Research Fellowship, Young Engineers of Indian National Academy of Engineering, Amar Dye Chem Award of IIChE, BRNS Young Scientist Award, and DST-BOYSCAST Fellowship. He is also an elected member of National Academy of Science, India.

Prof. Nishith Verma is a professor of chemical engineering at the Indian Institute of Technology Kanpur, India. His research interests are the development of carbon-based materials, especially carbon nanofibers and nanoparticles in adsorption, catalytic reactions, biomedical, drug delivery, sensors, microbial fuel cells and agricultural applications. He has also developed lattice Boltzmann method-based mathematical models for adsorption and flow systems. Prof. Verma has published more than 100 research articles in peer-reviewed international journals and has 8 Indian and US patents filed/granted. Prof. Verma is an Alexander Humboldt Fellow.

Contributors

Cristina Agabo García
University of Pablo de Olavide
Seville, Spain

Zuzana Benková
Polymer Institute
Slovak Academy of Sciences
Slovakia

Anil Boda
BARC
Mumbai, India

Robert Chang-Tang Chang
National Ilan University
Taiwan

Bor-Yann Chen
National Ilan University
Taiwan

M. Natália D. S. Cordeiro
University of Porto
Portugal

Ashish Kumar Singha Deb
BARC
Mumbai, India

Vivekanand Gaur
Columbus Industries
Ohio

Krishna M. Gupta
NUS
Singapore

Gassan Hodaifa
University of Pablo de Olavide
Seville, Spain

Jianwen Jiang
NUS
Singapore

Anitha Kommu
IIT Kanpur
India

Byeong-Kyu Lee
University of Ulsan
Korea

Mridula P. Menon
National Ilan University
Taiwan

Sk. Musharaf Ali
BARC
Mumbai, India

Santiago Rodriguez-Perez
University of Pablo de Olavide
Seville, Spain

Thi-Huong Pham
University of Ulsan
Korea

Pooja Sahu
BARC
Mumbai, India

Jayant K. Singh
IIT Kanpur
India

Arun Kumar Subramani
National Ilan University
Taiwan

Nishith Verma
IIT Kanpur
India

Xiao-Dan Xie
National Ilan University
Taiwan

Qiao-Jie Yu
National Ilan University
Taiwan

Ke-Fu Zhou
National Ilan University
Taiwan

1

Theory, Molecular, Mesoscopic Simulations, and Experimental Techniques of Aqueous Phase Adsorption

Jayant K. Singh and Nishith Verma

CONTENTS

1.1 Adsorption from Liquid Solution

In Chapter 1, we briefly present the fundamentals of adsorption theory. In addition, we provide an overview of simulation methods and experimental techniques that are commonly used to study the adsorption phenomena from liquid solution.

1.2 Thermodynamics of Surface Adsorption

Consider a mixture of two species such that there is a phase separation into phase α and phase β. The Gibbs-Duhem equation for the composite phase α + phase β + interface system can be written as:

$$-SdT + Vdp - N_1 d\mu_1 - N_2 d\mu_2 - Ad\gamma = 0 \tag{1.1}$$

where S, V, and N_i are the total entropy, volume, and number of moles of species i in the two-phase system, respectively, and μ_i is the chemical potential of component i. To maintain equilibrium between the phases, an isothermal change in pressure must be accompanied by changes in the chemical potentials that permit them to remain equal between the phases. These changes can be described by Gibbs-Duhem equations written separately for the two phases:

$$\begin{aligned} -S^\alpha dT + V^\alpha dp - N_1^\alpha d\mu_1 - N_2^\alpha d\mu_2 = 0 \\ -S^\beta dT + V^\beta dp - N_1^\beta d\mu_1 - N_2^\beta d\mu_2 = 0 \end{aligned} \tag{1.2}$$

with the phases indicated by the superscript (α or β). Alternatively, the above equation can be written as:

$$\begin{aligned} V^\alpha\left(-s^\alpha dT + dp - c_1^\alpha d\mu_1 - c_2^\alpha d\mu_2\right) = 0 \\ V^\beta\left(-s^\beta dT + dp - c_1^\beta d\mu_1 - c_2^\beta d\mu_2\right) = 0 \end{aligned} \tag{1.3}$$

where c_i is the concentration of species i in the bulk phase and s is the entropy density. By subtracting each equation in (1.3) from equation (1.1), and dividing by the area, we obtain a Gibbs-Duhem equation that amplifies the effects of the interface:

$$-\hat{s}dT + \tfrac{1}{A}(V - V^{\alpha} - V^{\beta})dp - \Gamma_1 d\mu_1 - \Gamma_2 d\mu_2 - d\gamma = 0 \quad (1.4)$$

where $\hat{s}$ is the surface excess entropy. The surface excess of species i is defined as:

$$\Gamma_i = \left(N_i - c_i^{\alpha}V^{\alpha} - c_i^{\beta}V^{\beta}\right)/A \quad (1.5)$$

At the level of detail where the interface is significant, the volumes V^{α} and V^{β} are ambiguous. The value of Γ_i depends on the position of the Gibbs dividing surface and the volume. Thus, an alternative approach for the calculation of surface excess amount is also preferred. The surface excess amount Γ_i, for adsorption from liquid phase can be written as

$$\Gamma_i = \left(N_i - c_i^{l}V_o^{l}\right)/A \quad (1.6)$$

where V_o^l is the volume of the liquid phase up to the interface and i = 1, 2 represent the component.

For a binary system, we can write:

$$\Gamma_1 = \left(N_1 - c_1^{l}V_o^{l}\right)/A \quad (1.7)$$

$$\Gamma_2 = \left(N_2 - c_2^{l}V_o^{l}\right)/A \quad (1.8)$$

Thus, we can write:

$$\Gamma = \frac{\left(N^{o} - c^{l}V_o^{l}\right)}{A} \quad (1.9)$$

where $\Gamma = \Gamma_1 + \Gamma_2$, $N^o = N_1 + N_2$, $c^l = c_1^l + c_2^l$.

Then the reduced excess amount for the component 1 is defined by elimination of V_o from the above equation.

1.2.1 Langmuir Isotherm

The correlation of one component adsorption on a solid surface is well investigated and reported for the gas phase. Several of such correlations such as Langmuir, Freundlich, Slips, Troth, and others can easily be extended to the liquid-phase adsorption.

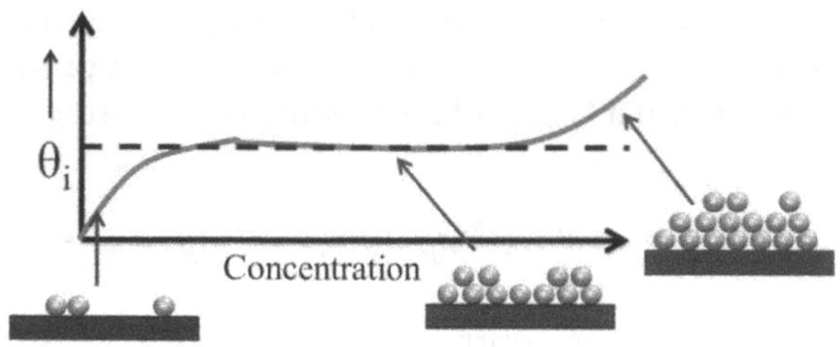

FIGURE 1.1
Representation of the adsorption process, where symbol (θ_i) is the fraction of the surface sites occupied. (From Lyubchik, S. et al., Comparison of the Thermodynamic Parameters Estimation for the Adsorption Process of the Metals from Liquid Phase on Activated Carbons, in *Thermodynamics – Interaction Studies – Solids, Liquids and Gases*, InTech, Rijeka, p 95–122, Ch. 4.

The Langmuir isotherm (1918) is one of the simplest models of physical adsorption [1]. This model assumes monolayer coverage, constant binding energy between surface and adsorbate, and no interaction among adsorbed molecules. The Langmuir isotherm is expressed as:

$$q_e = \frac{K_L \times q_{\max} C_e}{1 + K_L C_e} \tag{1.10}$$

where C_e is the equilibrium concentrations of the adsorptive in liquid phase, and q_e is the amount adsorbed on the adsorbent surface. Langmuir isotherm has two adjustable parameters viz., $q_{\max}$ and K_L, where $q_{\max}$ is the maximum adsorption capacity (monolayer coverage), usually expressed in terms of mmol of the adsorbate per (g) of adsorbent, and K_L is the constant of Langmuir isotherm. Figure 1.1 presents an illustration of monolayer coverage of adsorbed species on the adsorbent, and the corresponding type of adsorption isotherm.

1.2.2 Freundlich Isotherm

This is an empirical isotherm, which assumes that the amount adsorbed at equilibrium has a power law dependence on the concentration of the solute [1]. The Freundlich isotherm is expressed as:

$$q_e = K_F \times q_{\max} C_e^{1/n} \tag{1.11}$$

It may be considered to be a two-parameter isotherm, where K_F and n are the "adjustable parameters." These parameters are calculated from a "log-log plot" of q_e versus C_e, which should be a straight line of slope $1/n$ and intercept log K_F. The parameters K_F and $1/n$ are related to adsorption capacity and adsorption efficiency, respectively.

1.2.3 BET (Brunauer, Emmett and Teller) Isotherm

The Brunauer-Emmett-Teller equation (1938) considers multilayer adsorption behavior for many systems [1,2]. Figure 1.1 presents the adsorption isotherm and corresponding layer formation of adsorbed species on the adsorbent. The equation is given as

$$q_e = \frac{K_{BET} \times q_{\max} C_e}{(C_i - C_e) \times \left[1 + (K_{BET} - 1) \times (C_e / C_i) \right]} \tag{1.12}$$

where C_i is concentration of the solute when all layers are saturated, $q_{\max}$ is the maximum amount of solute adsorbed and K_{BET} is the parameter related to the binding intensity for all layers. This is one of the most popular equations for determining the specific surface area of a solid.

1.3 Computer Simulations for Adsorption Studies

Theoretical studies tend to be based around simple models: aiming to reduce the complexity of the problem but retaining the essential physics. Thus, it is often difficult to compare theoretical predictions with experimental results. Computer simulations have played a valuable role in bridging the gap between theory and experiment. Computer simulations can provide exact results for these simple models, and hence can be used as a test of theory. Computer modeling is used to provide insight and understanding of how complex systems behave beyond what theory and experiment could deliver separately [3]. We may test a theory by conducting a simulation using the same model. We may also test the model by comparing with experimental results. Furthermore, we may also carry out simulations of certain processes or phenomena on the computer that are difficult or impossible in the laboratory. Computer simulation also provides a route to the macroscopic details (equations of state, materials properties, etc.) from the microscopic detail (atomic masses, atomic charges, etc.). They also allow the straightforward elucidation of detail at the microscopic scale, which can be difficult (if not impossible) to probe from experiment. The simulations act as a bridge between microscopic and macroscopic length, and time scales as shown in Figure 1.2. The hidden detail behind bulk measurements can also be revealed. For instance, the structure of a fluid is easily found from simulation but somewhat more difficult from experiment. Additionally, it also provides a route to determining properties in situations out of the reach of experiment, such as high pressures or temperatures. However, in all these cases, predictions of properties are dependent on a good model of the interactions between molecules.

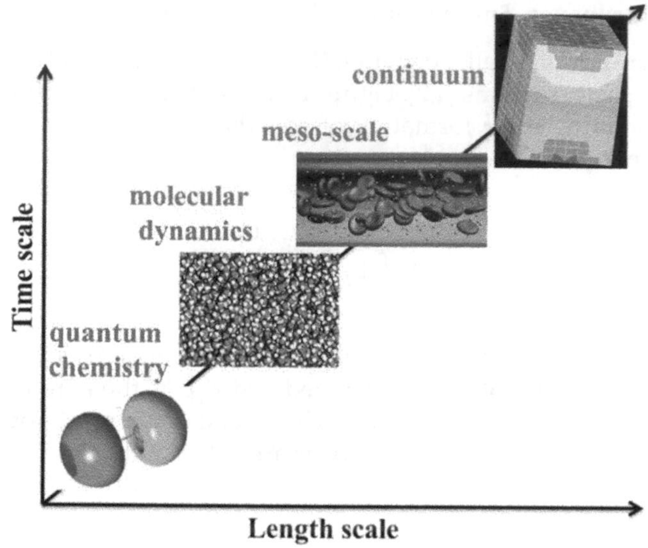

FIGURE 1.2
A schematic diagram of length and time scales accessible by simulation techniques.

In the studies of adsorption phenomena, computational studies can range from quantum chemical calculations to continuum simulations. In this book, we provide a brief description of simulation methodologies from molecular scale to mesoscopic scale.

1.3.1 Molecular Simulations

Molecular simulations at atomistic scale can broadly be divided into two categories in terms of methods: Monte Carlo simulation and molecular dynamics. These methods invariably use the statistical mechanical formalism, which connects thermophysical properties such as pressure, energy, and other macroscopic properties to the collection of microstates of systems of molecules, all having in common extensive properties, also called ensemble. A *microstate* of a system of molecules is a complete specification of all positions and momenta of all molecules. Most commonly, we encounter the total energy, the total volume, and/or the total number of molecules (of one or more species, if a mixture) as extensive properties.

Some examples of ensembles are:

- Micro-canonical ensemble
- Canonical ensemble
- Isothermal-Isobaric ensemble
- Grand-canonical ensemble.

1.3.2 Monte Carlo Simulations

Monte Carlo is a computer simulation that is based on repeated random sampling, i.e., using random numbers in scientific sampling. In general, this method uses random numbers to solve problems that are difficult to solve using other simple computational methods. The origin of Monte Carlo simulation technique lies in the early 1940s when Fermi, Ulam, von Neumann, Metropolis, and others used random numbers to examine problems in physics from stochastic perspective [4]. Monte Carlo simulation is used to study a wide range of scientific areas. In Monte Carlo simulations for molecular systems, configurations are generated using equilibrium probability distribution, π, based on a statistical mechanical ensemble. For example, a typical Monte Carlo simulation uses the following steps repetitively to perform sampling of phase space:

i. Generation of trial configuration by perturbing the original configuration. For example, a randomly selected particle of the existing system is perturbed by an amount (also referred as displacement move).
ii. The ratio of probabilities of going from a current configuration (old) to a next configuration (new) after a trial, as per the detailed balanced, is computed as:

$$\frac{P(old \rightarrow new)}{P(new \rightarrow old)} = \frac{\pi_{new}}{\pi_{old}} \tag{1.13}$$

For a displacement move, the above ratio is $\exp[-(U_{new} - U_{old})/kT]$, where U, k, and T are configurational energy, Boltzmann constant and temperature, respectively.
iii. If the trial is accepted, the new configuration is taken as the next state in the Markov chain; otherwise the original configuration is taken as the next state in the chain. The above is represented as:

$$P_{acc}(old \rightarrow new) = \min\left(1, \frac{\pi_{new}}{\pi_{old}}\right) \tag{1.14}$$

iv. The steps from i to iii are repeated for large number of steps in order to generate large number of configurations.
v. Interested properties are collected after every few 100 or suitable number of steps and average values are generated.

The Monte Carlo method is free from the restrictions of solving Newton's equation motion. Specific moves can be combined in a simulation for a greater flexibility in sampling important configurations.

1.3.3 Molecular Dynamics (MD) Simulation

For a set of N-particles in a system of volume V, the Newtonian motion can be expressed by two first-order differential equations as:

$$\dot{\mathrm{r}}_i = \frac{\mathrm{P}_i}{m_i} \tag{1.15}$$

$$\dot{\mathrm{p}}_i = -\frac{\partial U(\mathrm{r}^N)}{\partial \mathrm{r}_i} \tag{1.16}$$

The r_i and p_i are the position and momentum vectors of particle *i*, respectively. Here, U is the potential energy which depends on the positions of all the N-particles. In molecular dynamics simulation, we perform the integration of Newton's equation of motion for all the particles in a system and thus new configurations are being generated. Time averages are taken to evaluate average properties of interest. Molecular dynamics simulations are very similar to the real experiments. In a molecular dynamics simulations approach, we perform the following steps:

i. Preparation of initial system: Setting up the coordinates and assigning velocities of each particle of the system. It can be done either by placing the particles on a FCC lattice or using random number generator to place particles randomly in the simulation box. Initial velocities are usually generated according to the Boltzmann distribution for a given temperature.
ii. Force calculations: The forces on all particles at each time step, which are needed in the integration step.
iii. Integration: There are several algorithms that are designed to integrate equation of motion. The Verlet algorithm [5] is one of the simplest and also usually the best to integrate the equation of motion. The coordinate of the particle, around time t, is written in the following form using Taylor's expansion:

$$r(t+\Delta t) = r(t) + v(t)\Delta t + \frac{1}{2m} f(t)\Delta t^2 + \frac{1}{3!}\Delta t^3 . \dddot{r} + O(\Delta t^4) + \ldots \tag{1.17}$$

Similarly,

$$r(t-\Delta t) = r(t) - v(t)\Delta t + \frac{1}{2m} f(t)\Delta t^2 - \frac{1}{3!}\Delta t^3 . \dddot{r} + O(\Delta t^4) + \ldots \tag{1.18}$$

Addition of the above equations results:

$$r(t+\Delta t) = 2r(t) - r(t-\Delta t) + \frac{1}{2m} f(t)\Delta t^2 \tag{1.19}$$

Velocity of each particle can also be calculated from the above equations, as described below:

$$v(t) = \frac{r(t+\Delta t) - r(t-\Delta t)}{2\Delta t} + O(\Delta t^2) \tag{1.20}$$

In another algorithm, velocity Verlet algorithm [6,7], the velocities and positions are calculated at each time step.

The time average of any quantity calculated from the phase-space, generated using molecular dynamics simulations, is equivalent to the ensemble average under a significant number of sampling configurations. While NVE ensemble is a natural choice for molecular dynamic simulations, in practice we are interested in controlling the temperature and pressure for its experimental realization. Thus, other ensembles such as canonical (NVT) and isothermal-isobaric (NPT), and grand canonical (μVT) ensembles are more commonly used for real applications. To maintain a constant temperature or pressure of the system, the equation of motion employed in the molecular dynamics method is appropriately modified using different numerical techniques [8–10]. One of the popular molecular dynamics tools that is versatile and powerful in simulating adsorption from aqueous phase at molecular level is LAMMPS [11].

1.3.4 Average Properties

The trajectories obtained from the molecular dynamics simulations or Monte Carlo are used to calculate various properties of the system. The thermodynamics properties are evaluated using either integration over trajectories (MD) or using ensemble average (MC). In calculation of thermodynamical properties, adequate sampling is required in order to minimize the errors. It helps to minimize the fluctuations while computing structural and dynamical properties of the system. The trajectories can be visualized using VMD [12], which many is also useful in understanding the behavior and can also act as a check against an expected behavior of the system.

1.3.5 Density Profiles

Adsorption of liquid or impurities on a solid surface can usually be described by calculating density profiles along axis (or axes) of interest. A typical way to calculate density profile is by dividing the system into a certain number of bins and evaluating the number of particles accumulated in each bin at different frames based on their coordinates. The bin density is averaged over number of frames to obtain the density profiles along the axis of interest.

1.3.6 Hydrogen Bonding

Hydrogen bonding is a special type of dipole-dipole attraction between molecules. It results from the attractive force between the hydrogen atoms covalently attached to an electronegative atom (N, O, F) of a different molecule. The understanding of hydrogen bonding calculation using molecular dynamics is well documented by Ohmine and Tanaka [13]. The hydrogen bonding between any two water molecules are determined by the fulfillment of the following criteria [14,15]:

i. Distance between the oxygen atoms of two water molecules, R_{OO} should be less than 3.5 Å as shown in Figure 1.3(a), which is derived from radial distribution function of water molecules in molecular dynamics as well as from neutron and x-ray scattering data [16].
ii. The distance between the oxygen of the acceptor molecule and the hydrogen of the donor, $R_{OH} \leq 2.5$ (see Figure 1.3(a)). This distance is also derived from the first minimum of the RDF of water.
iii. The angle between the vector joining the intermolecular oxygen and O-H bond of acceptor molecule should be equal to 30°.

Figure 1.3(b) presents the hydrogen bonding formation between two water molecules by the fulfillment of above three conditions. Using different models of water molecules, the number of hydrogen bonding differ to some extent. Using TIP4P water model at 25° C, the average number of hydrogen bonds per molecule was found to be 3.59 [17], however, more recent study reported the value of 2.357 [18]. The structure and dynamics of liquid water is described using the instantaneous hydrogen bonding. Even in equilibrated water, the population operator of hydrogen bond fluctuates with time. The intermittent hydrogen bond fluctuation is characterized by the correlation function:

$$c(t) = \langle h(0)h(t) \rangle / \langle h \rangle \tag{1.21}$$

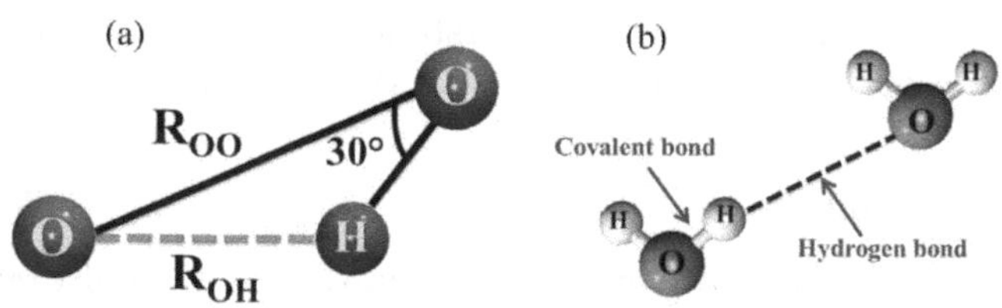

FIGURE 1.3
(a) Representation of hydrogen bonding criteria (not to scale) and (b) hydrogen bonding between two water molecules.

where h is the population operator of hydrogen bonds and $h(t)$ equals to unity if a water molecule is hydrogen bonded.

1.3.7 Diffusion Coefficient

The main advantage of MD simulation is the trajectory of particles/molecules of the system. Thus, it gives a route to calculate dynamical properties of the systems, i.e., diffusion coefficients, shear and bulk viscosity, and thermal conductivity. In this section, we discuss the computation of diffusion coefficient from MD simulations.

According Fick's law, the flux j of the diffusing species is proportional to the negative gradient in the concentration of that species. The proportionality constant, D, is referred as the diffusion coefficient.

$$u\rho(r,t) = \mathrm{D}\left[\nabla\rho(r,t)\right] \tag{1.22}$$

where ρ, u are the local density and velocity respectively which are functions of position r and t. Time evolution of the density of the system is described by:

$$\frac{\partial\rho(r,t)}{\partial t} = D\nabla^2\rho(r,t) \tag{1.23}$$

Using the boundary condition, $\rho(r,0) = \delta(r)$, the solution of the above equation yields an expression for $\rho(r,t)$. Our interest is to find the displacement of a particle in space and time. Ensemble average of displacement can be described as:

$$\langle r^2(t)\rangle = \int dr[r^2\rho(r,t)] \tag{1.24}$$

Now, the diffusion, which is displacement per unit time, can be described as:

$$\frac{\partial}{\partial t}\langle r^2(t)\rangle = \frac{\partial}{\partial t}\int dr\left[r^2\rho(r,t)\right] = D\int dr\left[r^2\nabla^2\rho(r,t)\right] \tag{1.25}$$

Solving above equation results Einstein equation of diffusion, as described below:

$$\frac{\partial\langle r^2(t)\rangle}{\partial t} = 2dD \tag{1.26}$$

Whereas D is the macroscopic transport coefficient, and d is the dimensionality of the system. We describe below two different approaches for calculating diffusion coefficients.

1.3.7.1 Einstein Relation

The Einstein relation is also called the mean square displacement (MSD) relation. In this method, the mean square displacement for every particle in the system are calculated and averaged over time. The mean square displacement is plotted as a function of time. The diffusion coefficients can be calculated using log-log plot of the MSDs as a function of time. In general, the MSD can be expressed in power law as $\langle r^2(t)\rangle \sim 2dDt^v$, where v is the anomaly parameter. For different values of v, the diffusion behavior can be characterized as

i. Sub-diffusion ($v < 1$), commonly found in case of polymer like macromolecular systems [19,20]
ii. Normal diffusion ($v = 1$)
iii. Super diffusion ($v > 1$)
iv. Ballistic diffusion ($v = 2$).

It is difficult to obtain all the regimes in a system; however, in Brownian granular systems, all four regimes can be observed [21].

1.3.7.2 Green-Kubo Relation

The diffusion coefficient can also be evaluated by the Green-Kubo relation [22,23], which is given by:

$$D = \frac{1}{3N}\int_0^\infty \left\langle \sum_{i=1}^{N} v_i(t)v_i(0) \right\rangle dt \tag{1.27}$$

The term $<v_i(t).\ v_i(0)>$ in the Green-Kubo relation is called velocity autocorrelation function (VACF), because it describes correlations between velocities of tagged particle at different times along an equilibrium trajectory. For classical systems, the Einstein relationship and the Green-Kubo relationship are strictly equivalent [23].

1.3.8 Residence Time

The residence time of molecule is an important parameter to evaluate in order to understand the binding ability of an absorbent. It is calculated using residence auto correlation function $C_R(t)$:

$$C_R(t) = \frac{\left\langle \sum_{i=1}^{N} \theta_i(t_0)\theta_i(t+t_0) \right\rangle}{\left\langle \sum_{i=1}^{N} \theta_i(t_0)\theta_i(t_0) \right\rangle} \tag{1.28}$$

$$C_R(t) = A\exp[-(t/\tau_s)] \tag{1.29}$$

where $\theta_i(0) = 1$ when the molecule i is found in the region of interest (could be within the first monolayer of a surface, or within the structure of a porous adsorbent at time $t = 0$). If an appended molecule continuously remains region of interest as the time 't' progresses, then $\theta_i(t) = 1$; $\theta_i(t) = 0$ when the molecule exit the region of interest. The $\theta_i(t)$ remains equal to 0 even if the molecule eventually returns inside the region. The residence time (τ_s) is evaluated by fitting an exponential form, as shown in equation (1.29), to the $C_R(t)$ values.

1.3.9 Free Energy Calculations

Free energy is arguably the most important general concept in physical chemistry. The free energies of molecular systems describe their tendencies to associate and react, and thus is an important property to evaluate in order to assess phase transition, adsorption, affinity of an ion to attach to a ligand and so forth. Hence, developing efficient methods to evaluate free-energy difference between two states is one of the important goals in molecular simulation community [24,25].

1.3.9.1 Basic Formulation of Free Energy Calculations

The statistical mechanical definition of free energy is in terms of the partition function, a sum of the Boltzmann weights of all the energy levels of the systems. However, only for the simplest model system can this free energy be represented by an analytical function. One can write a classical analog of the quantum mechanical partition function where the energy is viewed as a continuous function, rather than discrete. This is likely to be good approximation in most systems involving noncovalent interactions near room temperature. Unfortunately, the free energy represented in this way requires an integration over all 3N degrees of freedom, where N = number of atoms in the system. Thus, this is impractical in most cases. However, if one focuses on free energy difference then it can be represented as:

$$G_B - G_A = \Delta G = -RT\ln\left\langle e^{-\Delta H/RT} \right\rangle_A \tag{1.30}$$

where $\Delta H = H_B - H_A$ and $< >_A$ refers to an ensemble average over the system represented by Hamiltonian H_A. Equation (1.30) is the fundamental equation of free energy perturbation calculations [24,25]. If systems A and B differ in more than a trivial way, then equation (1.30) will not lead to a sensible free energy. One can, however, generalize the problem and describe the Hamiltonian $H(\lambda)$ as equation (1.31):

$$H(\lambda) = \lambda H_B + (1-\lambda)H_A \quad (1.31)$$

where λ can vary from 0 ($H = H_A$) to 1 ($H = H_B$). One can then generalize equation (1.30) as follows:

$$\Delta G = G_B - G_A = \sum_{\lambda=0}^{1} -RT \ln \langle e^{-\Delta H'/RT} \rangle_A \quad (1.32)$$

where $\Delta H' = H_{\lambda+\Delta\lambda} - H_\lambda$. Here, we break up the free energy calculation into windows, each one involving a small enough interval in λ to allow the free energy to be calculated accurately.

An alternative to free energy perturbation calculations is thermodynamic integration [24,25], where the free energy difference between two systems (one characterized by $H = H_A$ or $\lambda = 0$ in equation (1.31) and the other by $H = H_B$ or $\lambda = 1$ in equation (1.31)) can be represented as:

$$\Delta G = \int_{\lambda=0}^{\lambda=1} \left\langle \frac{\partial H}{\partial \lambda} \right\rangle_\lambda \mathrm{d}\lambda \quad (1.33)$$

The application of equation (1.33) requires one to evaluate the ensemble average of the derivative of the Hamiltonian with respect to λ, $<\partial H/\partial \lambda>_\lambda$ at various values of λ. One can then use numerical integration methods to calculate ΔG by equation (1.33).

The third commonly used method for free energy calculations is called umbrella sampling [24,25]. This method is developed by Torrie and Valleau in 1977 and since has been one of the major approaches for performing free-energy calculations along the whole reaction coordinates [26,27]. It is readily applied to both molecular dynamics and Monte Carlo simulations. This can either be aimed at in one simulation or in different simulations (windows), the distributions of which overlap. The effect of the bias potential to connect energetically separated regions in phase space gave rise to the name umbrella sampling.

1.3.9.2 Umbrella Sampling Method

In this section, the formalism of recovering unbiased free-energy differences from biased simulations will be discussed. The bias potential w_i of window *i* is an additional energy term, which depends only on the reaction coordinate:

$$E^b(r) = E^u(r) + w_i(\xi) \tag{1.34}$$

The superscript 'b' denotes biased quantities, whereas the superscript 'u' denotes unbiased quantities. Quantities without superscripts are always unbiased. In order to obtain the unbiased free energy $A_i(\xi)$, we need the unbiased distribution:

$$P_i^u(\xi) = \frac{\int \exp\left[-\beta E(r)\right]\delta\left[\xi'(r) - \xi\right]d^N r}{\int \exp\left[-\beta E(r)\right]d^N r} \tag{1.35}$$

Molecular dynamic simulations of the biased system provide the biased distribution along the reaction coordinate P_i^b. Assuming an ergodic system:

$$P_i^b(\xi) = \frac{\int \exp\left\{-\beta\left[E(r) + w_i(\xi'(r))\right]\right\}\delta[\xi'(r) - \xi]d^N r}{\int \exp\left\{-\beta\left[E(r) + w_i(\xi'(r))\right]\right\}d^N r} \tag{1.36}$$

Because the bias depends only on ξ and the integration in the numerator is performed over all degrees of freedom but ξ:

$$P_i^b(\xi) = \exp\left[-\beta w_i(\xi)\right] \times \frac{\int \exp\left[-\beta E(r)\right]\delta\left[\xi'(r) - \xi\right]d^N r}{\int \exp\left\{-\beta\left[E(r) + w_i(\xi'(r))\right]\right\}d^N r} \tag{1.37}$$

Using Equation (1.35) results in:

$$\begin{aligned} P_i^u(\xi) &= P_i^b(\xi)\exp[\beta w_i \xi] \times \frac{\int \exp\left\{-\beta\left[E(r) + w_i(\xi(r))\right]\right\}d^N r}{\int \exp\left[-\beta E(r)\right]d^N r} \\ &= P_i^b(\xi)\exp[\beta w_i \xi] \times \frac{\int \exp\left\{-\beta E(r)\exp\{\left[-\beta w_i[\xi(\vec{r})\right]\right\}d^N r}{\int \exp\left[-\beta E(r)\right]d^N r} \\ &= P_i^b(\xi)\exp\left[\beta w_i(\xi)\right]\left\langle \exp\left[-\beta w_i(\xi)\right]\right\rangle \end{aligned} \tag{1.38}$$

From equation (1.38), unbiased free energy $A_i(\xi)$ can be readily evaluated. $P_i^b(\xi)$ is obtained from a molecular dynamic simulation of the biased system, $w_i(\xi)$ is given analytically, and $F_i = -(1/\beta)\ln\langle\exp[-\beta w_i(\xi)]\rangle$ is independent of ξ:

$$A_i(\xi) = -(1/\beta)\ln P_i^b(\xi) - w_i(\xi) + F_i \tag{1.39}$$

This derivation is exact. No approximation enters apart from the assumption that the sampling in each window is sufficient. This is facilitated by an appropriate choice of umbrella potentials $w_i(\xi)$.

In umbrella sampling, the reaction coordinate is not constrained, but only restrained and pulled to a target value by a bias potential. Therefore, the full momentum space is sampled. Usually, the umbrella sampling is done in a series of windows, which are finally combined with weighted histogram analysis method (WHAM).

1.3.9.3 Weighted Histogram Analysis Method (WHAM)

Numerous methods have been proposed to unbias and recombine the results from umbrella sampling calculations, a promising one being the WHAM [28–30]. It aims to minimize the statistical error of unbiased distribution of function and this approach is routinely used to calculate the PMF along single coordinate:

$$P_i^u(\xi) = \sum_{i=1}^{Nw} n_i P_i^b(\xi) \times \left[\sum_{i=1}^{Nw} n_j e^{-\beta[w_i(\xi)-Fi]}\right]^{-1} \tag{1.40}$$

where N_w is the window number and n is the number of configurations for a biased simulation. The quantity F_i is considered as the dimensionless free energy corresponding to the ith window, and it can be determined by the following equation:

$$e^{-Fi/kBT} = \int d\xi e^{-\beta w_i(\xi)} P_i^u(\xi) \tag{1.41}$$

Because the distribution function itself depends on the set of constants F_i, the WHAM equations (1.40) and (1.41) must be solved self-consistently. In practice, this is achieved through an iteration procedure. Starting from an initial guess for the N_w free energy constants F_i, an estimate for the unbiased distribution is obtained from equation (1.40). The iteration cycle is repeated until both equations are satisfied.

The typical procedure followed for the calculations of potential of mean force (PMF) is illustrated below. The free energy is obtained from the PMF, extracted from a series of umbrella sampling simulations. A series of initial configurations is generated, each corresponding to a location wherein the molecule of interest (generally referred to as a "tagged molecule") is harmonically restrained at increasing center-of-mass (COM) distance from a reference surface using an umbrella biasing potential. This restraint allows the tagged molecule to sample the configurational space in a defined region along a reaction coordinate between the reference surface and the tagged molecule. Computing the probability distribution of the tagged molecule from the reference surface in each window, this relates to the PMF. The windows must allow for slight overlap of the tagged positions for proper reconstruction of the PMF curve. The following steps are used in the umbrella sampling procedure [30]:

i. Generate a series of configurations along a single degree of freedom (reaction coordinate)
ii. Extract frames from the trajectory in step 1 that correspond to the desired COM spacing
iii. Run umbrella sampling simulations on each configuration to restrain it within a window corresponding to the chosen COM distance
iv. Use the Weighted Histogram Analysis Method (WHAM) to extract the PMF and calculate free-energy different, ΔG.

To conduct umbrella sampling, one must generate a series of configurations along a reaction coordinate. Some of these configurations will serve as the starting configurations for the umbrella sampling windows, which are run in independent simulations. Here, we use a simple example for demonstrating the application of the approach of umbrella sampling. The Figure 1.3 illustrates these principles. The top image illustrates running the simulations by pulling the tagged molecule towards the surface, in order to generate a series of configurations along the reaction coordinate. In MD simulations, the harmonic term will add an extra force to the tagged molecule in its z-coordinate. These configurations are extracted after the simulation is complete (dashed arrows in between the top and middle images). The middle image corresponds to the independent simulations conducted within each sampling window, with the center of mass of the tagged molecule s restrained in that window by an umbrella biasing potential. Figure 1.4 shows the ideal result as a histogram of bins, with neighboring windows overlapping calculated from these simulations.

The most common analysis conducted for umbrella-sampling simulations is the extraction of the potential of mean force (PMF), which will yield the ΔG for the binding/unbinding process. The value of ΔG is simply the difference between the highest and lowest values of the PMF curve. A common method for estimating unbiased distribution function is Weighted Histogram Analysis Method (WHAM).

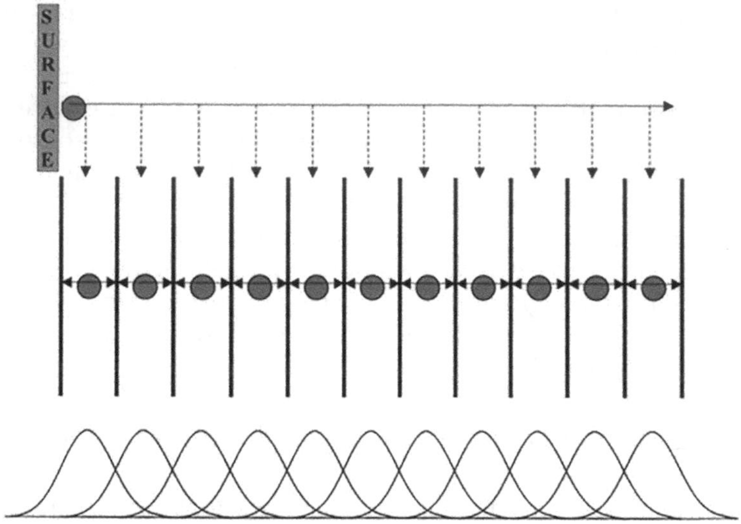

FIGURE 1.4
Representation of the umbrella sampling approach for evaluating the free-energy difference for bring a molecular from far to the surface of interest.

1.4 Mesoscopic Approach

Model simulation of flow field and associated concentration and temperature profiles in a packed bed adsorber are critical to predicting its performance. Although use of efficient adsorbents holds the key to achieving high separation and purification efficiency, prevailing process conditions in the bed may significantly impact the performance of the adsorbers. Besides being challenged with different solute concentration levels during a common operation, adsorbers are often subjected to varying flow rates of the liquid to be processed. In such cases, liquid velocity profiles within the bed, especially in the voids between the particles that are often too complex to be theoretically resolved. Adsorbers may also be subjected to non-isothermal conditions arising from, for example, exothermic heat of adsorption or elevated temperatures at which process liquid is delivered to the bed. Considering mathematical complexities involved in deriving analytical solutions to the Navier Stoke (NS) equation coupled with species and energy conservation equations, various numerical techniques including Finite Difference and Finite Element have been used to solve velocity, concentration, and temperature profiles in packed bed adsorbers.

Mesoscopic approach-based lattice Boltzmann methods (LBMs) are relatively newer. Yet the methods have become popular over the past decade for solving a wide range of fluid flow problems, especially for geometries having complex boundaries. Specific examples of successful applications of LBMs with a high degree of accuracy are the simulation of complex shallow water flow, cavity driven flow, and flow around circular cylinder and through bead packs [31]. With the advent of fast computational machines several LBM-based models have been developed, which consume relatively less CPU time. Computational implementation of such methods is also simple.

Historically, Frisch et al. were among the first to lay the foundation of LBM in the late 80s [32]. The primitive version of the model, known as "lattice gas model," considered a small number of 'fluid particles' instead of large number of individual molecules as considered in the molecular dynamic approach. A 'fluid particle' is considered to be a large group of molecules, which is much larger than a molecule but is considerably smaller than the length scale of the simulation. Therefore, the number of data to be stored in computer is relatively less. The main postulate of the model is that particles are assumed to move along the links of the regular underlying numerical grids, and the particle motion evolves over discrete time steps. Momentum conservation law is applied at each discrete time step. Considering that macroscopic properties of fluid are determined at lattice sites over discrete time steps, simulation using lattice gas model is relatively faster. The main feature of the model is that all collisions are assumed to occur at the same time. Therefore, the computer program may be implemented in parallel computing machines to reduce computational times. The primitive lattice gas models have had reasonably good success in simulating fluid flow. However, the model lacked Galilean invariance. Number of possible collisions considered at each time step was also computationally restrictive. Therefore, this model could not be widely used. However, it gave rise to what we know today as the LBM-based models. The LBM-based models, introduced in late nineties, are derived from the classical Boltzmann equation:

$$\frac{\partial f}{\partial t} + \vec{c}.\frac{\partial f}{\partial \vec{r}} + \vec{F}.\frac{\partial f}{\partial \vec{c}} = \Omega(f) \tag{1.42}$$

where $\Omega(f)$ is the collision function, $F(\vec{r},t)$ is the external body force at a coordinate r and time t, $\vec{c}(\vec{r},t)$ is the particle velocity, and $f(\vec{r},\vec{c},t)$ is the particle distribution function. Briefly, the particle distribution function is a statistical parameter from which the macroscopic properties of the fluid, such as velocity, density, and concentration, can be determined. The model is updated in the same manner as was done in the original lattice gas model. However, instead of considering individual particles to be traveling along the links, it is the distribution function which is calculated at the nodes where the links meet. The LBM-based models have indeed overcome all difficulties encountered in the lattice gas model. At the same time, it has also retained the advantages associated with the original version. At present, LBM-based models are perceived to be an efficient engineering tool for simulating complex fluid flow and its associated transport properties such as concentration and temperature. An elaborate description of the model is presented below.

1.4.1 Introduction to LBM

For a beginner, a simple example to understand the implementation of LBM-based models is the simulation of the two-dimensional (2-D) Poisuille flow of an incompressible fluid in a channel of rectangular cross-section, using the D2Q9 square lattice (2 dimensions, 9 directions). The model may be extended to solve convective-diffusion equation. As earlier stated, origin of LBM is traced to the kinetic theory of gases. Fluid density, momentum, and energy are determined from the particle number density distribution functions by considering the appropriate integrals. In theory, the computations appeared to be straightforward. However, in practice it was difficult to mathematically solve equation (1.42) because of the complicated form of the collision function. By proposing a simple collision function based on Bhatnagar, Gross and Krook's (BGK) approach, $\Omega(f) = -\frac{1}{\tau}\left(f_i - f_i^{eq}\right)$, where τ is the relaxation time which is of the order of time between collisions, and represents the rate at which the number density f_i deviates from the local equilibrium value f_i^{eq} due to collision, mathematical solution to the LBM-based models became considerably simpler [33]. Present versions of the models have become even more versatile by using a multiple-relaxation-time-based approach.

The mechanistic approach to using LBM-based models is the recovery of macroscopic flow behavior of the fluid from the mesoscopic flow representation of the particle movements. At the start of each time step, particles are assumed to collide at each node of the lattice. This stage is known as *collision*. Post-collision, each particle except rest particles travels in a straight line along the different lattice links to the next node. This stage is known as *streaming*. The particles arriving at the new nodes again collide at the beginning of the new time step, and thereafter, they undergo streaming. The process is repeated until the particles reach the equilibrium state. At each time step, the macroscopic quantities such as fluid density and velocities are calculated from the particle number density distribution function.

1.4.2 D2Q9 Square Lattice

As earlier mentioned, fluid particles are assumed to move along the lattice links. It is also assumed that only one particle can move along the link at any time, which allows particles to collide only at the nodes, and not on the links. A schematic presentation of the D2Q9 square lattice is shown below (Figure 1.5).

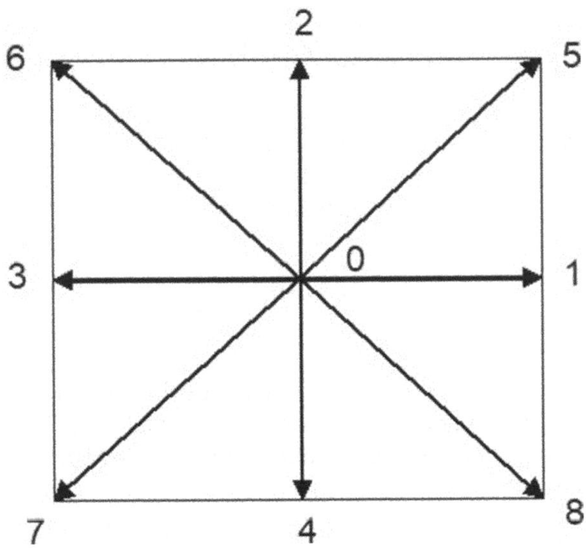

FIGURE 1.5
A schematics of the D2Q9 square lattice.

The particle velocities e_i along the different links i are given by:

$$\vec{ei} = \begin{cases} \cos\left(\frac{\Pi}{4}(i-1)\right)\hat{i} + \sin\left(\frac{\Pi}{4}(i=1)\right)\hat{j} & i = 1,2,3,4 \\ \sqrt{2}\left(\cos\left(\frac{\Pi}{4}(i-1)\right)\hat{i} + \sin\left(\frac{\Pi}{4}(i-1)\right)\hat{j}\right) & i = 5,6,7,8 \\ 0 & i = 0 \end{cases} \tag{1.43}$$

From the above-description, it is clear that the lattice allows only three types of particles: (1) those moving along the horizontal and vertical links of the square lattice (i = 1–4), (2) those moving along the diagonal directions of the square (i = 5–8), and (3) rest particles at the centre of the lattice (i = 0).

1.4.3 Model Development for the 2-D Flow of a Pure Fluid

The macroscopic properties (density ρ and velocity u) of the fluid flow are calculated from the particle density distribution function f as follows:

$$\sum_{i=0}^{n} f_i = \rho \tag{1.44}$$

$$\sum_{i=0}^{n} f_i \vec{e}_i = \rho \vec{u} \tag{1.45}$$

where n is the number (9) of links in the D2Q9 lattice. The BGK-approximated Boltzmann transport equation describes change in the number density considering both collision and streaming steps at each time step:

$$f_i(\vec{r} + \vec{e}_i, t+1) - f_i(\vec{r}, t) = -\frac{1}{\tau}\left(f_i - f_i^{eq}\right) \tag{1.46}$$

where r is the direction vector, τ is the relaxation time, and f_i^{eq} is the equilibrium distribution function. As recast below, equation (1.46) may be mathematically interpreted as the combination of two independent steps:

1) Streaming: $f_i(\vec{r}^{\,+}, t+1) = f_i(\vec{r}, t)$, and

2) Collision: $f_i(\vec{r} + \vec{e}_i, t+1) - f_i(\vec{r}^{\,+}, t+1) = -\frac{1}{\tau}\left(f_i - f_i^{eq}\right)$

where superscript (+) denotes the distribution function after streaming. After determining equilibrium distribution function discussed below, particle distribution functions are determined using appropriate boundary conditions at each time step.

1.4.4 Equilibrium Distribution Function

Collision function in the transport equation contains equilibrium distribution function f^{eq}, which is chosen such that it satisfies the conservation of mass, momentum, and higher order quantities such as stresses:

$$\sum_{i=0}^{8} f_i^{eq} = \rho \tag{1.47}$$

$$\sum_{i=0}^{8} f_i^{eq} \vec{e}_i = \rho \vec{u} \tag{1.48}$$

$$\sum_{i=0}^{8} f_i^{eq} e_{i\alpha} e_{i\beta} = P_{\alpha\beta} + \rho u_\alpha u_\beta \tag{1.49}$$

where P is the stress tensor. Further, f^{eq} is expressed in terms of dynamic collision invariants, namely, density, momentum, kinetic energy, and momentum flux tensor, as follows:

$$f_i^{eq}(\vec{u}) = f_i^{eq}(0)(1 + Au_\alpha e_{i\alpha} + Bu_\alpha u_\alpha + Cu_\alpha u_\beta e_{i\alpha} e_{i\beta}) \tag{1.50}$$

where:

$$f_i^{eq}(0) = \begin{cases} \overline{f}^0 & i = 0 \\ \overline{f}^{1*} & i = 1,2,3,4 \\ \overline{f}^{2*} & i = 5,6,7,8 \end{cases}$$

The coefficients A, B, C, $\overline{f}^0$, $\overline{f}^{1*}$, and $\overline{f}^{2*}$ are determined by subjecting the equilibrium distribution function to the isotropy and Galilean invariance conditions [34]:

$$f_i^{eq}(\vec{u}) = f_i^{eq}(0)\left(1 + 3u_\alpha e_{i\alpha} - \frac{3}{2}u_\alpha u_\alpha + \frac{9}{2}u_\alpha u_\beta e_{i\alpha} e_{i\beta}\right) \tag{1.51}$$

where:

$$f_i^{eq}(0) = \begin{cases} \frac{4}{9}\rho & i = 0 \\ \frac{1}{9}\rho & i = 1,2,3,4 \\ \frac{1}{36}\rho & i = 5,6,7,8 \end{cases} \tag{1.52}$$

Macroscopic conservation equations including Continuity and Navier-Stokes equations are shown to be recovered from the Boltzmann transport equation, using the Chapman-Enskog method:

$$\partial_t \rho + \partial_\alpha \rho u_\alpha = 0 \tag{1.53}$$

$$\partial_t \rho u_\alpha + \partial_\beta \rho u_\alpha u_\beta = -\partial_\alpha P_{\alpha\beta} + \nu \partial_\beta \partial_\beta \rho u_\alpha \tag{1.54}$$

The macroscopic kinematic diffusion coefficient ν is related to the LBM relaxation time as:

$$\nu = \frac{2\tau - 1}{6} \tag{1.55}$$

Readers may refer the referenced studies for the mathematical details of the Chapman-Enskog method [35,36].

1.4.5 Boundary Conditions

Implementation of boundary conditions is briefly described here for the Poiseuille flow problem. Figure 1.6 describes different types of boundaries and unknown distribution functions (underlined) for the simulation of 2-D fluid flow, using the D2Q9 lattice:

i. Top and bottom walls excluding corners
ii. Inlet and outlet planes excluding corners
iii. Four corners.

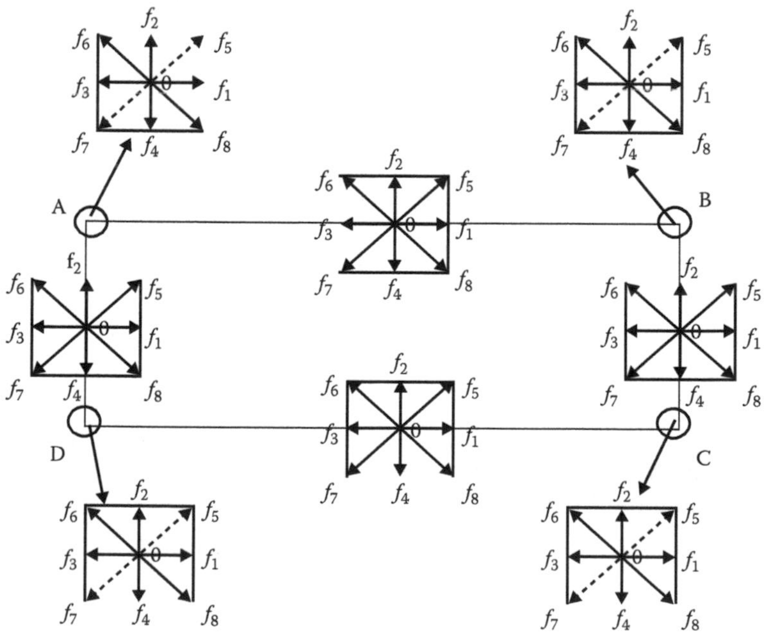

FIGURE 1.6
Setting up of LBM boundary conditions, functions underlined are unknown. (From Manjhi, N., Lattice Boltzmann Modeling for the Velocity and Concentration Profiles in a Packed Bed, M. Tech. Thesis, Indian Institute of Technology Kanpur, Kanpur (India), 2006.)

Standard bounce-back rule is commonly used to determine particle distribution functions at the wall boundary nodes [37]. According to the rule, the distribution functions on the links interior to the boundary nodes are numerically set equal to those on the opposite exterior links. For example, at the wall parallel to the direction $\vec{e}_1$, the distribution function approaching the boundary via link $\vec{e}_2$ (interior to the wall) is set equal to that on the opposite link $\vec{e}_4$ (exterior to the wall). Readers should be able to easily show that the standard bounce back condition produces no slip-fluid velocity at the solid walls.

An improved version (high accuracy) of the LBM boundary conditions was presented by Zou et al. [38], based on the 'non-equilibrium' approach. In this approach, the bounce back rule is applied to the non-equilibrium component of the distribution functions normal to the boundary of the walls, and also, to inlet and outlet sections of the channel. Re-refer Figure 1.6. Boundary conditions for a *stationary top wall* may be applied in the following way:

At any time step of computations, $f_0, f_1, f_2, f_3, f_5, f_6$ are known from the previous time step because these distribution functions originate from within the computational domain. Fluid velocity u_x and u_y are set to zero, considering no slip velocity at the walls. Applying the conservation of mass at the boundary nodes, the unknown distribution functions are expressed in terms of known distribution functions: $f_4 + f_7 + f_8 = \rho - (f_0 + f_1 + f_2 + f_3 + f_5 + f_6)$. Conservation of x-momentum yields: $f_4 + f_7 + f_8 = f_2 + f_5 + f_6$, whereas the conservation of y-momentum yields: $f_7 - f_8 = (f_1 + f_5) - (f_3 + f_6)$. Finally, equating the non-equilibrium component of the functions normal to the wall yields: $f_2 - f_2^{eq} = f^4 - f_{eq}^4$. Therefore, four unknown equilibrium distribution functions (underlined in the figure), namely, f_4, f_7, f_8, and ρ are calculated from as many equations. The similar procedure is also adopted for the *bottom wall*, *inlet*, and *outlet planes*. In the planes, the fluid velocity is, however, prescribed.

Four corner nodes of the channel require special treatment, considering that each node contains pairs of links which do not point into the flow filed, as illustrated by the dashed arrows for node *A* in Figure 1.6. These are called "buried" links, because their populations are confined to the wall, and they do not stream or bring in mass into the flow field [39]. Nevertheless, they participate in collision, and their populations are pre-defined. To determine all distribution functions at the corners, two additional equations are required. For example, at the bottom inlet corners, bounce back rule is applied on the non-equilibrium component of the distribution functions normal to wall and inlet, as follows:

$$f_1 - f_1^{eq} = f_3 - f_3^{eq} \text{ (normal to inlet) and } f_2 - f_2^{eq} = f_4 - f_4^{eq} \text{ (normal to wall)}$$

Readers may like to derive the boundary conditions for the Poisuille flow in the channel, as an exercise, and check their results with the following Table 1.1.

TABLE 1.1

LBM Boundary Conditions for D2Q9 Square Lattice

Cases	Unknowns	Equations	Results
Case 1 Wall below	f_2, f_5, f_6, ρ	(B.1), (B.2), (B.3), (B.4)	$f_2 = f_4$ $f_5 = f_7 - \frac{(f_1 - f_3)}{2}$ $f_6 = f_8 + \frac{(f_1 - f_3)}{2}$
Case 2 Wall above	f_4, f_7, f_8, ρ	(B.1), (B.2), (B.3), (B.4)	$f_4 = f_2$ $f_7 = f_5 - \frac{(f_1 - f_3)}{2}$ $f_8 = f_6 + \frac{(f_1 - f_3)}{2}$
Case 3 Inlet	f_1, f_5, f_8, u_x	(B.1), (B.2), (B.3), (B.5)	$u_x = 1 - \frac{(f_0 + f_2 + f_4) + 2(f_3 + f_6 + f_7)}{\rho_{in}}$ $f_1 = f_3 + \frac{2}{3}\rho_{in} u_x$ $f_8 = f_6 + \frac{f_2 - f_4}{2} + \frac{1}{6}\rho_{in} u_x$ $f_5 = \rho_{in} - \sum_{i=0, i \neq 5}^{8} f_i$
Case 4 Outlet	f_3, f_6, f_7, u_x	(B.1), (B.2), (B.3), (B.5)	$u_x = 1 + \frac{(f_0 + f_2 + f_4) + 2(f_3 + f_6 + f_7)}{\rho_{out}}$ $f_3 = f_1 - \frac{2}{3}\rho_{in} u_x$ $f_7 = f_5 + \frac{f_2 - f_4}{2} + \frac{1}{6}\rho_{out} u_x$ $f_5 = \rho_{out} - \sum_{i=0, i \neq 6}^{8} f_i$
Case 5 Bottom inlet corner	f_1, f_2, f_5, f_6, f_8	(B.1), (B.2), (B.3), (B.4), (B.5)	$f_1 = f_3$ $f_2 = f_4$ $f_5 = f_7$ $f_6 = f_8 = \frac{\rho_{in} - \left(f_0 + 2(f_3 + f_4 + f_7)\right)}{2}$

(Continued)

TABLE 1.1 (CONTINUED)

LBM Boundary Conditions for D2Q9 Square Lattice

Cases	Unknowns	Equations	Results
Case 6 Top inlet corner	f_1, f_4, f_5, f_7, f_8	(B.1), (B.2), (B.3), (B.4), (B.5)	$f_1 = f_3$ $f_4 = f_2$ $f_8 = f_6$ $f_7 = f_5 = \dfrac{\rho_{in} - \left(f_0 + 2(f_2 + f_3 + f_6)\right)}{2}$
Case 7 Bottom outlet corner	f_2, f_3, f_5, f_6, f_7	(B.1), (B.2), (B.3), (B.4), (B.5)	$f_2 = f_4$ $f_3 = f_1$ $f_6 = f_8$ $f_7 = f_5 = \dfrac{\rho_{out} - \left(f_0 + 2(f_1 + f_4 + f_8)\right)}{2}$
Case 8 Top outlet corner	f_3, f_4, f_6, f_7, f_8	(B.1), (B.2), (B.3), (B.4), (B.5)	$f_3 = f_1$ $f_4 = f_2$ $f_6 = f_8 = \dfrac{\rho_{out} - \left(f_0 + 2(f_1 + f_2 + f_5)\right)}{2}$ $f_7 = f_5$

Source: Manjhi, N., Lattice Boltzmann Modeling for the Velocity and Concentration Profiles in a Packed Bed, M. Tech. Thesis, Indian Institute of Technology Kanpur, Kanpur (India), 2006.

The equations used for determining unknown distribution functions at each boundary are as follows:

Conservation of mass: $f_0 + f_1 + f_2 + f_3 + f_4 + f_5 + f_6 + f_7 + f_8 = \rho$ (B.1)

Conservation of momentum in x-direction: $f_1 - f_3 + f_5 - f_6 - f_7 + f_8 = \rho u_x$ (B.2)

Conservation of momentum in y-direction: $f_2 - f_4 + f_5 + f_6 - f_7 - f_8 = \rho u_y$ (B.3)

Bounce back rule at the non-equilibrium component of the distribution function normal to the wall: $f_2 - f_2^{eq} = f_4 - f_4^{eq}$ (B.4)

Bounce back rule at the non-equilibrium component of the distribution function normal to the wall: $f_1 - f_1^{eq} = f_3 - f_3^{eq}$ (B.5)

Model validation may be performed by comparing the simulation results with the analytical solution to the Poiseuille flow problem. Figure 1.7 describes the LBM results for the steady-state 2D velocity profiles of a pure gas (N_2) in a horizontal channel of rectangular cross-section. The simulation was carried out for 64 × 131 nos of x (length) and y (width) nodes, with step sizes Δx and Δy set to unity. Steady-state velocity distribution was assumed when:

$$\frac{\sum_i\sum_i \left|u_x(i,j,t+1)-u_x(i,j,t)\right|+\left|u_y(i,j,t+1)-u_y(i,j,t)\right|}{\sum_i\sum_i \left|u_x(i,j,t)-u_y(i,j,t)\right|} < 10^{-12}.$$

The upstream and downstream densities in the channel were assumed to be 2.7001 and 2.6999, respectively. Density difference across the tube length was chosen to be small so that the steady-state velocity solution could be compared with the corresponding analytical solution to the Poisuille flow problem for an incompressible fluid. Initial conditions were set at the equilibrium distribution values corresponding to the constant gas density (ρ) of

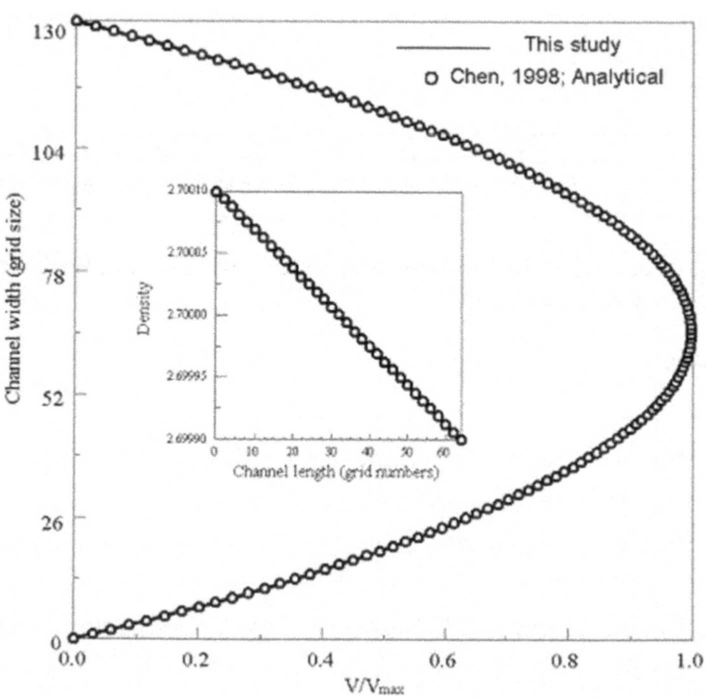

FIGURE 1.7

LBM simulation of Poiseuille flow in the channel (Grids = 64 × 131, τ = 2.0, Reynolds number = 0.43), and the model validation. (From Agarwal, S. et al., *Heat and Mass Transfer*, 41: 843–854, 2005.)

2.7, which is the average mean density, and zero fluid velocity. The value of τ was set at 2.0. As observed from the figure, the model results are in excellent agreement with the analytical solution for the velocity profiles, $u_x = u_o(1-y^2/W^2)$, where W is the half width of the channel, $u_o = W^2/(2\rho\nu)(\Delta p/\Delta x)$, and also, with the published data. The simulation results (Figure 1.7) show a linear pressure gradient across the channel, which is consistent with the theory. As expected, variation in the fluid density across the width of the channel was found to be insignificant.

1.4.6 LBM-Based Model for Adsorption in Packed Beds

Before we describe the development of LBM-based models for a packed bed adsorber, let us understand the approach to developing the model for the simple case of a binary component fluid flowing without phase change through an empty tube, using a d2q9 square lattice. The approach is similar to that described earlier for the flow of a pure fluid. The difference is with respect to the distribution functions. In the present case, there are two distribution functions, f for the combined (total) density (ρ) of two components (solute and carrier fluid) in the binary mixture, and d for the difference ($\Delta\rho$) between the densities of two components. Therefore, knowing the two quantities, ρ and $\Delta\rho$, density distributions of the individual species, solute and carrier may be determined. As per the lattice Boltzmann approach, the transport equation for ρ remains identically the same as equation (1.46), considering collision and streaming. The constraints on the equilibrium distribution f_i^0 also remain the same as those described by equations (1.47–1.49). The BGK approximation for d is defined in the similar fashion as that for pure fluid:

$$\mathbf{d}_i(\mathbf{r}+\mathbf{e}_i, t+1) - d_i(\mathbf{r}, t) = \frac{-1}{\tau_d}\left(d_i - d_i^0\right) \tag{1.56}$$

where d_i^0 is the local equilibrium distribution function for $\Delta\rho$, and τ_d is the relaxation time to reach local equilibrium d_i^0, and is related to the lattice diffusion coefficient, D as $D = (\tau_d - ½)$. The constraints on d_i^0 are as follows:

$$\sum_i d_i^0 e_{i\alpha} = \rho u_\alpha \quad \textit{and} \quad \sum_i d_i^0 e_{i\alpha} e_{i\beta} = \Gamma\Delta\mu\delta_{\alpha\beta} + \Delta\rho u_\alpha u_\beta \tag{1.57}$$

where Ѓ is the mobility and $\Delta\mu$ is the chemical potential difference between two components. Assuming dilute solution and small Mach number, mobility and chemical potential may be equated with diffusivity and density (or concentration) as: D = ЃRT and $\Delta\mu = \Delta\rho RT$. The recovery of the following convective-diffusion macroscopic equation for ideal solution follows directly by applying the Chapman-Enskog expansion procedure on equation (1.56) with the constraints given by equation (1.57):

$$\partial_t \Delta\rho + \partial_\alpha \Delta\rho u_\alpha = D\theta \partial_\beta \partial_\beta \Delta\rho \tag{1.58}$$

where $\theta = (\tau_d - ½)$. A suitable choice of the two equilibrium distribution functions, f_i^i and d_i^0 for a binary miscible fluid on a d2q9 square lattice is defined as per the procedure developed by Swift et al. [36]. In this case, the respective equilibrium distribution functions for ρ and $\Delta\rho$ satisfy the constraints (1.47–1.49) and (1.57) and are defined in the similar form as that of the pure fluid:

$$f_i^0(u) = f_i[0]\left[T_i + Au.e_i - Bu.u + C(e_i.u)^2\right], \text{ and}$$

$$d_i^0(u) = d_i[0]\left[t_i + au.e_i - bu.u + c(e_i.u)^2\right] \tag{1.59}$$

Numerical values of all coefficients (A, B, C, a, b, c) are determined to be the same as for equation (1.51), except the following coefficients:

$$f_i^0(0) = 4\rho/9 \text{ for } i = 0, \rho/9 \text{ for } i = 1-4, \text{and } \rho/36 \text{ for } i = 5-8.$$

$$d_i^0(0) = 4\Delta\rho/9 \text{ for } i = 0, \Delta\rho/9 \text{ for } i = 1-4, \text{and } \Delta\rho/36 \text{ for } i = 5-8. \tag{1.60}$$

$T_i = 9/4(1 - 5/3T)$ and $t_i = 9/4(1 - 5/3D)$, where T and D are temperature and diffusion coefficients, respectively. Model simulations may be performed by adjusting values of θ or τ_d.

1.4.7 Adsorption Breakthrough Analysis

Let us focus on the development of a simple 1D LBM-based model for packed columns. Such models assume radially flat velocity profiles in the column. Refer to Figure 1.8a. Initially, a pure liquid is assumed to flow at a constant flowrate Q (velocity = u) through a tubular column of length L and inside diameter d. The column is uniformly packed with *small* adsorbent particles (tube to particle diameter ratio > 10). At a certain time, a step change in the liquid concentration to C_l is introduced at the inlet to the column.

The sigmoidal (S-shaped) type of the unsteady-state concentration profile measured at the outlet of the column, shown in Figure 1.8b, is the breakthrough curve. Assuming isothermal condition, the 1D unsteady-state concentration profiles of the solute in the column in such a scenario can be

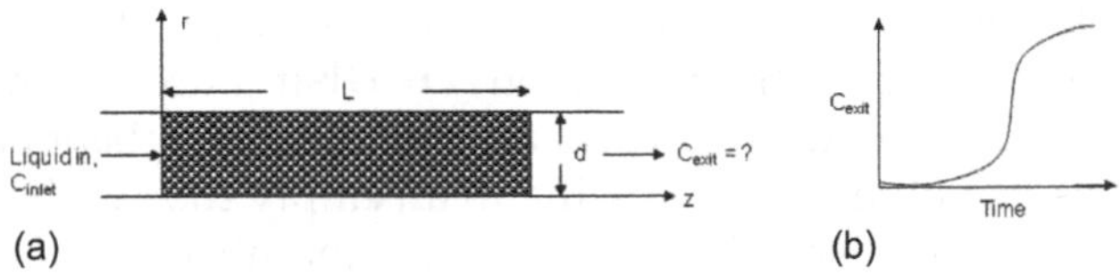

FIGURE 1.8
(a) Adsorption in packed bed. (b) Breakthrough curve. (From Agarwal, S. et al., *Heat and Mass Transfer*, 41: 843–854, 2005.)

described by two species balances, one (C_l) in the liquid phase and the other (C_s) on the solid surface:

$$\textit{Liquid phase}: \quad \partial_t C_l + \partial_\alpha C_l u_\alpha = D\nabla^2 C_l - a\partial_t C_s(1-\varepsilon) \tag{1.61}$$

where D is the axial dispersion coefficient, a is the specific adsorption surface area of the adsorbent, and ε is the bed porosity. Particle (film) mass transfer coefficient is assumed to be large. Also, radial concentration gradient in the column is assumed to be insignificant considering that radial peclet number is large because of the small diameter of the tubular adsorber. Assuming first order adsorption and desorption rates, species balance on the surface of the adsorbent particles is derived as:

$$\textit{Surface}: \quad \partial_t C_s = k_a C_l - k_d C_s \tag{1.62}$$

where k_a and k_d are adsorption and desorption rate constants, respectively. If we assume instantaneous equilibrium of the surface with the liquid phase, equation (1.62) may be recast as $\partial_t C_s = \partial_t C_l(dC_l/dC_l)$, where the term in the parenthesis on the right hand side of the equation is nothing but the slope (m) of the adsorption isotherm, determined under equilibrium conditions from equation (1.62). Equation (1.62) allows the linear as well as non-linear dependence of the rate on the surface concentration Cs via the desorption rate constant, kd. In other words, assuming $k_d = k_d(C_s)$, one may explain both types of adsorption isotherm, linear as well as non-linear. In either case, equation (1.62) is simplified as the following:

$$\partial_t C_l + \partial_\alpha C_l u_\alpha = D\partial_\alpha \partial_\alpha C_l - am\partial_t C_l(1-\varepsilon) \tag{1.63}$$

By arranging the terms in the above equation, the modified liquid phase species balance equation can be cast as a diffusion-convection equation as follows:

$$\partial_t C_l + \partial_\alpha C_l u_\alpha^1 = D^1 \partial_\alpha \partial_\alpha C_l \tag{1.64}$$

where the velocity and dispersion coefficient are modified by the constant factor, $1 + am(1-\varepsilon)$. It is clear from equation (1.64) that concentration of the liquid in a narrow tubular adsorber packed with small adsorbent particles can be modelled as that of the liquid flowing in an empty tube without reaction. However, velocity and dispersion coefficient are modified (see the analogous convective-diffusion macroscopic equation 1.58). Therefore, velocity in the momentum conservation equation is also set at u_{α}^{1} for consistency with the definition of velocity in equation (1.64). If the velocity is assumed to be independent of the radial location (assuming plug flow), u_{α}^{1} becomes a function of α only, as in the case for C_l. Seeking solution to the macroscopic equation (1.64) completes the theoretical analysis for 1D concentration breakthrough in a packed bed adsorber.

1.4.8 Boundary Conditions

As discussed in the preceding section, binary fluid flow may be described by two distribution functions, f and d. The former distribution function is used for the combined density of two components ρ and the latter is used for the difference between densities of two components, $\Delta\rho$. Therefore, the same boundary conditions (Table 1.1) are imposed on both distribution functions at each boundary site.

Chapter 6 presents a detailed analysis of 3D LBM-based models discussed in literature for adsorption and their salient features.

1.5 Experimental Techniques

The experimental methodology to measure the aqueous phase adsorption capacity of a solid under batch or static conditions is to first allow the adsorbent material to equilibrate with the solute in the aqueous solution at a constant temperature. A simple species-specific balance in the aqueous and solid phase can then be used to understand the interplay of different experimental conditions and accurately determine equilibrium loading of the solute in the solid adsorbents:

$$q = \frac{V(C_o - C_e)}{w} \tag{1.65}$$

where q (mg/g) is the equilibrium loading of the adsorbate or solute in the solid phase, C_0 is the initial aqueous phase solute concentration (mg/L), C_e is the aqueous phase solute concentration (mg/L) at equilibrium, and V (L) is the volume of the solution in contact with the adsorbent of weight w (g).

It is clear from equation (1.65) that q vs. C_e represents a single data point on the adsorption isotherm curve for the liquid-solid system under study. The entire isotherm curve may be constructed by using different amounts of adsorbent keeping the other two experimental variables, namely, C_0 and V to be constant. Alternatively, different volumes of the solution can be used keeping C_0 and w to be constant. Yet in another scenario, different initial concentrations of the solutes in the solution can be used keeping V and w to be constant. Under either experimental condition, a unique isotherm or q vs. C_e relationship will represent the system at the constant temperature. Readers must note that equation (1.65) does not describe the rate of adsorption or the time the system will take to equilibrate, nor does it describe the type of adsorbent whether the material is porous or highly porous or non-porous with only the external surface contributing to adsorption.

Precautions must be taken to perform the batch experiments. Unlike gas-solid adsorption, majority of aqueous phase adsorption show negligible heat of adsorption and the process may be considered to be isothermal. Yet maintaining a constant adsorption temperature is necessary during the experiments because equilibrium solute loading may be sensitive to small variation in the ambient or surrounding atmospheric temperature. It is important to mention here that it is not uncommon for many adsorption systems to reach equilibrium in more than 12 h. In such cases, a thermally non-insulated system will see a significant swing in the laboratory or room temperature. Consequently, the measurements may yield erroneous or inaccurate data. Therefore, it is strongly recommended that adsorption experiments be performed in a constant temperature bath or thermally insulated enclosure with the provision of adjusting/setting adsorption temperatures over a wide range (less than room temperature to ~80 °C).

Constant and uniform concentration in the test solution is another necessary condition that must be maintained during the batch adsorption tests. Such condition can be maintained by uniformly stirring the solution at a constant rate. Technically, adsorption must not be limited by the interparticle mass transfer rate and an approximately constant solute concentration should exist in the solution. Also, small (micron) sized or adsorbent particles or powdery materials tend to flocculate, and relatively larger particles may settle in the solution. Stirring helps to keep the particles suspended and uniformly dispersed in the solution. Generally, internal stirring is not desired because the adsorbent particles may collide with the stirrer blades and break. In some cases, particles may adhere to the stirrer surface. Externally mechanical (motorized) shaking of containers, viz. glass flasks mounted with clamp holders, is the most commonly used means to create stirred conditions in the solution. Such equipments (shakers) are commercially available.

A few special types (beads and fabrics) of adsorbents require a careful design of the stirring arrangement for the test solutions mixed with the adsorbent materials. In a recent study, Yadav et al. [41] prepared the iron nanoparticles-dispersed phenolic precursor-based carbon beads (approximately 0.8 mm size). Carbon nanofibers (CNFs) were grown on the carbon microbeads, using catalytic chemical vapor deposition. The prepared materials were tested as catalysts for catalytic wet air oxidation (CWAO) of phenol using a batch reactor. However, comparative adsorption tests were performed on the materials to determine the amount of phenol removed from water, contributed from the adsorption of phenol on the material surface. Because of the potential danger of the CNFs leached from the beads under vigorous stirred conditions, an especially configured impeller cum catalyst/adsorbent holder was used which held the CNFs-containing carbon beads (Figure 1.9). The beads were placed in the wire-meshed cylindrical baskets mounted on the shaft of the impeller. Such especially configured impeller served dual role of material holder as well as stirrer without damaging the CNFs grown over the beads under stirred conditions.

Gupta et al. [42] developed an aluminum-impregnated hierarchal web of carbon fibers for the removal of dissolved fluoride ions in water by adsorption. Micron-size activated carbon fibers (ACF) in the fabric form were used as a substrate to grow carbon nanofibers (CNF) by CVD. The micro-nano carbon fibers (ACF/CNF) thus prepared were impregnated with aluminum and tested for the adsorption of fluoride ions under batch conditions. Figure 1.10 is the schematics of the especially designed and developed

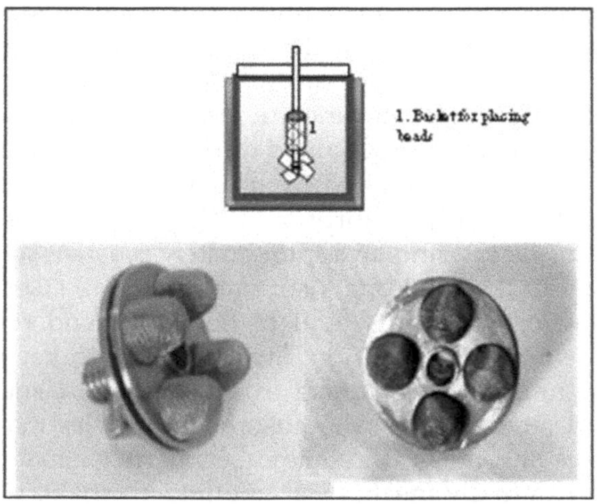

FIGURE 1.9
A specially configured reactor for bead positioning. (From Yadav, A. et al., *Journal of Environmental Chemical Engineering*, 4: 1504–1513, 2016.)

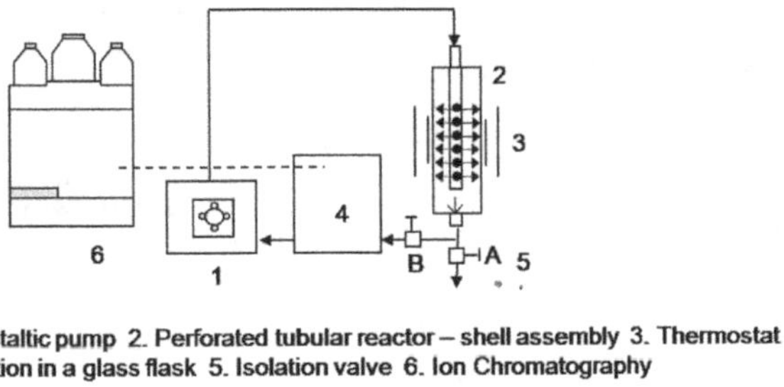

FIGURE 1.10
Schematics of the experimental setup used for the adsorption studies on the ACF (fabric) adsorbents. (From Gupta, A. et al., *Industrial and Engineering Chemistry Research*, 48: 9697–9707, 2009.)

experimental set-up used in the study for the aqueous phase adsorption using a fabric adsorbent material. A cylindrical tube (*I.D.* = 2.5 cm, *O.D.* = 2.8 cm, *L* = 6 cm) was mounted inside a cylindrical shell (*I.D.* = 4.0 cm, *L* = 10 cm) with provisions made for the water inlet and outlet. One end of the tube was closed. The outer surface of the tube was perforated with holes of diameter 0.1 mm at center-to-center distances of 0.4 cm. The ACF adsorbent was wrapped over the perforated section of the tube. The test solution in a glass container was continuously re-circulated to the inlet of the tube, using a peristaltic pump as shown in the schematics.

We have earlier mentioned that batch adsorption tests can be performed under any of three scenarios: keep two of three variables (V, C_o and w) constant and vary third variable. Measure C_e after sufficiently long time when the equilibrium is attained, i.e., solution concentration does not change with time. Considering that adsorption is a reversible process, accompanied by desorption, precautions should be taken to use sufficient quantity of solution vis-à-vis that of adsorbent such that neither solution is depleted with the solute, nor the surface is saturated with the solute, and a true dynamic equilibrium is attained during the tests. In other words, excess quantities of adsorbent-dose or solution must be avoided. Readers should, however, note that we are interested in measuring specific loading (mg/g) of the solute and not the actual loading for constructing isotherm. Therefore, q_e depends on C_e alone at the constant temperature. Very often several combinations of adsorbent dose and solution volume are used to determine the optimized relative amounts of V and w, or V and C_o, or w or C_o. Verma et al. [43] used similar exercise to determine the optimized amounts of adsorbent-dose to remove Cr(VI) from synthetic test water at 30 °C. The red dotted circle shown in Figure 1.11 indicates the region of experimental conditions that may be used to perform batch adsorption tests and construct the corresponding isotherm. In such a region, the percentage removal or equilibrium loading is neither large nor small.

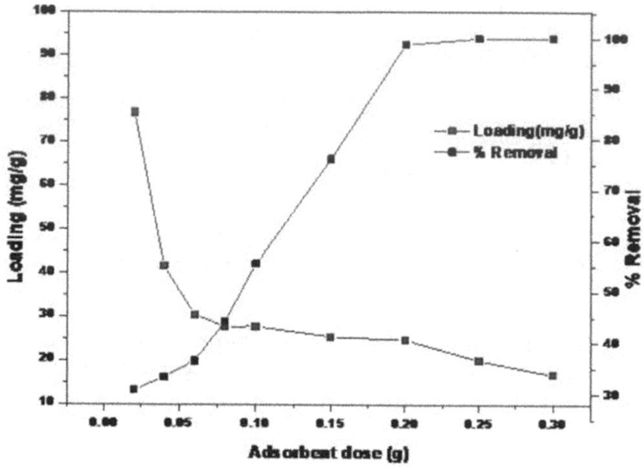

FIGURE 1.11
Effect of adsorbent-dose on the adsorption of Cr(VI) (temperature = 30 °C, initial concentration = 100 mg/L, time = 24 h). (From Verma, N. K. et al., *Green Processing and Synthesis*, 4: 37–46, 2015.)

1.5.1 Equilibrium Adsorption Loading from Flow Study

Although batch adsorption tests are routinely performed to determine the equilibrium loading of a solute and the adsorption capacity of the material, adsorption tests can also be performed under flow or dynamic conditions to determine the equilibrium data. Re-refer to Figure 1.12 for the typical breakthrough concentration profiles measured at the exit of the column for different solute (fluoride ions in this case) concentrations [42].

In theory, steady-state condition never exists during adsorption, except towards the later part of the experiment when the aqueous phase concentration at the exit of the adsorber equals approximately to that at the inlet of the adsorber. By making the specie balance for the solute, it is easy to show that the total uptake of the solute (mg) by the adsorbent (w g), from the instance the solute is injected in the flowing stream to the inlet of the column until it is saturated with the solute, can be determined as follows:

$$Uptake(mg) = Q(C_{in}T - \int_0^T C_{exit}\, dt) \tag{1.66}$$

where Q is the liquid flowrate (liters per min); C_{in} and C_{exit} are the solute concentrations in the liquid at the inlet and outlet of the column, respectively, and T is the total time (min) of adsorption until the bed is saturated with the solute in the influent stream. Readers should again convince themselves that theoretically, the specific capacity (mg/g) of the material calculated in such (flow) case must be the same as that measured under batch conditions, if the adsorption

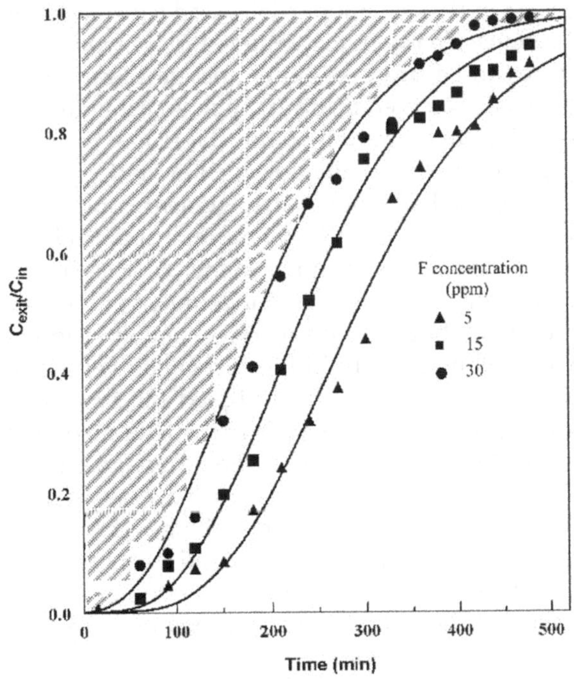

FIGURE 1.12
Breakthrough data for different inlet fluoride concentrations (w = 3 g, Q = 0.01 liters per min, T = 303 K). (From Gupta, A. et al., *Industrial and Engineering Chemistry Research*, 48: 9697–9707, 2009.)

temperature is the same under two conditions, and the final equilibrium concentration C_e of batch test equals the inlet concentration C_{in} of flow test. Readers should also be able to note that the expression in the closed parenthesis of the right hand side of equation (1.66) can also be evaluated by numerically integrating the area (hatched) above the breakthrough curve shown in Figure 1.12.

Indeed, flow studies, performed using a packed bed column of adsorbents, are prone to experimental errors. The errors are induced from maldistribution or channeling of liquid flow in the packed bed. Therefore, precautions must be taken to ensure that the adsorbent materials are uniformly packed in the column and incoming liquid to the column is uniformly dispersed over the packed bed. If the column diameter is relatively large, a shower-like arrangement is recommended for the liquid over the bed.

In recent times, several types of novel micro-nano particles (~100–500 nm) have been synthesized as potential adsorbents for wastewater treatment. Such materials have relatively larger adsorption capacities owing to their large specific surface areas. It is extremely difficult, however, to use micro-nano particles in a packed bed column under flow conditions because such particles are likely to be entrained in the flowing stream. Therefore, Hosseini et al. [44] used the packed bed of sand particles as the contactor between an aqueous suspension of Fe/Cu nanoparticles and a nitrate solution for nitrate reduction applications. Zhang et al. [45] used glass wool as the support for silica nanoparticles to separate gallium and germanium in water under flow conditions in a packed bed column. Such an arrangement considerably reduced the possibility of entrainment, but the risk of the support-materials themselves being entrained with the flow still existed.

Saraswat et al. [46] used a packing arrangement consisting of a thin film composite (TFC) reverse osmosis (RO) membrane and stacked layers of polymeric materials as a support for the micro-nano particles in the column. Such an arrangement not only prevented the entrainment of the adsorbent materials and support-particles, but also minimized channeling and maldistribution of flow through the column. Figure 1.13 is a schematics of the Perspex-column (length = 60 mm, I.D. = 6 mm) specially designed and fabricated for the breakthrough analysis. As shown, the column contained the adsorbents (micro-nano particles) sandwiched between stacked layers of polymeric materials having graded (decreasing) sizes (0.8, 0.5 and 0.05 mm-average size). The top layers consisted of 0.05 mm and 0.5 mm-sized adsorbents. This way, there existed a gradient of different bed porosities in the stacked layers and the flow distribution of the liquid to be treated was uniform in the column. The test data in the study [46] showed that the equilibrium loadings measured from batch and flow tests were within ±15% difference.

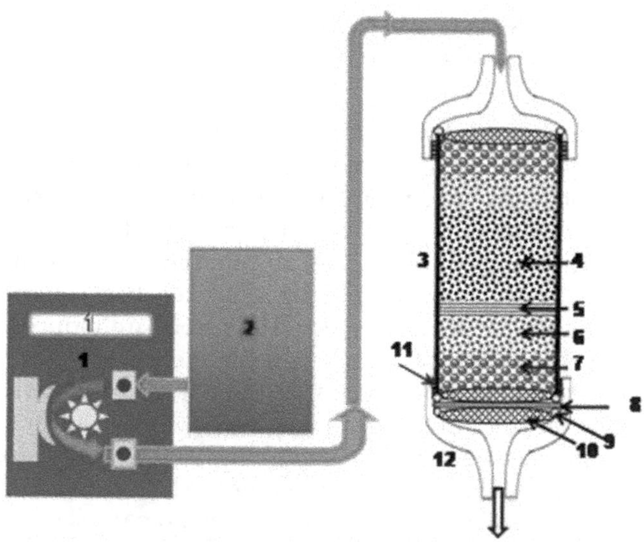

FIGURE 1.13
Schematics of the packed bed column used for dynamic (flow) study of the adsorption of vitamin B_{12} by micro-nanoparticles: 1. Peristaltic pump, 2. VB_{12} in container, 3. Perpsex column, 4-6. adsorbents (~200 nm-0.05 mm-0.5 mm), 7. adsorbents (0.8 mm), 8. TFC RO-membrane, 9. O-rings, 10. SS mesh, 11. Thread-fitting, 12. End-stub. (From Saraswat, R. et al., *Chemical Engineering Journal*, 197: 250–260, 2012.)

References

1. Dutta, B. K. (2009) *Principles of Mass Transfer and Separation Processes*, PHI Learning Private Limited: New Delhi.
2. Lyubchik, S., Lyubchik, A., Lygina, O., Lyubchik, S., Fonseca, I. Comparison of the Thermodynamic Parameters Estimation for the Adsorption Process of the Metals from Liquid Phase on Activated Carbons. In *Thermodynamics - Interaction Studies - Solids, Liquids and Gases*, InTech: Rijeka, p 95–122, Ch. 4.
3. Allen, M. P., Tildesley, D. J. (1987) *Computer Simulation of Liquids*, Oxford University Press: Walton Street, Oxford.
4. Metropolis, N., Rosenbluth, A. W., Rosenbluth, M. N., Teller, A. H. (1953) Equation of State Calculations by Fast Computing Machines. *The Journal of Chemical Physics* 21: 1087–1092.
5. Verlet, L. (1967) Computer "Experiments" On Classical Fluids. I. Thermodynamical Properties of Lennard-Jones Molecules. *Physical Review* 159: 98–103.
6. Swope, W. C., Andersen, H. C., Berens, P. H., Wilson, K. R. (1982) A Computer Simulation Method for the Calculation of Equilibrium Constants for the Formation of Physical Clusters of Molecules: Application to Small Water Clusters. *The Journal of Chemical Physics* 76: 637–649.
7. Rapaport, D. C. (2004) *The Art of Molecular Dynamics Simulation*, Cambridge University Press: New York.
8. Nosé, S. (1984) A Unified Formulation of the Constant Temperature Molecular Dynamics Methods. *The Journal of Chemical Physics* 81: 511–519.
9. Hoover, W. G. (1985) Canonical Dynamics: Equilibrium Phase-Space Distributions. *Physical Review A* 31: 1695–1697.
10. Hoover, W. G. (1986) Constant-Pressure Equations of Motion. *Physical Review A* 34: 2499–2500.
11. Plimpton, S. (1995) Fast Parallel Algorithms for Short-Range Molecular Dynamics. *Journal of Computational Physics* 117: 1–19.
12. Humphrey, W., Dalke, A., Schulten, K. (1996) Vmd: Visual Molecular Dynamics. *Journal of Molecular Graphics* 14: 33–38.
13. Ohmine, I., Tanaka, H. (1993) Fluctuation, Relaxations, and Hydration in Liquid Water. Hydrogen-Bond Rearrangement Dynamics. *Chemical Reviews* 93: 2545–2566.
14. Luzar, A., Chandler, D. (1996) Hydrogen-Bond Kinetics in Liquid Water. *Nature* 379: 55–57.
15. Luzar, A., Chandler, D. (1996) Effect of Environment on Hydrogen Bond Dynamics in Liquid Water. *Physical Review Letters* 76: 928–931.
16. Ferrario, M., Haughney, M., McDonald, I. R., Klein, M. L. (1990) Molecular-Dynamics Simulation of Aqueous Mixtures: Methanol, Acetone, and Ammonia. *The Journal of Chemical Physics* 93: 5156–5166.
17. Jorgensen, W. L., Madura, J. D. (1985) Temperature and Size Dependence for Monte Carlo Simulations of Tip4p Water. *Molecular Physics* 56: 1381–1392.
18. Zielkiewicz, J. (2005) Structural Properties of Water: Comparison of the Spc, Spce, Tip4p, and Tip5p Models of Water. *The Journal of Chemical Physics* 123: 104501.

19. Desai, T. G., Keblinski, P., Kumar, S. K., Granick, S. (2006) Molecular-Dynamics Simulations of the Transport Properties of a Single Polymer Chain in Two Dimensions. *The Journal of Chemical Physics* 124: 084904.
20. Azuma, R., Takayama, H. (1999) Diffusion of Single Long Polymers in Fixed and Low Density Matrix of Obstacles Confined to Two Dimensions. *The Journal of Chemical Physics* 111: 8666–8671.
21. Bodrova, A., Dubey, A. K., Puri, S., Brilliantov, N. (2012) Intermediate Regimes in Granular Brownian Motion: Superdiffusion and Subdiffusion. *Physical Review Letters* 109: 178001.
22. McQuarrie, D. A. (2000) *Statistical Mechanics*, University Science Books: Virginia.
23. Frenkel, D., Smit, B. (2001) *Understanding Molecular Simulation: From Algorithms to Applications*, Academic Press: Orlando.
24. Kollman, P. (1993) Free Energy Calculations: Applications to Chemical and Biochemical Phenomena. *Chemical Reviews* 93: 2395–2417.
25. Christophe, C., Andrew, P. (2007) *Free Energy Calculations Theory and Applications in Chemistry and Biology*, Springer Verlag: Berlin Heidelberg.
26. Torrie, G. M., Valleau, J. P. (1974) Monte Carlo Free Energy Estimates Using Non-Boltzmann Sampling: Application to the Sub-Critical Lennard-Jones Fluid. *Chemical Physics Letters* 28: 578–581.
27. Torrie, G. M., Valleau, J. P. (1977) Nonphysical Sampling Distributions in Monte Carlo Free-Energy Estimation: Umbrella Sampling. *Journal of Computational Physics* 23: 187–199.
28. Kumar, S., Rosenberg, J. M., Bouzida, D., Swendsen, R. H., Kollman, P. A. (1992) The Weighted Histogram Analysis Method for Free-Energy Calculations on Biomolecules. I. The Method. *Journal of Computational Chemistry* 13: 1011–1021.
29. Souaille, M., Roux, B. (2001) Extension to the Weighted Histogram Analysis Method: Combining Umbrella Sampling with Free Energy Calculations. *Computer Physics Communications* 135: 40–57.
30. Roux, B. (1995) The Calculation of the Potential of Mean Force Using Computer Simulations. *Computer Physics Communications* 91: 275–282.
31. Succi, S. (2001) *The Lattice Boltzmann Equation for Fluid Dynamics and Beyond*, Oxford Press: Oxford.
32. Frisch, U., Hasslacher, B., Pomeau, Y. (1986) Lattice-Gas Automata for the Navier-Stokes Equation. *Physical Review Letters* 56: 505–1508.
33. Bhatnagar, P. L., Gross, E. P., Krook, M. (1954) A Model for Collision Processes in Gases. I. Small Amplitude Processes in Charged and Neutral One-Component Systems. *Physical Review* 94: 511–525.
34. Manjhi, N. (2006) Lattice Boltzmann Modeling for the Velocity and Concentration Profiles in a Packed Bed, M. Tech. Thesis, Indian Institute of Technology Kanpur, Kanpur (India).
35. Chen, S., Doolen, G. D. (1998) Lattice Boltzmann Method for Fluid Flows. *Annual Review of Fluid Mechanics* 30: 329–364.
36. Swift, M. R., Orlandini, E., Osborn, W. R., Yeomans, J. M. (1996) Lattice Boltzmann Simulation of Liquid-Gas and Binary Fluid Systems. *Physical Review E* 54: 5041–5046.
37. Ziegler, D. P. (1993) Boundary Conditions for Lattice Boltzmann Simulations. *Journal of Statistical Physics* 71: 1171–1177.

38. Zou, Q., Hou S., Chen S., Doolen G. D. (1995) Improved Incompressible Lattice Boltzmann Model for Time-Independent Flows. *Journal of Statistical Physics* 81: 35–38.
39. Maier, R. S., Bernard, R. S., Grunau, D. W. (1996) Boundary Conditions for the Lattice Boltzmann Method. *Physics of Fluids* 8: 1788–1795.
40. Agarwal, S., Verma, N., Mewes, D. (2005) 1D Lattice Boltzmann Model for Adsorption Breakthrough. *Heat and Mass Transfer* 41: 843–854.
41. Yadav, A., Teja, A. K., Verma, N. (2016) Removal of phenol from water by catalytic wet air oxidation using carbon bead-supported iron nanoparticle-containing carbon nanofibers in an especially configured reactor. *Journal of Environmental Chemical Engineering* 4: 1504–1513.
42. Gupta, A., Deva, D., Sharma, A., Verma, N. (2009) Adsorptive Removal of Fluoride by Micro-Nano Hierarchal Web of Activated Carbon Fibers. *Industrial and Engineering Chemistry Research* 48: 9697–9707.
43. Verma, N. K., Khare, P., Verma, N. (2015) Synthesis of iron-doped resorcinol formaldehyde-based aerogels for the removal of Cr(VI) from water. *Green Processing and Synthesis* 4: 37–46.
44. Mossa, S. F., Hosseini, Ataie-Ashtiani, B., Kholghi, M. (2011) Nitrate Reduction by Nano-Fe/Cu Particles in Packed Column, *Desalination* 276: 214–221.
45. Zhang, L., Guo, X., Li, H., Yuan, Z., Liu, X., Xu, T. (2011) Separation of Trace Amounts of Ga and Ge in Aqueous Solution Using Nano-Particles Micro-Column, *Talanta* 85: 2463–2469.
46. Saraswat, R., Talreja, N., Deva, D., Sankararamakrishnan, N., Sharma, A., Verma, N. (2012) Development of Novel in Situ Nickel-Doped, Phenolic Resin-Based Micro-Nano-Activated Carbon Adsorbents for the Removal of Vitamin B-12. *Chemical Engineering Journal* 197: 250–260.

2

Graphene Nanopores-Based Separation of Impurities from Aqueous Medium

Anitha Kommu and Jayant K. Singh

CONTENTS

2.1 Introduction

Pollution of water resources because of indiscriminate disposal of various organic and inorganic pollutants is a worldwide environmental concern. The pollutants are released into the environment from several industries such as metallurgy, tannery, mining, chemical manufacturers, paper, and pesticides, to name a few [1,2]. Discharge of wastewater from industries causes serious environmental problem and threat to human and aquatic lives because of its toxic, non-degradable, persistent and accumulative characteristics [3,4]. Typical symptoms of copper poisoning include anemia, liver, and kidney damage, as well as stomach pain [5–7]. High concentrations of lead entering into human body can cause abdominal pain, headache, chronic nephritis of the kidney, brain damage, and central nervous system disorders [8]. Long term exposure of inorganic salts and organic pollutants may cause serious damage to liver, kidney, lung cancer, reduction in hemoglobin formation, hypertension, and itching [9–11]. The textile manufacturing and dyeing industries utilize large quantities of several types of dyes and release the dye pollutants into environment as waste water effluents. Most dyes are highly toxic, bio-degradable, and even carcinogenic to both microbial populations and mammalian animals [12]. Therefore, it is necessary to remove organic and inorganic materials from the industrial effluents before discharged into environment. Extraction of toxic gas and removal of aqueous pollutants in a safe and effective fashion are technically challenging. There are several conventional techniques that are being used for the elimination of hazardous pollutants, which include chemical precipitation, membrane filtration, ion exchange, adsorption, electro chemical treatment, evaporation, and solvent extraction technologies as shown in schematic Figure 2.1. However, these methods have some limitations such as limited adsorption capacity or removal efficiencies, small selectivity, and long equilibrium time. Decreasing the energy requirement and infrastructure costs of existing technologies remains a challenge. To improve the effective removal of organic and inorganic pollutants from wastewater by taking advantage of new materials and techniques has always been an interesting subject of research. In recent years, different types of nanostructured materials have been developed to overcome these principal drawbacks. We have summarized below the nanostructured materials, which have been applied for removal of the heavy metals, organics, and biological impurities from wastewater effluents.

Polymer and nanocomposite materials have drawn significant attention from many researchers for purification of seawater and wastewater containing high concentration impurities because of their excellent selectivity, permeability, and chemical resistance over a wider range of pH and temperature. The different types of inorganic nanomaterials such as zeolite, ZnO, SiO_2, TiO_2, metal–organic frameworks, and nanoclay materials have been utilized for the preparation of nanocomposite membrane to enhance the physicochemical properties and are claimed to be efficient in terms of its efficiency, permeability and selectivity [13]. Among them, carbon nanotube and graphene-based nanomaterials are considered to be the most promising and prominent contenders for the advancement of membrane separation process towards the wastewater treatment [14]. In particular, carbon-based nanomaterials have received considerable attention because of their novel properties such as easy accessibility, excellent mechanical properties, biocompatibility, and environmental

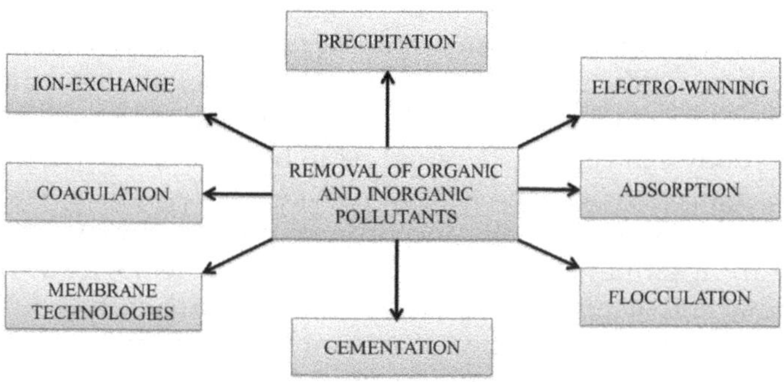

FIGURE 2.1
Conventional methods for the removal of organic pollutants and heavy metal ions.

friendliness [15–19]. To date, research has focused on the ability of carbon nanotubes to serve as nanostructure material for the separation of impurities from aqueous medium [20–23]. Recent studies have also attempted to improve adsorption capacity, selectivity, and water permeability by molecular sieve action of polymers [24–27], ceramics [28–31], and biological molecules [32–34]. Although these materials are theoretically promising, their practical applications are not yet fully explored. Membranes fabricated from materials such as zeolites and carbon nanotubes are difficult to shape in a cost effective and scalable manner and prove somewhat ineffective with regard to the exclusion of salt ions, as well as presenting a lower adsorption of metal ion, low water flux [35–38], hence illustrating the need for ultrathin, low cost membranes. Rise in demand for environmental solutions and the urgent need to tackle the water crisis has led to the endless discovery of new classes of nanomaterials.

Carbon based nanomaterials possess exceptional surface properties, structural characteristics, large surface area, high tensile strength, low thickness, charge density, and mechanical strength [39]. Therefore, such materials play a major role in purification and desalination process. Advances in molecular simulation studies of graphene-based materials improve the interactions and transportation of water molecules through exfoliated graphene and its derivatives. Additionally, the large surface area can provide a broad spectrum of functionalization and surface modification. Moreover, it forms hybrid nanocomposite and complex with various nanoparticles. Numerous studies have confirmed that nanoporous graphene-based membrane materials provide required surface properties such as high hydraulic permeation rates and salt rejection, selectivity and lower transport resistance. Mahmoud et al. [23] stated that graphene nanosheets have the sufficient strength to withstand high pressure and the excellent physicochemical characteristic for effective desalination process in spite of its insignificant thickness. There remains a challenge to attain scalable and cost-effective preparation of graphene and its functionalized derivatives for membrane applications even though vast theoretical assurance.

In this chapter, we focus on the current applications of graphene-based material as adsorbents and desalination membranes for the removal of metal ions and organic contaminants. We then present the recent developments in applications of graphene-based materials for water treatment. The chapter provides an insight into the removal of metal ions and organic pollutants using the graphene based materials that can play a valuable role in the water purification.

2.2 Separation of Metal Ions from Aqueous Solution

2.2.1 Graphene-Based Membranes

Graphene is a carbon-based material that has recently received attention as a potentially selective material for water desalination and purification membranes. Graphene is a two dimensional sheet that consists of a hexagonal honeycomb lattice of covalently bonded carbon atoms and it has many superiorities over conventional materials [40,41]. The most fascinating properties of graphene are high mechanical strength, thermal and electrical conductivity, high surface functionality, and large surface area [42–44]. These properties suggest that graphene is suitable for several applications such as transistor fabrication, super capacitors, improved batteries, solar cells, and water purification [45]. One such potential application is the use of graphene sheets as membranes for the separation of inorganic salts and small organic molecules. Despite its negligible thickness, membranes made of graphene are considered to be impermeable to gases and liquids [46]. Nanopores in graphene sheet are formed by 'knocking out' carbon atoms from the matrix, which is initially examined through a series of theoretical studies [47,48]. In this section, we present a brief overview of recent experiments, computer simulations, and theoretical models developed to investigate water transport through different types of nanoporous graphene (NPG) membranes. This will help to identify the potential applications of nanoporous and functionalized graphene membranes and the advantage they may offer over polymer-based commercial filtration membranes. Size of nanopores can suitability be tuned to allow only the passage of water molecules through NPG, while not allowing ions and unwanted substances to pass through the membrane shown in Figure 2.2a. Several simulations studies identified NPG membranes with artificial nanopores to have great promise for nanofiltration and RO based seawater desalination, gas separation, and selective ion passage [49]. Water flux and salt rejection through NPG is greatly dependent on pore size, applied pressure, and the chemistry of the pores, as well as how the pores are functionalized [48,50]. Cohen-Tanugi and Grossman [48] investigated graphene pores functionalized with hydroxyl groups and hydrogen atoms (Figure 2.2b) to study how the desalination dynamics change with pore size and applied hydrostatic pressure. Their investigation revealed that, in addition to pore size, desalination performance is also sensitive to pore chemistry. Hydrophilic hydroxyl groups roughly double the rate at which the water permeates the graphene membrane [39]. Hydrogen terminated nanpores show better water selectivity while functionalization with hydroxyl groups enhance the rate of water transport (Figure 2.3a). Simulation results also show that the transport of water through these nanoporous membranes could reach up to 66 L/cm^2/day/MPa with greater than 99% salt rejection. In contrast, water transport through a conventional RO membrane approximately reaches 0.01–0.05 L/cm^2/day/MPa with similar salt rejection. Various studies showed that salt rejection and water permeability using nanopores based membrane are 2–3 orders of magnitude greater than that measured in the commercialized membranes [23,48] (Figure 2.3c). These values revealed the great potential for the utilization of functionalized NPG as a high-permeability desalination membrane. Consequently, the ability to tune the selectivity of synthetic nanopores in graphene membranes

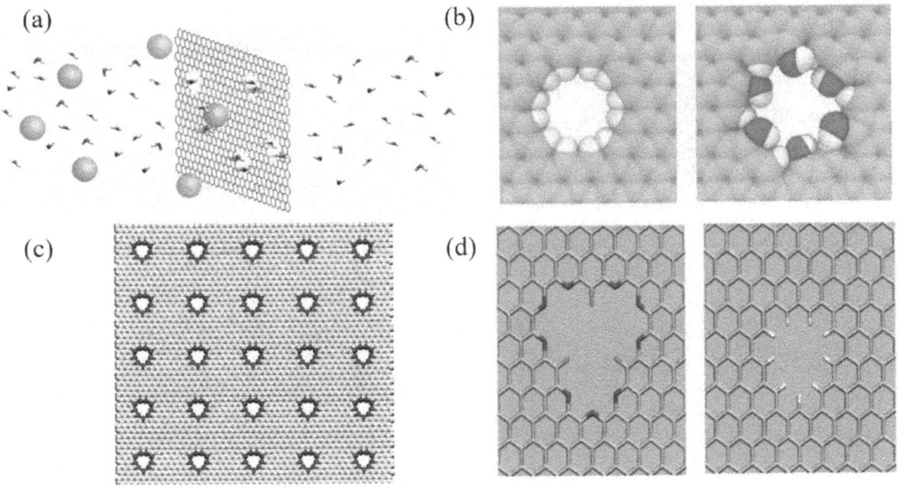

FIGURE 2.2
(a) Passage of ions and water molecules through nanoporous graphene sheet. (b) Nanoporous graphene functionalized with hydrogen and hydroxyl groups for various pore sizes. (c) Nitrogen functionalized nanoporous graphene sheet (dark gray represents nitrogen atom). (d) Fluorine-nitrogen-terminated (left) nanopore to facilitate the passage of cations and hydrogen-terminated (right) nanopore to favor the passage of anions. (a. Reprinted by permission from Macmillan Publishers Ltd. *Desalination* (Aghigh, A. et al., *Desalination*, 365 : 389–397, 2015.), copyright 2012; b. Reprinted with permission from Cohen-Tanugi, D., Grossman, J. C., *Nano Letters*, 12 : 3602–3608, 2012. Copyright 2012 American Chemical Society; d. Reprinted with permission from Sint, K. et al., *Journal of the American Chemical Society*, 130 : 16448–16449, 2008. Copyright 2008 American Chemical Society.)

further promotes the use of graphene in desalination technologies [39]. Rejection of ions is found to involve steric effects, and electrostatic interactions between charged species and the pores [51,52]. Hydrophilic functionalized pores allow more water flux but less salt rejection than the hydrophobic functionalized pores [48]. Thus, the ion rejection can be tuned by functionalization of the pores. Sint et al. [51] predicted that functionalized nanopore in graphene monolayer could serve as ionic sieves of high selectivity and transparency using molecular dynamics (MD) simulations. Gai et al. [53,54] designed the graphene pores with ion etching and decorated with negatively charged atoms like nitrogen, fluorine, and positively charged hydrogen atoms (see Figure 2.2d). Pore selectivity was influenced by the electrostatic interactions between the ions and functional groups positioned at the edge of the nanopore. Functionalizing the edge of some pores with nitrogen, fluorine, and hydrogen acts as ionic sieves with high selectivity. For example, the F-N functionalized nanopore permits the passage of Li^+, Na^+, and K^+ ions, yet Cl^- and Br^- ions penetrated the H-pore. The nanopores whose edge atoms are terminated with hydroxyl group can double the water flux through them [48] by designing three biomimetic Na^+- or K^+-selective graphene nanopores,

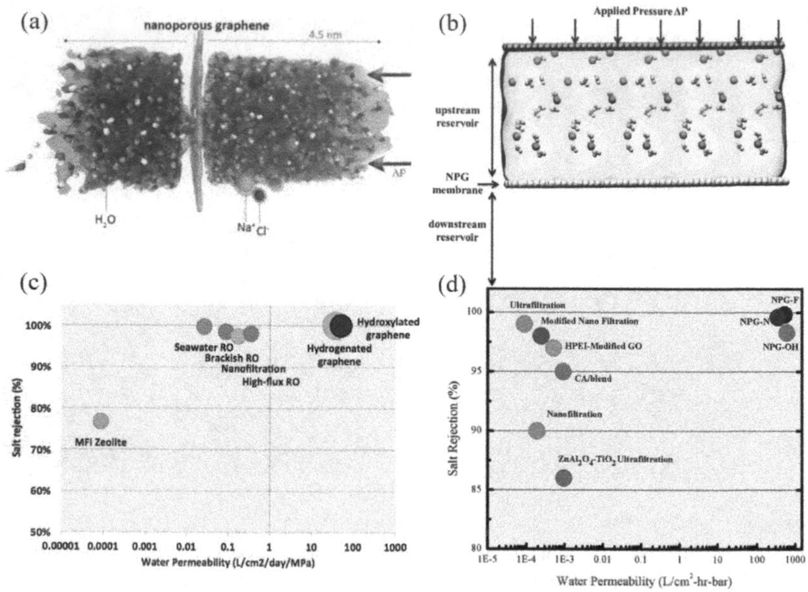

FIGURE 2.3
(a) Nanoporous graphene sheet can effectively filter NaCl salt from water. (b) Overall view of the computational system, water (light gray surface), metal ions (dark gray color), nitrate ions (white and light gray color), and pressure applied on piston (gray color). (c) Salt rejection versus water permeability for the nanoporous graphene and other separation materials. (d) Comparison of water permeability and heavy metal ions rejection of functionalized NPG with the existing various technologies reported in literature. (a,c. Reprinted with permission from Cohen-Tanugi, D., Grossman, J. C., *Nano Letters*, 12 : 3602–3608, 2012. Copyright 2012 American Chemical Society.)

based on MD simulations. The simulation results show that a nanopore containing four carbonyl groups favors the conduction of K^+ over Na^+ and a nanopore functionalized by four negatively charged carboxylate groups selectively binds Na^+ but selectivity transports K^+ over Na^+. The ion selectivity of the smaller diameter pores containing three carboxylate groups can be tuned by varying the intensity of the applied voltage bias. Under lower voltage bias, the three carboxylate groups of graphene nanopore transports ions in a single-file manner and approves Na^+ ions. When comparing carboxylated, aminated, and hydroxylated pores, Konatham et al. [52] concluded that hydroxylated pores may be optimal for nanoporous graphene separation, because they offer strong free energy barriers for ions passage at both low and moderate ionic strength (0.025 and 0.25M); at the same time, the selectivity of charged functionalized pores decreases significantly as ionic strength increases [52]. We have investigated the heavy metal ion separation performance using NPG membranes with variable pore size, pore chemistry, and applied hydrostatic pressure [55]. Compare the nitrogen, fluorine, and hydroxyl (–OH) functionalized graphene nanopores (see Figure 2.2c) on the ability to reject $Pb(NO_3)_2$, $Cu(NO_3)_2$, $Cd(NO_3)_2$, $Co(NO_3)_2$ and $Zn(NO_3)_2$ using water permeability, salt rejection, and free-energy profiles. The MD simulations are performed using 0.5 M concentration for each metal ion using their nitrate salts in an aqueous solution. A representative snapshot is shown in Figure 2.3b. First, an upstream reservoir is initially filled with a salt water of 0.5 M concentration. The second portion of the box is made up of an initially empty downstream reservoir. The physical rigid piston made of a single layer of graphene in the *xy* plane is positioned initially at end of the solution containing reservoir, and subsequently allowed to push the salt water towards the NPG membrane at a prescribed external force.

This study shows that the NPG functionalized with N (NPG-N) shows higher salt rejection with intermediate permeability compared to NPG functionalized with F (NPG-F) and OH (NPG-OH). NPG-OH shows higher water permeability with lower salt rejection compared to NPG-N and NPG-F. However, NPG-F shows lowest permeability compared to the other two NPGs considered in this study. The energy barrier for the water molecule is higher for NPG-OH followed by NPG-N and NPG-F, which shows that the water flux will be more for NPG-F than the other two NPGs considered in this study.

The water flux and PMF results obtained for water using MD simulations are also in agreement with the DFT calculations. Moreover, the permeability of the water is 4–5 orders more than the existing technologies as shown in Figure 2.3d. For the case of NPG-OH, we have observed very high permeability with ~90% salt rejection percentage compared to NPG-N and NPG-F. On the other hand, NPG-N has intermediate permeability with ~100% salt rejection. However, NPG-F shows the low permeability compared to the other two NPGs considered in this study. Overall, the enhanced water permeability and salt rejection using functionalized NPG can offer important advantages over existing technologies [56,57]. The findings thus suggested that ion selectivity can be optimized by varying the pore size, shape, and number of functional ligands attached to the nanopores present in the graphene membrane. While enhancing water flux may not dramatically improve the RO performance because of the thermodynamics limits associated with salt removal, such enhancement is often encouraging because it helps achieve high throughput desalination of water. Other functionalization schemes have also been shown to enable high rejection rate of heavy metal ions and transition-metal ions [55,58] and to more effectively exclude chloride and nitrate ions [52]. Subsequent to the prominent examination of graphene for desalination, numerous theoretical studies have exposed the potential superiority of graphene membranes to state-of-the-art polymer-based filtration [48,51,59].

2.2.2 Graphene Oxide

Graphene sheets can be modified with various functional groups as well as integrate with other classes of inorganic nanoparticles to change its physicochemical properties. The modified graphene sheets have exciting opportunities in various applications including desalination. The chemical modification or surface functionality of these graphene materials make significant changes in physicochemical properties and improve the permeation and selectivity of the resultant membrane [60–62]. In recent years, several attempts have been made in the synthesis, chemical modification, and surface functionalization graphene sheet with hydroxyl-, amine-, and carboxylate-terminated groups for the enhancement of separation of impurities from wastewater. The graphene monolayers can easily be functionalized by several modification methods to facilitate the physicochemical behavior and surface properties.

Over the last few years, graphene oxide (GO), considered as the oxidized graphene, is being modified with functional groups such as epoxy, hydroxyl, and carboxyl functional groups on the surface. The oxygen-containing functional groups on the GO surface have remarkable hydrophilic character, large surface area, and chemical reactivity, resulting in high adsorption capacity. The graphene oxide surfaces have high surface area value of ~2620 m^2/g, which indicates that graphene oxide nanosheets should have high sorption capacity in the preconcentration of heavy metal ions from large volumes of aqueous solutions [63]. The functional groups containing oxygen atoms have a lone electron pair and they can efficiently bind the metal ions to form a metal complex by sharing an electron pair. Recently published papers have shown that GO has impressive adsorption properties toward metal ions [63–67]. Yang et al. [68] showed that GO had an impressive absorption capacity for Cu^{2+}, which is approximately 10 times greater than that of activated carbon. The GO shows maximum sorption capacities towards Pb^{2+}, Cd^{2+}, Co^{2+}, Eu^{3+}, and U^{4+} ions in aqueous solutions [63,66,67,69]. The abundant oxygen-containing functional groups on the surfaces of graphene oxide nanosheets played an important role on Cd(II) and Co(II) sorption. The maximum sorption capacities of Cd(II) and Co(II) on graphene oxide nanosheets at pH 6.0 and T = 303 K are reported to be 106.3 and 68.2 mg/g, respectively, which are higher than any other reported values. The more important is oxygen-containing functional group on the surfaces of graphene oxide nanosheets, which makes the adjacent oxygen atoms available to bind the metal ions [63]. Graphene oxide nanosheets may be suitable materials for ex-situ environmental remediation of heavy metal ions [63]. Composites and derivatives of GO increase the adsorption capacity and enhance the selectivity toward metal ions [70–72]. Several studies also reported that GO can be suitable materials for the removal of metal ions from large volumes of aqueous solutions in environmental pollution cleanup [63,66,67]. Excellent adsorption of GO and strong tendency for creation of GO-metal complexes to precipitate open the path to the removal of pollutants from wastewater. Moreover, graphene and graphene-based materials have recently been applied in solid-phase extraction and solid-phase microextraction [73]. GO can be grafted to polymeric chains that have reactive species such as hydroxyls and amines, that is, poly (ethylene glycol), polylysine, polyallyalmine, and poly(vinyl alcohol).

These materials combine the properties of their parts; the polymeric part offers dispersibility in certain solvents, mechanical strengthening, and several morphological characteristics, whereas graphene contributes to the electrical conductivity, chemical reactivity, and reinforcement of the mechanical properties [74–76]. Recently, it has been reported that covalently grafted polymers on the GO is an effective material for the removal of metal ions from industrial waste. The polymer grafted GO surfaces show high adsorption capacities of Cr^{4+}, Cu^{2+}, Cd^{2+}, and Ni^{2+} in aqueous solutions [77,78]. Starburst polyamidoamine (PAMAM) dendrimers have demonstrated binding affinities for metal ions [79]. By tethering dendrimers to the solid support, promising novel adsorbents can possess the mechanical and thermal stability of the support, and the strong affinity and high loading capacities of dendrimers [79]. Dendrimers such as PAMAM are frequently studied because of their versatility in a variety of terminal groups, including carboxylic acids. One of the main mechanisms of adsorption is the coordination of metal cations via electrostatic binding [79]. Mamadou et al. [80] found that the Cu(II) binding capacities of the EDA core PAMAM dendrimers were much larger and more sensitive to solution pH than those of linear polymers with amino groups, and the metal ion laden dendrimers can be regenerated by decreasing the solution pH to 4.0, thus enabling the recovery of the bound Cu(II) ions and recycling of the dendrimers [81]. The Graphene oxide/polyamidoamine dendrimers (GO-PAMAMs) were prepared via a "grafting-from" strategy (see Figure 2.4a) by Yang Yuan et al. [82]. The total adsorption capacity of

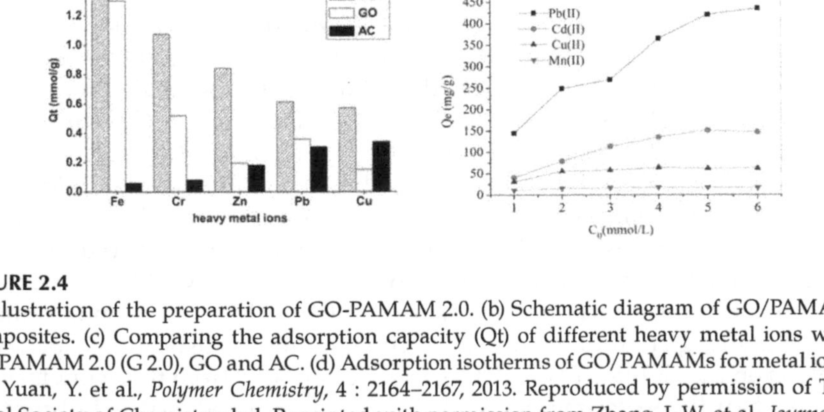

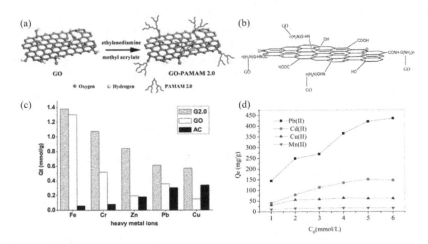

FIGURE 2.4
(a) Illustration of the preparation of GO-PAMAM 2.0. (b) Schematic diagram of GO/PAMAM Composites. (c) Comparing the adsorption capacity (Qt) of different heavy metal ions with GO-PAMAM 2.0 (G 2.0), GO and AC. (d) Adsorption isotherms of GO/PAMAMs for metal ions. (a,c. Yuan, Y. et al., *Polymer Chemistry*, 4 : 2164–2167, 2013. Reproduced by permission of The Royal Society of Chemistry; b,d. Reprinted with permission from Zhang, J.-W. et al., *Journal of Membrane Science*, 450 : 197–206, 2014. Copyright 2014 American Chemical Society.)

these heavy metal ions can reach 1.0007 mmol/g (see Figure 2.4c). Zhang et al. [81] prepared GO-PAMAMs via a "grafting to" method (see Figure 2.4b) and also compared them to the grafting-from strategy. The GO-PAMAMs were prepared more easily and present more efficient adsorption properties for heavy metal ions as shown in Figure 2.4d. The adsorption behavior of GO-PAMAMs for heavy metal ions in water solution was studied by changing the concentration of heavy metal ions, pH values, and temperature. The adsorption capacities of the GO/PAMAMs for the heavy metal ions were highly pH dependent [81].

Adsorption of GO/PAMAMs for Cu(II) or Mn(II) was a typical monomolecular layer. The maximum adsorption capacity of GO/PAMAMs for Pb(II) is determined to be 568.18 mg/g, which is higher than that for Cd(II), Cu(II), or Mn(II). Adsorption reaches the equilibrium state within 60 min. All of the above results indicate that GO/PAMAMs have excellent adsorption properties for removing heavy metal ions from wastewater. However, not much is known about the molecular interaction of ions with PAMAM dendrimers [81]. In our recent work [83], we focused towards understanding the molecular mechanisms involved in the adsorption of metal ions on the graphene, graphene oxide, and polyamidoamine (PAMAM) dendrimer with different terminal groups grafted on the graphene and graphene oxide surfaces (see Figure 2.5a). In particular, this presented a computational study of the Pb(II) ion adsorption behavior on the graphene and graphene oxide modified surfaces. In this work, using the base materials listed in Table 2.1, eight different surfaces are generated viz., graphene (GS), GS-PAMAM, GS-PAMAM-COO^-, GS-PAMAM-OH, graphene oxide (GO), GO-PAMAM, GO-PAMAM-COO^-, and GO-PAMAM-OH. The PAMAM dendrimer grafted on the graphene sheet considered in this work is of the G3 generation shown in Figure 2.5c. An initial system for a molecular dynamics simulation study is generated by taking the above optimized dendrimer grafted on the GO and GS surfaces (adsorbent) solvated in an aqueous concentrations of $Pb(NO_3)_2$ ionic salts. An illustration of such a system is shown in Figure 2.5b.

Adsorption of Pb^{2+} ions on the GO–PAMAM-COO^- surface is significantly more than the other surfaces for all five concentrations considered in this work. The highest monolayer adsorption capacities for Pb^{2+} ions on the GO–PAMAM-COO^-, GO–PAMAM-OH, GO–PAMAM, and GO surfaces are found to be 1523.1, 1224.6, 1100.3, 1043.93 mg/g, respectively (see Figure 2.5d). The monolayer adsorption capacities for Pb^{2+} ions on the GS–PAMAM-COO^-, GS–PAMAM-OH, GS–PAMAM, and GS surfaces are found to be 1050.1, 975.33, 862.4, 795.33 mg/g, respectively. Recent experimental work studied the removal of heavy metal ions from aqueous solutions using GO-PAMAM adsorbent.

The maximum adsorption capacity of GO-PAMAM for Pb^{2+} ion is found to be 568.18 mg/g at around 6 mmol/L. The maximum adsorption capacity of Pb^{2+} ion on the GO-PAMAM-COO^- surface calculated from this work is ~62% higher than the experimental work. It should be noted that pH of most of the wastewater streams is 6.0, whereas we have performed simulations for

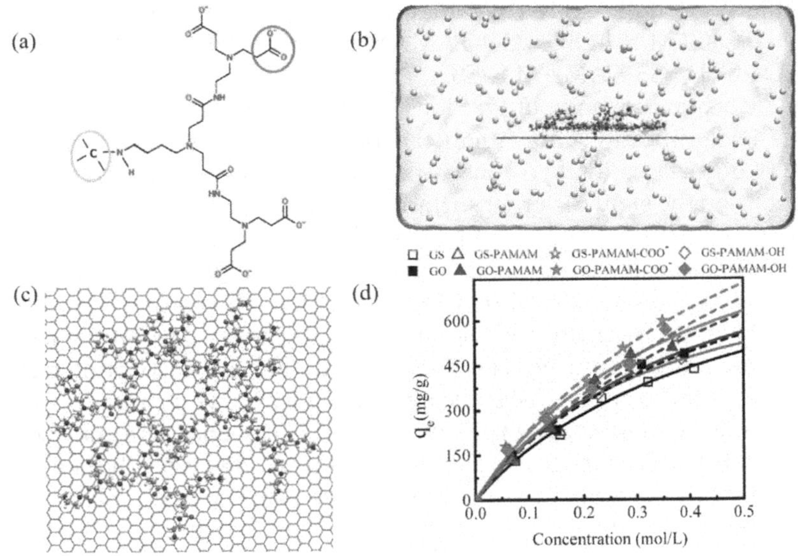

FIGURE 2.5
(a) Schematic diagram for a G0.5 PAMAM-COO^- grafted onto graphene sheet where dark gray circle group represents a carboxyl terminal group and light gray circle represents a carbon atom within the graphene sheet (GS). (b) Representative simulations system GO-PAMAM surface solvated in an ionic solution. Gray color represents the GS, light gray, dark gray, black and white color represents the carbon atoms of PAMAM dendrimer, oxygen atoms of functional groups and PAMAM, nitrogen atoms of PAMAM and hydrogen atoms of functional groups and PAMAM. Water represented as the light gray surface, nitrate ions represented as white and light gray, and metal ions are shown in light gray. (c) GO-PAMAM surface. (d) Langmuir adsorption isotherm showing the variation of total amount of metal ion adsorbed (q_e) with the concentration (mol/L) for adsorption of Pb^{2+} ions onto the eight different surfaces. Open symbols for different terminal groups (-NH_2, -COO^- and -OH) of dendrimer grafted to GS surface and filled symbols for different terminal groups (-NH2, -COO^- and -OH) of dendrimer grafted to GO surface. (Reprinted with permission from Kommu, A. et al., *The Journal of Physical Chemistry A*, 121 : 9320–9329, 2017. Copyright 2017 American Chemical Society.)

TABLE 2.1
Summary of the Materials Considered

Material	Description
GS	Graphene Sheet
GO	Graphene Oxide
PAMAM	Dendrimer with NH_2 terminal groups
PAMAM-COO^-	Dendrimer with COO^- terminal groups
PAMAM-OH	Dendrimer with OH terminal groups

the neutral system at a pH of 7. Nevertheless, qualitatively the experimental findings of the order of adsorption of Pb^{2+} metal ion on the different surfaces GO–PAMAM > GO > GS is akin to that seen in this work [81]. The maximum adsorption capacities of Pb^{2+} ion on different surfaces calculated from the Langmuir isotherm equation follow the order GO–PAMAM-COO^- > GO–PAMAM-OH > GO–PAMAM > GO > GS–PAMAM-COO^- > GS–PAMAM-OH > GS–PAMAM > GS.

The adsorption capacity of Pb^{2+} ions on the dendrimer grafted on the GO surface is significantly more than the bare GO, bare GS, and dendrimer grafted GS surfaces for five concentrations. The adsorption mechanism of metal ions on the eight different surfaces is discussed with the help of microscopic interactions between metal ion and solid surfaces. We also examined the self-diffusion coefficient and residence time of Pb^{2+} ions near the surfaces. Interestingly, we found that the interaction between the Pb^{2+} ion and dendrimer plays a significant role in enhancing their association with the dendrimer grafted surfaces. The results show that the adsorption capacity of the Pb^{2+} ion is improved significantly using carboxyl terminal groups of dendrimer grafted on graphene oxide surface. This section summarized and compared the difference between the graphene oxide and polymer grafted graphene oxide surfaces in removing various heavy metal ions present in wastewater.

2.3 Separation of Organic Compounds

In this section, we focus on the separation of organic compounds existing in various aqueous streams from multiple sources that are currently one of the most important environmental concerns. The organic solvents and dyes discharged by the industries are primary pollutants of water resources. It is necessary to remove these organic compounds from contaminated wastewater prior to its discharge to the environment to protect the aquatic lives and human beings. Thus, removal of such pollutants from contaminated water has become a subject of intense research [84,85]. Among these techniques, the adsorption method is considered to be one of the simplest and most attractive methods for separating organic pollutants from wastewater. The ideal adsorbent material should exhibit high gravimetric capacity, easy separation from cleaned water and easy cleaning for long-term cycling. The most common adsorbents are activated carbon [86], zeolites [87] and natural fibers [88], and other recent refined materials, including graphene capsules [89], collagen nanocomposites [90], polyurethane sponge [91], polyurethane and iron oxides composites [92], MnO_2 nanowires [93], and graphene hydrogels [94]. These adsorbents have been used for removal of organic solvents such as alcohols, aromatic compounds, and dyes from aqueous phase.

Ethanol is one of the common organic pollutants in various industries, which cause adverse effects on environment and human health such as oxidative damage of brain, liver, stomach etc. [95,96]. Therefore, it is necessary to separate ethanol from aqueous solution efficiently. Numerous materials have been studied as potential membranes such as zeolites [97,98], modified polydimethylsiloxane/polystyrene blended IPN pervaporation membranes [99] for the separation of ethanol-water mixtures [100]. Polymer or polymer-based hybrid membranes have also been widely used in ethanol/water separation owing to their relatively low cost and ease of processing [101]. Nevertheless, plasticization and swelling significantly decrease the performance of the polymer membranes and alter their permeability and selectivity [101]. Membranes based on zeolites such as NaA zeolites [98,102] show excellent performance for separation of water from ethanol compared to that based on polymers [103] etc. However, these processes have significant disadvantages, which have low efficiency, unsatisfactory regeneration, and cycling ability. Another class of porous materials named metal–organic frameworks (MOFs) have also been tested for the separation of alcohol/water mixtures, but their performances are not up to industrial benchmarks. Thus, development of new and advanced materials is necessary to overcome these principal drawbacks and to achieve the targets of high selectivity. In this direction, two-dimensional materials have attracted much recent attention because of their unique properties and high-surface area [103,104]. In recent years, the most prominent member of the family of layered materials has been graphene, which serves as a building block for few-layered graphene and graphite as well as for single and multi-walled carbon nanotubes [15]. Experiments are performed to study the adsorption behavior of ethanol-water within slit pores composed of graphene layers [105]. The hydrophobic nature of graphene surface induces preferential ethanol adsorption within the slit pore.

Most of the studies on ethanol-water mixture have focused on the performances of the adsorbents, especially in terms of the adsorbed number of molecules, without much of molecular insight into the separation behavior. In contrast, few works based on molecular dynamics simulations have focused on the properties of alcohol-water mixtures confined in nanopores. For example, molecular simulations studies have shown the competitive binding/adsorption between alcohols and water inside single walled carbon nanotubes (SWCNT) bundles/membranes [106]. Yang et al. used Monte Carlo and molecular dynamics techniques to study the preferential adsorption and diffusion of ethanol-water system on silicate crystal [102]. The structural and physiochemical behavior of alcohol-water mixture are different from that of pure components because of its hydrogen bonding effect. Several studies demonstrated that ethanol-water mixture could undergo phase separation under hydrophilic nanopores [107]. Zhao et al. [108] studied the effect of pore width and composition of ethanol-water mixtures confined within slit-shaped graphene nanopores using MD simulations (see Figure 2.6a). They found that ethanol molecules were preferentially adsorbed on the inner surface of a pore wall and formed an adsorbed ethanol layer. Essentially, the adsorbed alcohol layer can be equivalent to a new "interface" for the water molecules covering it. The diffusion coefficients for confined pure ethanol molecules are substantially less than that of bulk phase, indicating more

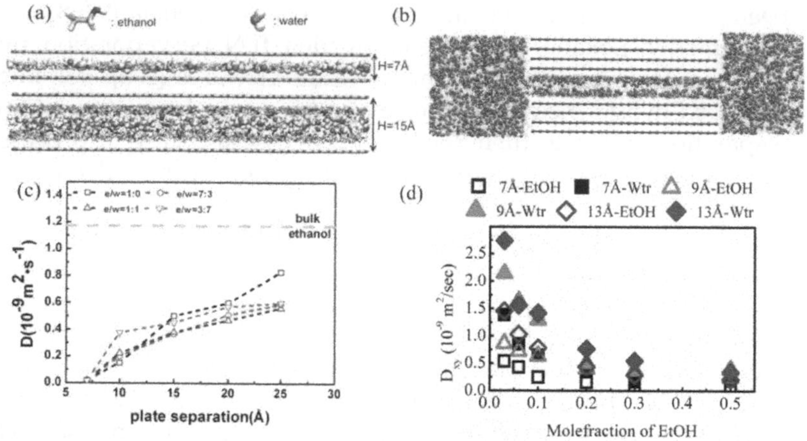

FIGURE 2.6
(a) Representative of simulation systems. The light gray walls represent the graphene sheets. (b) Representative simulation snapshot for the water/ethanol mixture in contact with the 13 Å graphene pore. Dark gray and white spheres represent oxygen, hydrogen atoms, and gray spheres represent carbon atoms of ethanol molecules, respectively. Light gray color represents water molecules. Gray color spheres represent the carbon atoms of graphene surface, respectively. (c) Diffusion coefficients of ethanol with various ethanol/water compositions confined in 7–25 Å graphene nanochannels. Diffusion coefficient of ethanol in bulk phase represented by light gray dash line. (d) Diffusion coefficient for water and ethanol molecules within three slit-shaped pores on the graphene (GS) surface as a function of ethanol mole fraction. (a,c. Reprinted with permission from Zhao, M., Yang, X. *The Journal of Physical Chemistry C*, 119 : 21664–21673, 2015. Copyright 2015 American Chemical Society; b,d. Reprinted with permission from Kommu, A., Singh, J. K., *The Journal of Physical Chemistry C*, 121 : 7867–7880, 2017. Copyright 2017 American Chemical Society.)

enhanced confinement effect as shown in Figure 2.6c. Phan et al. [107] have investigated the sorption, structure and dynamics of the ethanol-water mixtures confined in alumina pores. Structural and dynamical properties of ethanol and water under confinement as opposed to in the bulk to understand the transport mechanisms in the pores. To quantify the potential application of porous alumina pores as membrane materials for producing anhydrous ethanol. The investigation of the dynamical properties of the confined mixtures indicates that water diffuses through the narrow pores faster than ethanol. The combination of structural and dynamical results reported here for water–ethanol mixtures suggests that alumina-based porous materials could be used as permselective membranes for the removal of water from aqueous ethanol solutions. Furthermore, how the adsorption behavior of ethanol-water mixtures is different from that of graphene-based slit pores is not known. To this end, we have performed MD simulations to investigate the selectivity of ethanol-water in slit-shaped graphene pores [109]. We have reported the details concerning structural (i.e., density profiles, hydrogen bonding, and molecular orientation) and dynamical properties (i.e. self-diffusion coefficients and residence time) of ethanol-water confined in slit pores. In this work, the MD simulations are conducted for ethanol molecules dissolved in explicit water. Figure 2.6b presents a schematic picture of the simulation box containing slit pore made up of graphene surfaces (walls). In order to compare the properties of confined water and ethanol molecules within different pore widths of graphene sheets, we considered three pore sizes (7, 9 and 13 Å) under identical conditions. The suitability of the sheets for the separation of ethanol-water mixture is investigated by studying the adsorption and structural behavior of ethanol-water mixtures in slit pores with variable width (7 to 13 Å) using molecular dynamics simulations. The selectivity of ethanol is found to depend on the pore-width and nature of the pore walls. The selectivity of ethanol is highest for 9 Å pore and lowest for 7 Å pore. The investigation of the structural properties of the confined ethanol and water molecules within different pore widths are used to elucidate the adsorption behavior. Results from this study showed that the adsorption of ethanol molecules inside the graphene pore increases with increasing mole fractions of ethanol-water mixture. By comparing the selectivity of ethanol molecules within the different slit pores, 9 Å slit pore shows the highest efficiency of ethanol-water separation compared to the 7 and 13 Å pores. The combination of structural and dynamical properties results reported here for ethanol-water system suggests that pore size 9 Å of graphene surface are useful for the separation of ethanol from ethanol-water system (see Figure 2.6d). Based on the above results, it can be concluded that molecular sieving plays an important role in lower pores, whereas surface-fluid interaction is a governing factor for larger pore sizes. In a recent work, Joshi et al. [110] using MD simulations have confirmed that the graphene-based materials with well-defined pore sizes can be used for filtration and separation technologies for the extraction of valuable solutes from complex mixtures. The graphene capillary acts as molecular sieves, blocking all solutes with hydrated radii larger than the capillary size. Results show that the selectivity of ethanol decreases with increase in mole fraction of ethanol. The selectivity of ethanol is higher for 9 Å slit pore compared to 7 and 13 Å pores. The results confirm that the ethanol molecules easily permeate through the 9 Å slit pore compared to the 7 Å. However, the ethanol molecules allowed into the graphene pores obviously depend on the width of the pore. Therefore, the pore width is the crucial parameter to dictate the selectivity of ethanol. This suggests that pore 9 Å of graphene surface is promising for ethanol/water separation.

2.4 Potential of Mean Force (PMF)

Several different methods have been implemented for calculation of the free energy of various systems. Here we discuss the commonly employed technique, namely, umbrella sampling. Torrie and Valleau develop this method in 1977. The method has been one of the major approaches for performing simulations along predetermined reaction coordinates. It is readily applied to both molecular dynamics and Monte Carlo simulations. This procedure enables the calculation of the potential of mean force (PMF), which governs the elementary microscopic steps of metal ions/water permeation through the pore. The PMF simulation also been accepted as an effective tool in studying the microscopic mechanism of various adsorption processes. The energy barrier calculated from PMF profiles can shed further light on the energetics of metal ion rejection across pore. The method is described in detail in Chapter 1. To calculate the PMF $W(z)$, we choose a reaction coordinate z along which we perform umbrella sampling with harmonic biasing with harmonic biasing potentials. The umbrella sampling is performed with window size of $\Delta z = 0.05$ nm and the harmonic biasing potentials in each window is chosen to be $0.5K_z\,(z\text{-}z_0)^2$, where K_z is the harmonic force constant and z_0 is the location of the center of window i. The yielded umbrella histograms were then unbiased and combined using the weighted histogram analysis method [111] (WHAM) to form a complete PMF ($W(z)$) profile using a tolerance factor of 10^{-6}. He et al. [59] explored the ion selectivity of bioinspired graphene nanopores by calculating PMF, that is, the free energy landscape of the ions passing through the biological channels, which is shown in Figure 2.7a. Their PMF calculations revealed that Na^+ and K^+ pass through four carbonyl (4CO) and four negatively charged carboxylate (4COO) nanopores, respectively. The Na^+ encounter energy barrier of ~2.9 kJ/mol higher than the K^+ for the nanopore with four carbonyl groups, which mimics the KcsA selectivity filter, thus showing the selectively conduct of K^+ over Na^+. A nanopore functionalized with four negatively charged carboxylate groups, which mimics the selectivity filter of the NavAb Na^+ channel, selectively binds Na^+ but transports K^+ over Na^+, due to its stronger affinity (by 4.3 kJ/mol) for Na^+. Konatham et al. [52] studied the penetration of ion and water across the NPG nanopores by calculating potential mean force along the direction perpendicular to the NPG pore. The graphene nanopore functionalized with hydroxyl functional groups encounter large free energy barriers for Cl^- ions (19 kcal/mol) while passing through the membrane at both low and moderate feed concentrations (see Figure 2.7b). On the other hand, the PMF results obtained for pores functionalizing with COO^- groups indicate an effective free-energy barrier of ~11 kcal/mol for Cl^- ions as they attempt to diffuse across the pore at higher ionic strength. In the Figure 2.7c, we observe that PMF of Cl^- ions across the GS pore with varying diameters of 14.5, 10.5, and 7.5 Å nanopores. PMF profiles suggest that Cl^- ions can diffuse through the pristine 14.5 Å pore (black line) without the need to overcome significant free-energy barriers. The authors found that hydroxyl functionalized nanopore shows larger energy barrier for Cl^- passage at even higher salt concentrations, indicating that steric effects are mostly responsible for the salt rejection across the pore even in the presence of concentration polarization. Xue et al. [112] studied the water and ion rejection across the

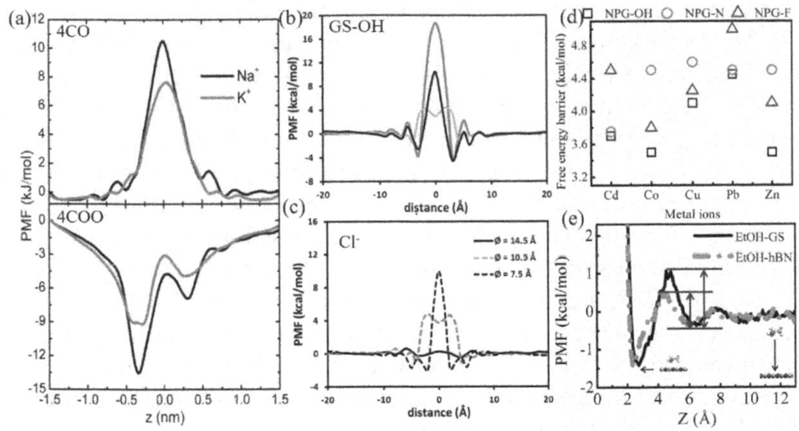

FIGURE 2.7
(a) One-dimensional PMF for Na^+ and K^+ traversing through (A) 4CO and (B) 4COO nanopores. (b) One-dimensional PMF along the direction perpendicular to the hydroxyl functionalized GS pores for Cl^- ion at 0.025 (light gray) and 0.25 M NaCl (black). The gray colored curve represents the result obtained on the pristine pore with of diameter 10.5 Å. (c) One-dimensional potential of mean force profiles obtained along the direction perpendicular to the GS pores for Cl^- ion in 0.25 M NaCl solutions. Pristine GS pores of diameter 14.5 (black continuous line), 10.5 (light gray dashed line), and 7.5 Å (black dashed line). (d) Comparing the free energy barrier of different metal ions passing through the hydroxyl, nitrogen, and fluorine functionalized NPG. (e) Potential of mean force (PMF) as a function of distance between the one ethanol molecule and graphene (GS) and hBN surfaces. The graphene sheet is at z = 0. (a. Reprinted with permission from He, Z. et al., *ACS Nano*, 7 : 10148–10157, 2013. Copyright 2013 American Chemical Society; b,c. Reprinted with permission from Konatham, D. et al., *Langmuir*, 29 : 11884–11897, 2013. Copyright 2013 American Chemical Society; d. Reprinted with permission from Kommu, A., Singh, J. K., *The Journal of Physical Chemistry C*, 121 : 7867–7880, 2017. Copyright 2017 American Chemical Society.)

α-graphyne nanopores by calculating the PMF free energies using umbrella sampling. The pristine graphyne nanopore shows an energy barrier for water 2 kcal/mol that exhibits higher water permeability through the graphene pores. The energy barriers for different ions (Na^+, Cl^-, Mg^{2+}, K^+, and Ca^{2+}) passing through the α-graphyne pores were also calculated. The PMF calculations indicated that the energy barrier opposing the passage for Na^+, K^+, Cl^- ions is around ~10 kcal/mol. The energy barrier for divalent ions conduction through the α-graphyne pores is around ~60 kcal/mol. Therefore, the pristine graphyne membranes can transport water molecules rapidly across the nanopore. In our recent work [55], we quantify the effect of functionalizing pores with nitrogen, fluorine, and hydroxyl (–OH) on the ability to reject $Pb(NO_3)_2$, $Cu(NO_3)_2$, $Cd(NO_3)_2$, $Co(NO_3)_2$, and $Zn(NO_3)_2$ using PMF profiles. The PMF profiles for ions and water were calculated across NPG-OH, NPG-N, and NPG-F in different ionic solutions. The PMF is evaluated in the direction perpendicular to the membrane pore using umbrella sampling. Based on the PMF profile of a metal ion and water molecule in ionic solutions for

NPG-OH pore, the energy barriers of water molecule passing across the pore are 3.0, 3.4, 3.4, 3.0, and 2.7 kcal mol^{-1} in the presence Cd^{2+}, Co^{2+}, Cu^{2+}, Pb^{2+}, and Zn^{2+} ion, respectively. The energy barrier values for different metal ions at center of the pore are 3.7, 3.5, 4.5, 4.5, and 3.5 kcal mol^{-1} for Cd^{2+}, Co^{2+}, Cu^{2+}, Pb^{2+}, and Zn^{2+}, respectively. The higher the energy barrier is, the higher is the resistance to pass through the NPG pore. Hence, increase/decrease in the energy barrier should increase/decrease the salt rejection ability of the NPG pore (Figure 2.7d). The PMF profiles for ions and water molecule across NPG-N, the energy barriers for the water molecule at the center of the pore (0 Å) in Cd^{2+}, Co^{2+}, Cu^{2+}, Pb^{2+}, and Zn^{2+} ionic solutions are 4.1, 3.1, 3.1, 3.5, and 3.0 kcal mol^{-1}, respectively. On the other hand, the energy barrier at the pore, associated with the ion movement in water Cd^{2+}, Co^{2+}, Cu^{2+}, Pb^{2+}, and Zn^{2+} ions is 2.0, 4.5, 4.6, 3.7, and 4.5 kcal mol^{-1}, respectively. The above analysis shows that for NPG-N the Co^{2+}, Cu^{2+}, and Zn^{2+} ion display higher energy barrier in comparison to that seen for other metal ions in the water. The behavior, position of energy barrier, and energy minima observed in PMF profiles of water and ions are qualitatively akin to that seen for NPG-N and NPG-OH. However, the order of increasing energy barrier for ions and water is different. For ions, the increasing order for the energy barrier is Pb^{2+}, Cd^{2+}, Co^{2+}, Cu^{2+}, and Zn^{2+}. In the case of water, the increasing order for energy barriers is Pb^{2+}, Zn^{2+}, Cd^{2+}, Co^{2+}, and Cu^{2+}. In general, it is evident from the above that the energy barrier for water molecules is lower than that experienced by metal ions. Hence, water molecules pass through the pore with relative ease. In particular, the water flux is lowest in the presence of Pb^{2+} ionic solution, which is in contrast to the behavior seen in NPG-N and NPG-OH. This behavior corroborates the PMF characteristics observed in different functionalized NPGs.

The PMF simulations have also been accepted as an effective tool in studying the microscopic mechanism of various adsorption processes. We have recently explored the underlying mechanism of the demonstrated adsorption and selectivity of ethanol and water molecules on graphene and hBN surfaces. Figure 2.7e shows the potential of mean force (PMF) based free energies for water and ethanol molecules. The separation of ethanol from water within pores is primarily because of the free-energy difference among the ethanol and water in pores. Thus, the PMF calculation can provide an insight into the characteristics of water and ethanol within hBN and graphene pores. PMF method has been widely used as an effective tool in studying microscopic mechanism of various adsorption processes. For example, Sun et al. [113] explored the adsorption behavior of 13 different peptides on a hydrophobic surface using PMF. Kerisit et al. [114] studied the adsorption of solvent and impurity ions on a mineral surface using PMF. Xu et al. [115] explored the adsorption free energy of single surfactant on solid surface with the PMF profiles, which has been successfully used to study the macromolecule transfer in complex interfacial regions. This study revealed the microscopic interaction mechanisms determined for the solute adsorption on solid surfaces. Free energy profiles of metal ions and surfactants adsorbed on the surface are correlated with solvent density. We have also calculated PMF as a function of distance between one molecule (water or ethanol) and the surface (graphene and hBN) using umbrella sampling [109]. In Figure 2.7e, two energy barriers are 0.34 and 1.5 kcal/mol, corresponding to the transfer of one ethanol molecule from the bulk to the graphene surface. Similarly, two energy barriers, 2.40 and 0.40 kcal/mol, are observed for the case of ethanol molecule transfer from the graphene surface to the bulk. In the case of hBN surface, two energy barriers are 0.37 and 1.00 kcal/mol, for one ethanol molecule transfer from the bulk to the hBN surface. The corresponding energy barriers for the reverse process, ethanol molecule transfer from the hBN surface to the bulk, are 2.00 and 0.40 kcal/mol. Each free energy barrier represents a distinct adsorption/desorption process step. The lower energy barrier represents that ethanol molecule can easily adsorb on the hBN surface. This indicates that the ethanol is preferentially adsorbed on a hBN surface compared to a graphene surface.

2.5 Summary

In this chapter, we described various graphene-based materials that are used as adsorbent materials for removal of various aquatic pollutants, such as divalent metal ions and other organic pollutants. It is evident that graphene-based materials are promising alternatives to activated carbon and other adsorbent materials that are currently being considered for wastewater treatment. The simplicity, flexibility, and high adsorption capacity of nanostructured materials make it more attractive for the selective removal of toxic pollutants from industrial wastewaters. Therefore, these materials have potential to be one of the most versatile and reliable materials for future wastewater treatment applications. The effective use and experimental demonstration of these materials for the removal of organic and inorganic pollutants from wastewater still remain a challenge. We believe that future work will prove essential in bridging the gap between the computational understanding and experimental progress. However, more work is still required before these membranes become commercially viable, including fundamental understanding of transport processes, cost reduction, and pore size optimization. Further, the prospects of reduced capital and operating costs should encourage the development of these new materials.

Acknowledgments

This work is funded by BRNS, the department of atomic energy, India.

References

1. Ali, I., Gupta, V. K. (2007) Advances in Water Treatment by Adsorption Technology. *Nature Protocols* 1 : 2661–2667.
2. Gupta, V. K., Carrott, P. J. M., Ribeiro Carrott, M. M. L., Suhas. (2009) Low-Cost Adsorbents: Growing Approach to Wastewater Treatment—A Review. *Critical Reviews in Environmental Science and Technology* 39 : 783–842.
3. Hultberg, M., Isaksson, A., Andersson, A., Hultberg, B. (2007) Traces of Copper Ions Deplete Glutathione in Human Hepatoma Cell Cultures with Low Cysteine Content. *Chemico-Biological Interactions* 167 : 56–62.
4. Shen, X., Lee, K., König, R. (2001) Effects of Heavy Metal Ions on Resting and Antigen-Activated CD4+ T Cells. *Toxicology* 169 : 67–80.
5. Mubarak, N. M., Sahu, J. N., Abdullah, E. C., Jayakumar, N. S. (2014) Removal of Heavy Metals from Wastewater Using Carbon Nanotubes. *Separation & Purification Reviews* 43 : 311–338.
6. Gao, H.-W., Chen, F.-F., Chen, L., Zeng, T., Pan, L.-T., Li, J.-H., Luo, H.-F. (2007) A Novel Detection Approach Based on Chromophore-Decolorizing with Free Radical and Application to Photometric Determination of Copper with Acid Chrome Dark Blue. *Analytica Chimica Acta* 587 : 52–59.
7. Zhao, X.-T., Zeng, T., Li, X.-Y., Hu, Z. J., Gao, H.-W., Xie, Z. (2012) Modeling and Mechanism of the Adsorption of Copper Ion onto Natural Bamboo Sawdust. *Carbohydrate Polymers* 89 : 185–192.
8. Korn, M. d. G. A., de Andrade, J. B., de Jesus, D. S., Lemos, V. A., Bandeira, M. L. S. F., dos Santos, W. N. L., Bezerra, M. A., Amorim, F. A. C., Souza, A. S., Ferreira, S. L. C. (2006) Separation and Preconcentration Procedures for the Determination of Lead Using Spectrometric Techniques: A Review. *Talanta* 69 : 16–24.
9. Zaichick, S., Zaichick, V., Karandashev, V., Nosenko, S. (2011) Accumulation of Rare Earth Elements in Human Bone Within the Lifespan. *Metallomics* 3 : 186–194.
10. Porru, S., Placidi, D., Quarta, C., Sabbioni, E., Pietra, R., Fortaner, S. (2001) The Potencial Role of Rare Earths in the Pathogenesis of Interstitial Lung Disease: A Case Report of Movie Projectionist as Investigated by Neutron Activation Analysis. *Journal of Trace Elements in Medicine and Biology* 14 : 232–236.
11. Zhang, H., Feng, J., Zhu, W., Liu, C., Xu, S., Shao, P., Wu, D., Yang, W., Gu, J. (2000) Chronic Toxicity of Rare-Earth Elements on Human Beings. *Biological Trace Element Research* 73 : 1–17.
12. Doulati Ardejani, F., Badii, K., Yousefi Limaee, N., Mahmoodi, N. M., Arami, M., Shafaei, S. Z., Mirhabibi, A. R. (2007) Numerical Modelling and Laboratory Studies on the Removal of Direct Red 23 and Direct Red 80 Dyes from Textile Effluents Using Orange Peel, a Low-Cost Adsorbent. *Dyes and Pigments* 73 : 178–185.
13. Qadir, D., Mukhtar, H., Keong, L. K. (2017) Mixed Matrix Membranes for Water Purification Applications. *Separation & Purification Reviews* 46 : 62–80.
14. Ong, C. S., Goh, P. S., Lau, W. J., Misdan, N., Ismail, A. F. (2016) Nanomaterials for Biofouling and Scaling Mitigation of Thin Film Composite Membrane: A Review. *Desalination* 393 : 2–15.
15. Iijima, S. (1991) Helical Microtubules of Graphitic Carbon. *Nature* 354 : 56–58.
16. Dresselhaus, M. S., Dresselhaus, G., Saito, R. (1995) Physics of Carbon Nanotubes. *Carbon* 33 : 883–891.

17. Niyogi, S., Hamon, M. A., Hu, H., Zhao, B., Bhowmik, P., Sen, R., Itkis, M. E., Haddon, R. C. (2002) Chemistry of Single-Walled Carbon Nanotubes. *Accounts of Chemical Research* 35 : 1105–1113.
18. Ugarte, D., Châtelain, A., de Heer, W. A. (1996) Nanocapillarity and Chemistry in Carbon Nanotubes. *Science* 274 : 1897–1899.
19. Bahr, J. L., Tour, J. M. (2002) Covalent Chemistry of Single-Wall Carbon Nanotubes. *Journal of Materials Chemistry* 12 : 1952–1958.
20. Humplik, T., Lee, J., O'Hern, S. C., Fellman, B. A., Baig, M. A., Hassan, S. F., Atieh, M. A., Rahman, F., Laoui, T., Karnik, R., Wang, E. N. (2011) Nanostructured Materials for Water Desalination. *Nanotechnology* 22 : 292001.
21. Cohen-Tanugi, D., McGovern, R. K., Dave, S. H., Lienhard, J. H., Grossman, J. C. (2014) Quantifying the Potential of Ultra-Permeable Membranes for Water Desalination. *Energy & Environmental Science* 7, 1134–1141.
22. Fane, A. G., Wang, R., Hu, M. X. (2015) Synthetic Membranes for Water Purification: Status and Future. *Angewandte Chemie International Edition* 54 : 3368–3386.
23. Mahmoud, K. A., Mansoor, B., Mansour, A., Khraisheh, M. (2015) Functional Graphene Nanosheets: The Next Generation Membranes for Water Desalination. *Desalination* 356 : 208–225.
24. Lau, W. J., Ismail, A. F., Misdan, N., Kassim, M. A. (2012) A Recent Progress in Thin Film Composite Membrane: A Review. *Desalination* 287 : 190–199.
25. Geise, G. M., Paul, D. R., Freeman, B. D. (2014) Fundamental Water and Salt Transport Properties of Polymeric Materials. *Progress in Polymer Science* 39 : 1–42.
26. Dulebohn, J., Ahmadiannamini, P., Wang, T., Kim, S.-S., Pinnavaia, T. J., Tarabara, V. V. (2014) Polymer Mesocomposites: Ultrafiltration Membrane Materials with Enhanced Permeability, Selectivity and Fouling Resistance. *Journal of Membrane Science* 453 : 478–488.
27. Dong, H., Zhao, L., Zhang, L., Chen, H., Gao, C., Winston Ho, W. S. (2015) High-Flux Reverse Osmosis Membranes Incorporated with NaY Zeolite Nanoparticles for Brackish Water Desalination. *Journal of Membrane Science* 476 : 373–383.
28. Safarpour, M., Khataee, A., Vatanpour, V. (2015) Thin Film Nanocomposite Reverse Osmosis Membrane Modified by Reduced Graphene Oxide/TiO2 with Improved Desalination Performance. *Journal of Membrane Science* 489 : 43–54.
29. Zhang, J.-W., Fang, H., Wang, J.-W., Hao, L.-Y., Xu, X., Chen, C.-S. (2014) Preparation and Characterization of Silicon Nitride Hollow Fiber Membranes for Seawater Desalination. *Journal of Membrane Science* 450 : 197–206.
30. Garofalo, A., Donato, L., Drioli, E., Criscuoli, A., Carnevale, M. C., Alharbi, O., Aljlil, S. A., Algieri, C. (2014) Supported MFI Zeolite Membranes by Cross Flow Filtration for Water Treatment. *Separation and Purification Technology* 137 : 28–35.
31. Kujawa, J., Cerneaux, S., Koter, S., Kujawski, W. (2014) Highly Efficient Hydrophobic Titania Ceramic Membranes for Water Desalination. *ACS Applied Materials & Interfaces* 6 : 14223–14230.
32. Li, X., Chou, S., Wang, R., Shi, L., Fang, W., Chaitra, G., Tang, C. Y., Torres, J., Hu, X., Fane, A. G. (2015) Nature Gives the Best Solution for Desalination: Aquaporin-Based Hollow Fiber Composite Membrane with Superior Performance. *Journal of Membrane Science* 494 : 68–77.
33. Tang, C. Y., Zhao, Y., Wang, R., Hélix-Nielsen, C., Fane, A. G. (2013) Desalination by Biomimetic Aquaporin Membranes: Review of Status and Prospects. *Desalination* 308 : 34–40.

34. Pendergast, M. M., Hoek, E. M. V. (2011) A Review of Water Treatment Membrane Nanotechnologies. *Energy & Environmental Science* 4 : 1946–1971.
35. Fornasiero, F., In, J. B., Kim, S., Park, H. G., Wang, Y., Grigoropoulos, C. P., Noy, A., Bakajin, O. (2010) pH-Tunable Ion Selectivity in Carbon Nanotube Pores. *Langmuir* 26 : 14848–14853.
36. Hu, Z., Chen, Y., Jiang, J. (2011) Zeolitic Imidazolate Framework-8 as a Reverse Osmosis Membrane for Water Desalination: Insight from Molecular Simulation. *The Journal of Chemical Physics* 134 : 134705.
37. Cho, C. H., Oh, K. Y., Kim, S. K., Yeo, J. G., Sharma, P. (2011) Pervaporative Seawater Desalination Using NaA Zeolite Membrane: Mechanisms of High Water Flux and High Salt Rejection. *Journal of Membrane Science* 371 : 226–238.
38. Lai, L., Shao, J., Ge, Q., Wang, Z., Yan, Y. (2012) The Preparation of Zeolite NaA Membranes on the Inner Surface of Hollow Fiber Supports. *Journal of Membrane Science* 409–410 : 318–328.
39. Dervin, S., Dionysiou, D. D., Pillai, S. C. (2016) 2D Nanostructures for Water Purification: Graphene and Beyond. *Nanoscale* 8 : 15115–15131.
40. Geim, A. K. (2009) Graphene: Status and Prospects. *Science* 324, 1530.
41. Allen, M. J., Tung, V. C., Kaner, R. B. (2010) Honeycomb Carbon: A Review of Graphene. *Chemical Reviews* 110 : 132–145.
42. Wang, H., Yuan, X., Wu, Y., Huang, H., Peng, X., Zeng, G., Zhong, H., Liang, J., Ren, M. (2013) Graphene-Based Materials: Fabrication, Characterization and Application for the Decontamination of Wastewater and Wastegas and Hydrogen Storage/Generation. *Advances in Colloid and Interface Science* 195–196 : 19–40.
43. Geim, A. K., Novoselov, K. S. (2007) The Rise of Graphene. *Nature Materials* 6, 183–191.
44. Chowdhury, S., Balasubramanian, R. (2014) Recent Advances in the Use of Graphene-Family Nanoadsorbents for Removal of Toxic Pollutants from Wastewater. *Advances in Colloid and Interface Science* 204 : 35–56.
45. Aghigh, A., Alizadeh, V., Wong, H. Y., Islam, M. S., Amin, N., Zaman, M. (2015) Recent Advances in Utilization of Graphene for Filtration and Desalination of Water: A Review. *Desalination* 365 : 389–397.
46. Bunch, J. S., Verbridge, S. S., Alden, J. S., van der Zande, A. M., Parpia, J. M., Craighead, H. G., McEuen, P. L. (2008) Impermeable Atomic Membranes from Graphene Sheets. *Nano Letters* 8 : 2458–2462.
47. Jiang, D.-e., Cooper, V. R., Dai, S. (2009) Porous Graphene as the Ultimate Membrane for Gas Separation. *Nano Letters* 9 : 4019–4024.
48. Cohen-Tanugi, D., Grossman, J. C. (2012) Water Desalination across Nanoporous Graphene. *Nano Letters* 12 : 3602–3608.
49. Berry, V., (2013) Impermeability of Graphene and Its Applications. *Carbon* 62, 1–10.
50. Suk, M. E., Aluru, N. R. (2010) Water Transport through Ultrathin Graphene. *The Journal of Physical Chemistry Letters* 1 : 1590–1594.
51. Sint, K., Wang, B., Král, P. (2008) Selective Ion Passage through Functionalized Graphene Nanopores. *Journal of the American Chemical Society* 130 : 16448–16449.
52. Konatham, D., Yu, J., Ho, T. A., Striolo, A. (2013) Simulation Insights for Graphene-Based Water Desalination Membranes. *Langmuir* 29 : 11884–11897.
53. Gai, J.-G., Gong, X.-L., Wang, W.-W., Zhang, X., Kang, W.-L. (2014) An Ultrafast Water Transport Forward Osmosis Membrane: Porous Graphene. *Journal of Materials Chemistry A* 2 : 4023–4028.

54. Gai, J.-G., Gong, X.-L. (2014) Zero Internal Concentration Polarization FO Membrane: Functionalized Graphene. *Journal of Materials Chemistry A* 2 : 425–429.
55. Kommu, A., Namsani, S., Singh, J. K. (2016) Removal of Heavy Metal Ions Using Functionalized Graphene Membranes: A Molecular Dynamics Study. *RSC Advances* 6 : 63190–63199.
56. Fu, F., Wang, Q. (2011) Removal of Heavy Metal Ions from Wastewaters: A Review. *Journal of Environmental Management* 92 : 407–418.
57. Barakat, M. A. (2011) New Trends in Removing Heavy Metals from Industrial Wastewater. *Arabian Journal of Chemistry* 4 : 361–377.
58. Sun, P., Liu, H., Wang, K., Zhong, M., Wu, D., Zhu, H. (2014) Selective Ion Transport through Functionalized Graphene Membranes Based on Delicate Ion–Graphene Interactions. *The Journal of Physical Chemistry C* 118 : 19396–19401.
59. He, Z., Zhou, J., Lu, X., Corry, B. (2013) Bioinspired Graphene Nanopores with Voltage-Tunable Ion Selectivity for Na+ and K+. *ACS Nano* 7 : 10148–10157.
60. Liu, G., Jin, W., Xu, N. (2015) Graphene-Based Membranes. *Chemical Society Reviews* 44 : 5016–5030.
61. You, Y., Sahajwalla, V., Yoshimura, M., Joshi, R. K. (2016) Graphene and Graphene Oxide for Desalination. *Nanoscale* 8 : 117–119.
62. Chang, H., Wu, H. (2013) Graphene-Based Nanocomposites: Preparation, Functionalization, and Energy and Environmental Applications. *Energy & Environmental Science* 6 : 3483–3507.
63. Zhao, G., Li, J., Ren, X., Chen, C., Wang, X. (2011) Few-Layered Graphene Oxide Nanosheets as Superior Sorbents for Heavy Metal Ion Pollution Management. *Environmental Science & Technology* 45 : 10454–10462.
64. Zhao, G., Ren, X., Gao, X., Tan, X., Li, J., Chen, C., Huang, Y., Wang, X. (2011) Removal of Pb(ii) Ions from Aqueous Solutions on Few-Layered Graphene Oxide Nanosheets. *Dalton Transactions* 40 : 10945–10952.
65. Sitko, R., Turek, E., Zawisza, B., Malicka, E., Talik, E., Heimann, J., Gagor, A., Feist, B., Wrzalik, R. (2013) Adsorption of Divalent Metal Ions from Aqueous Solutions using Graphene Oxide. *Dalton Transactions* 42 : 5682–5689.
66. Sun, Y., Wang, Q., Chen, C., Tan, X., Wang, X. (2012) Interaction between Eu(III) and Graphene Oxide Nanosheets Investigated by Batch and Extended X-ray Absorption Fine Structure Spectroscopy and by Modeling Techniques. *Environmental Science & Technology* 46 : 6020–6027.
67. Zhao, G., Wen, T., Yang, X., Yang, S., Liao, J., Hu, J., Shao, D., Wang, X. (2012) Preconcentration of U(vi) Ions on Few-Layered Graphene Oxide Nanosheets from Aqueous Solutions. *Dalton Transactions* 41 : 6182–6188.
68. Yang, S.-T., Chang, Y., Wang, H., Liu, G., Chen, S., Wang, Y., Liu, Y., Cao, A. (2010) Folding/Aggregation of Graphene Oxide and Its Application in Cu2+ Removal. *Journal of Colloid and Interface Science* 351 : 122–127.
69. Sitko, R., Zawisza, B., Malicka, E. (2012) Modification of Carbon Nanotubes for Preconcentration, Separation and Determination of Trace-Metal Ions. *TrAC Trends in Analytical Chemistry* 37 : 22–31.
70. Madadrang, C. J., Kim, H. Y., Gao, G., Wang, N., Zhu, J., Feng, H., Gorring, M., Kasner, M. L., Hou, S. (2012) Adsorption Behavior of EDTA-Graphene Oxide for Pb (II) Removal. *ACS Applied Materials & Interfaces* 4 : 1186–1193.

71. Sitko, R., Janik, P., Feist, B., Talik, E., Gagor, A. (2014) Suspended Aminosilanized Graphene Oxide Nanosheets for Selective Preconcentration of Lead Ions and Ultrasensitive Determination by Electrothermal Atomic Absorption Spectrometry. *ACS Applied Materials & Interfaces* 6 : 20144–20153.
72. Sitko, R., Janik, P., Zawisza, B., Talik, E., Margui, E., Queralt, I. (2015) Green Approach for Ultratrace Determination of Divalent Metal Ions and Arsenic Species Using Total-Reflection X-ray Fluorescence Spectrometry and Mercapto-Modified Graphene Oxide Nanosheets as a Novel Adsorbent. *Analytical Chemistry* 87 : 3535–3542.
73. Liu, Q., Shi, J., Jiang, G. (2012) Application of Graphene in Analytical Sample Preparation. *TrAC Trends in Analytical Chemistry* 37 : 1–11.
74. Lin, Y., Jin, J., Song, M. (2011) Preparation and Characterisation of Covalent Polymer Functionalized Graphene Oxide. *Journal of Materials Chemistry* 21 : 3455–3461.
75. Goncalves, G., Marques, P. A. A. P., Barros-Timmons, A., Bdkin, I., Singh, M. K., Emami, N., Gracio, J. (2010) Graphene Oxide Modified with PMMA via ATRP as a Reinforcement Filler. *Journal of Materials Chemistry* 20 : 9927–9934.
76. Ramanathan, T., Abdala, A. A., Stankovich, S., Dikin, D. A., Herrera-Alonso, M., Piner, R. D., Adamson, D. H., Schniepp, H. C., Chen, X., Ruoff, R. S., Nguyen, S. T., Aksay, I. A., Prud'Homme, R. K., Brinson, L. C. (2008) Functionalized Graphene Sheets for Polymer Nanocomposites. *Nature Nanotechnology* 3 : 327–331.
77. Tan, P., Sun, J., Hu, Y., Fang, Z., Bi, Q., Chen, Y., Cheng, J. (2015) Adsorption of Cu2+, Cd2+ and Ni2+ from Aqueous Single Metal Solutions on Graphene Oxide Membranes. *Journal of Hazardous Materials* 297 : 251–260.
78. Mukherjee, R., Bhunia, P., De, S. (2016) Impact of Graphene Oxide on Removal of Heavy Metals Using Mixed Matrix Membrane. *Chemical Engineering Journal* 292 : 284–297.
79. Chong, L., Dutt, M. (2015) Design of PAMAM-COO Dendron-Grafted Surfaces to Promote Pb(II) Ion Adsorption. *Physical Chemistry Chemical Physics* 17 : 10615–10623.
80. Diallo, M. S., Christie, S., Swaminathan, P., Johnson, J. H., Goddard, W. A. (2005) Dendrimer Enhanced Ultrafiltration. 1. Recovery of Cu(II) from Aqueous Solutions Using PAMAM Dendrimers with Ethylene Diamine Core and Terminal NH2 Groups. *Environmental Science & Technology* 39 : 1366–1377.
81. Zhang, F., Wang, B., He, S., Man, R. (2014) Preparation of Graphene-Oxide/Polyamidoamine Dendrimers and Their Adsorption Properties toward Some Heavy Metal Ions. *Journal of Chemical & Engineering Data* 59 : 1719–1726.
82. Yuan, Y., Zhang, G., Li, Y., Zhang, G., Zhang, F., Fan, X. (2013) Poly(amidoamine) Modified Graphene Oxide as an Efficient Adsorbent for Heavy Metal Ions. *Polymer Chemistry* 4 : 2164–2167.
83. Kommu, A., Velachi, V., Cordeiro, M. N. S., Singh, J. K. (2017) Removal of Pb(II) Ion Using PAMAM Dendrimer Grafted Graphene and Graphene Oxide Surfaces: A Molecular Dynamics Study. *The Journal of Physical Chemistry A* 121 : 9320–9329.
84. Shannon, M. A., Bohn, P. W., Elimelech, M., Georgiadis, J. G., Mariñas, B. J., Mayes, A. M. (2008) Science and Technology for Water Purification in the Coming Deecades. *Nature* 452 : 301–310.

85. Zhang, F., Zhang, W. B., Shi, Z., Wang, D., Jin, J., Jiang, L. (2013) Nanowire-Haired Inorganic Membranes with Superhydrophilicity and Underwater Ultralow Adhesive Superoleophobicity for High-Efficiency Oil/Water Separation. *Advanced Materials* 25 : 4192–4198.
86. Bayat, A., Aghamiri, S. F., Moheb, A., Vakili-Nezhaad, G. R. (2005) Oil Spill Cleanup from Sea Water by Sorbent Materials. *Chemical Engineering & Technology* 28 : 1525–1528.
87. Adebajo, M. O., Frost, R. L., Kloprogge, J. T., Carmody, O., Kokot, S. (2003) Porous Materials for Oil Spill Cleanup: A Review of Synthesis and Absorbing Properties. *Journal of Porous Materials* 10 : 159–170.
88. Deschamps, G., Caruel, H., Borredon, M.-E., Bonnin, C., Vignoles, C. (2003) Oil Removal from Water by Selective Sorption on Hydrophobic Cotton Fibers. 1. Study of Sorption Properties and Comparison with Other Cotton Fiber-Based Sorbents. *Environmental Science & Technology* 37: 1013–1015.
89. Sohn, K., Joo Na, Y., Chang, H., Roh, K.-M., Dong Jang, H., Huang, J. (2012) Oil Absorbing Graphene Capsules by Capillary Molding. *Chemical Communications* 48 : 5968–5970.
90. Thanikaivelan, P., Narayanan, N. T., Pradhan, B. K., Ajayan, P. M. (2012) Collagen Based Magnetic Nanocomposites for Oil Removal Applications. *Scientific Reports* 2 : 230.
91. Zhu, Q., Pan, Q., Liu, F. (2011) Facile Removal and Collection of Oils from Water Surfaces through Superhydrophobic and Superoleophilic Sponges. *The Journal of Physical Chemistry C* 115 : 17464–17470.
92. Calcagnile, P., Fragouli, D., Bayer, I. S., Anyfantis, G. C., Martiradonna, L., Cozzoli, P. D., Cingolani, R., Athanassiou, A. (2012) Magnetically Driven Floating Foams for the Removal of Oil Contaminants from Water. *ACS Nano* 6 : 5413–5419.
93. Yuan, J., Liu, X., Akbulut, O., Hu, J., Suib, S. L., Kong, J., Stellacci, F. (2008) Superwetting Nanowire Membranes for Selective Absorption. *Nature Nanotechnology* 3 : 332–336.
94. Cong, H.-P., Ren, X.-C., Wang, P., Yu, S.-H. (2012) Macroscopic Multifunctional Graphene-Based Hydrogels and Aerogels by a Metal Ion Induced Self-Assembly Process. *ACS Nano* 6 : 2693–2703.
95. Bondy, S. C., Guo, S. X. (1994) Effect of Ethanol Treatment on Indices of Cumulative Oxidative Stress. *European Journal of Pharmacology: Environmental Toxicology and Pharmacology* 270 : 349–355.
96. Faisal, M., Khan, S. B., Rahman, M. M., Jamal, A., Umar, A. (2011) Ethanol Chemi-sensor: Evaluation of Structural, Optical and Sensing Properties of CuO Nanosheets. *Materials Letters* 65 : 1400–1403.
97. Zhang, K., Lively, R. P., Noel, J. D., Dose, M. E., McCool, B. A., Chance, R. R., Koros, W. J. (2012) Adsorption of Water and Ethanol in MFI-Type Zeolites. *Langmuir* 28 : 8664–8673.
98. Simo, M., Sivashanmugam, S., Brown, C. J., Hlavacek, V. (2009) Adsorption/Desorption of Water and Ethanol on 3A Zeolite in Near-Adiabatic Fixed Bed. *Industrial & Engineering Chemistry Research* 48 : 9247–9260.
99. Ahmed, I., Pa, N. F. C., Nawawi, M. G. M., Rahman, W. A. W. A. (2011) Modified Polydimethylsiloxane/Polystyrene Blended IPN Pervaporation Membrane for Ethanol/Water Separation. *Journal of Applied Polymer Science* 122 : 2666–2679.

100. Lai, C.-L., Liou, R.-M., Chen, S.-H., Shih, C.-Y., Chang, J. S., Huang, C.-H., Hung, M.-Y., Lee, K.-R. (2011) Dehydration of Ethanol/Water Mixture by Asymmetric Ion-Exchange Membranes. *Desalination* 266 : 17–24.
101. Shi, Q., He, Z., Gupta, K. M., Wang, Y., Lu, R. (2017) Efficient Ethanol/Water Separation via Functionalized Nanoporous Graphene Membranes: Insights from Molecular Dynamics Study. *Journal of Materials Science* 52 : 173–184.
102. Yang, J. Z., Liu, Q. L., Wang, H. T. (2007) Analyzing Adsorption and Diffusion Behaviors of Ethanol/Water through Silicalite Membranes by Molecular Simulation. *Journal of Membrane Science* 291 : 1–9.
103. Chen, S.-H., Yu, K.-C., Lin, S.-S., Chang, D.-J., Liou, R. M. (2001) Pervaporation Separation of Water/Ethanol Mixture by Sulfonated Polysulfone Membrane. *Journal of Membrane Science* 183 : 29–36.
104. Terrones, M., Botello-Méndez, A. R., Campos-Delgado, J., López-Urías, F., Vega-Cantú, Y. I., Rodríguez-Macías, F. J., Elías, A. L., Muñoz-Sandoval, E., Cano-Márquez, A. G., Charlier, J.-C., Terrones, H. (2010) Graphene and Graphite Nanoribbons: Morphology, Properties, Synthesis, Defects and Applications. *Nano Today* 5 : 351–372.
105. Severin, N., Sokolov, I. M., Rabe, J. P. (2014) Dynamics of Ethanol and Water Mixtures Observed in a Self-Adjusting Molecularly Thin Slit Pore. *Langmuir* 30 : 3455–3459.
106. Tian, X., Yang, Z., Zhou, B., Xiu, P., Tu, Y. (2013) Alcohol-Induced Drying of Carbon Nanotubes and Its Implications for Alcohol/Water Separation: A Molecular Dynamics Study. *The Journal of Chemical Physics* 138 : 204711.
107. Phan, A., Cole, D. R., Striolo, A. (2014) Preferential Adsorption from Liquid Water–Ethanol Mixtures in Alumina Pores. *Langmuir* 30 : 8066–8077.
108. Zhao, M., Yang, X. (2015) Segregation Structures and Miscellaneous Diffusions for Ethanol/Water Mixtures in Graphene-Based Nanoscale Pores. *The Journal of Physical Chemistry C* 119 : 21664–21673.
109. Kommu, A., Singh, J. K. (2017) Separation of Ethanol and Water Using Graphene and Hexagonal Boron Nitride Slit Pores: A Molecular Dynamics Study. *The Journal of Physical Chemistry C* 121 : 7867–7880.
110. Joshi, R. K., Carbone, P., Wang, F. C., Kravets, V. G., Su, Y., Grigorieva, I. V., Wu, H. A., Geim, A. K., Nair, R. R. (2014) Precise and Ultrafast Molecular Sieving Through Graphene Oxide Membranes. *Science* 343 : 752–754.
111. Souaille, M., Roux, B. (2001) Extension to the Weighted Histogram Analysis Method: Combining Umbrella Sampling with Free Energy Calculations. *Computer Physics Communications* 135 : 40–57.
112. Xue, M., Qiu, H., Guo, W. (2013) Exceptionally Fast Water Desalination at Complete Salt Rejection by Pristine Graphyne Monolayers. *Nanotechnology* 24 : 505720–505727.
113. Sun, Y., Dominy, B. N., Latour, R. A. (2007) Comparison of Solvation-Effect Methods for the Simulation of Peptide Interactions with a Hydrophobic Surface. *Journal of Computational Chemistry* 28 : 1883–1892.
114. Kerisit, S., Parker, S. C., (2004) Free Energy of Adsorption of Water and Metal Ions on the {10$\bar{1}$4} Calcite Surface. *Journal of the American Chemical Society* 126 : 10152–10161.
115. Xu, Z., Yang, X., Yang, Z. (2008) On the Mechanism of Surfactant Adsorption on Solid Surfaces: Free-Energy Investigations. *The Journal of Physical Chemistry B* 112 : 13802–13811.

3

Computational Chemistry Assisted Simulation for Metal Ion Separation in the Aqueous-Organic Biphasic Systems

Sk. Musharaf Ali, Anil Boda, Ashish Kumar Singha Deb, and Pooja Sahu

CONTENTS

3.1 Introduction

To meet the large concentrated energy needs for a developing country, the only sustainable energy resource available to the entire world in a longer-term time frame is nuclear energy [1], where uranium is used as a fuel in the nuclear reactor to generate the electricity. The uranium from primary resources meets only 65 per cent of world reactor requirements and the rest of the requirement can be met by secondary sources of supply. Uranium in the form of U_3O_8 can be recovered from a secondary source like "fertilizer phosphoric acid" by solvent extraction. The importance of nuclear energy was identified at the very beginning of the Indian atomic energy program. A three-stage nuclear power program, based on a closed nuclear fuel cycle, was planned to fulfill the growing energy demands in the country. The success of the three-stage nuclear energy program depends on efficient nuclear fuel reprocessing. The usually practiced nuclear fuel reprocessing can be classified in two types: namely, open and closed cycles. In the open cycle, the entire waste will be disposed of after being subjected to proper waste treatment. This results in huge underutilization of the energy potential of uranium. On the other hand, in the closed cycle, chemical separation of U-238 and Pu-239 are carried out, followed by radioactive fission products separation, sorted out according to their half-lives and activity appropriately, and disposed of with minimum environmental disturbance. As a part of long term energy strategy, India had chosen a closed cycle in view of its phased expansion of nuclear power generation extending through the second and third stages. The schematic of different nuclear fuel cycle reprocessing is displayed in Figure 3.1. But it is a well-known fact that there are some disadvantages in using nuclear energy owing to the concern about the handling and disposal

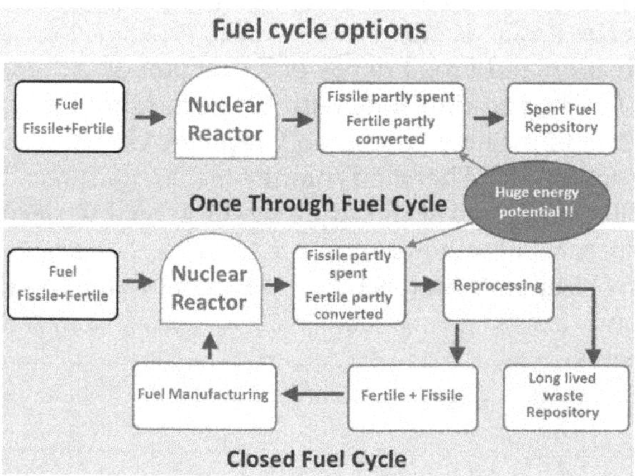

FIGURE 3.1
Schematic of nuclear fuel cycle.

of radiotoxic waste [2]. There is a continuing effort from experimentalists [3–11] and computational chemists [12–26] to design a variety of ligands for the safe removal and recovery of radiotoxic waste.

The spent fuel generated in nuclear reactors consists mainly (>98.5%) of uranium and short-lived fission products, which do not pose a long-term hazard. However, approximately 1 wt % of the spent fuel is composed of plutonium and the minor actinides (Am, Cm, Np), which are highly radiotoxic. Plutonium is the major contributor to the long-term radio toxicity of the spent nuclear fuel. Liquid-Liquid extraction is the proven technology for the nuclear waste reprocessing and partitioning. In the PUREX process, tri-n-butylphosphate (TBP) extracts uranium and plutonium that can be reused as Mixed Oxide "MOX" fuel in the reactors. The left over waste solutions contain minor actinides (MAs), mainly Am^{3+} and Cm^{3+} that are the main contributors to the long-term radio toxicity [27–30]. If the MAs are removed, the compulsory storage time of the remaining waste solution can be reduced significantly to a few hundred years from several thousand years [31,32] (Figure 3.2). One way to reduce the radio toxicity of the waste solution is partitioning and transmutation (P&T) of the long-lived minor actinides.

The P&T requires separation of actinides (An) from lanthanides (Ln). This separation is necessary since the lanthanides have higher neutron capture cross sectional areas than the actinides and thus absorb neutrons [33] during the transmutation process [34,35]. Additionally, during target preparation for transmutation, lanthanides do not form solid solutions with MAs and segregates in separate phases with the tendency to grow under thermal treatments. Because of the chemical similarities between lanthanides and

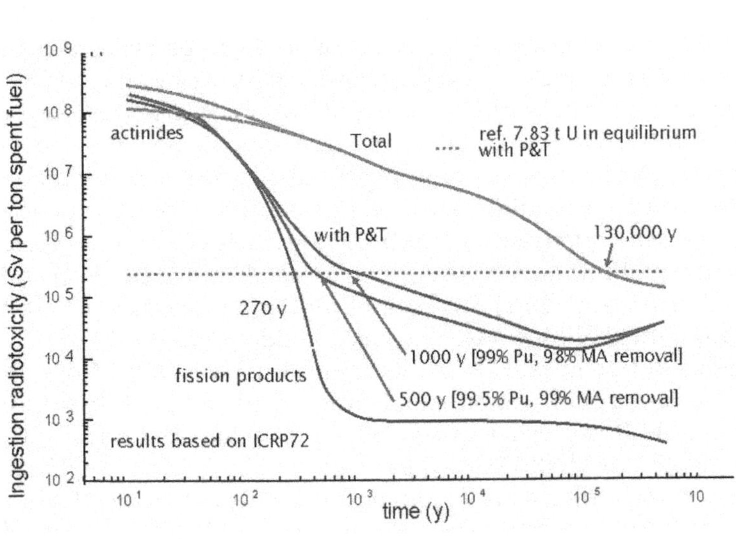

FIGURE 3.2
Partitioning of minor actinides and radio toxicity of nuclear waste.

actinides, the separation of the radioactive minor actinides from the lanthanides is a major challenge in nuclear waste reprocessing [36].

Understanding the separation of metal ions from the aqueous phase to the organic phase is of great relevance from the basic scientific knowledge as well as to the development of advanced, especially in nuclear waste reprocessing. In that direction, computational chemistry comprising Quantum electronic structure calculations (QESC) and molecular dynamics (MD) simulations technique has been emerged as a strong complimentary tool to the experiments by providing microscopic pictures at the molecular level. Even virtual experiments can be conducted using computational chemistry tools alone and the results can be used to reduce the cost and time of performing the experiments. Many a time when it is difficult to perform the experiments, computational simulation can provide a breakthrough. In view of the robustness of QESC and MD simulations, the present chapter is devoted to the application of QESC and MD simulations in the microscopic understanding of the ligand-aided metal ion transfer processes and thus its assistance in nuclear waste management.

3.2 Complexation of U(VI) towards N,N-Dihexyl-2-ethylhexanamide (DH2EHA)

There is a continuing effort to develop an alternate of the much-used TBP in the nuclear industry. One such alternate is N,Ndialkyl aliphatic amides, which have received particular attention in the back-end of nuclear fuel cycle [37–39]. The diamides have been found to be better as compared to TBP, because (i) they are easy to synthesize and have low aqueous solubility, (ii) the innocuous nature of chemical and radiolytic degradation products, (iii) have low secondary waste volume generations, and (iv) their extraction property can be tuned by tailoring the alkyl groups. In addition, fission product extraction was quite low as compared to TBP [37–41], which has been exploited for separation of uranium from rare earths. In order to corroborate the experimental findings, Vikas Kumar et al. [42] have performed density functional theoretical (DFT) calculations for the complexes of U(VI), Nd(III), and Yb(III) with N,N-dihexyl-2- ethyhexyl amide (DH2EHA).

Structures of free DH2EHA and its complexes with UO_2^{2+}, Nd^{3+}, and Yb^{3+} ions in the presence of nitrate anion was calculated with hybrid density functional B3LYP [43] using split-valence plus polarization (SVP) basis set as implemented in TURBOMOLE package [44,45]. Earlier, the B3LYP functional was shown to be working quite well for the actinides [46,47], though some times BP86 performs better [48] than the B3LYP. The scalar relativistic effective core potentials (ECP) were considered for UO_2^{2+}, Nd^{3+}, and Yb^{3+} ions, where 60 electrons were kept in the core of U whereas 28 electrons were kept in the core of Nd and Yb respectively [49]. The quintet and doublet spin state was used for Nd and Yb complexes during the optimization and energy calculation. The free energy was computed at 298.15K. The B3LYP functional was shown to be quite successful in predicting the thermodynamics of actinides [50,51]. The water and dodecane phase was taken care by the use of conductor-like screening model (COSMO) [52] with a dielectric constant of 80 and 2 respectively. The complexation reaction was simulated as:

$$UO_2(NO_3)_{2(water)} + 2DH2EHA_{(dodecane)} = UO_2(NO_3)_2.2DH2EHA_{(dodecane)} \quad (3.1)$$

$$M(NO_3)_{3(water)} + xDH2EHA_{(dodecane)} = M(NO_3)_3.xDH2EHA_{(dodecane),}$$
$$M = Nd, Yb; x = 2,3 \quad (3.2)$$

The free energy of extraction, ΔG_{ext}, for the above complexation reaction was evaluated using standard thermodynamic methods [22,23,53].

3.2.1 Structure and Structural Parameters of Various Chemical Species

By and large, the extraction of metal ions are carried out in aqueous acidic solution, where, the metal ions remain coordinated with the nitrate or mixed nitrate and aqua molecules in the first sphere of coordination. In view of this, the first sphere coordination of metal ion was optimized with nitrate ions as well as with mixed nitrate and aqua molecules. In the case of uranyl ion, two nitrate ions were found to be in bidentate mode, whereas 3 nitrate ions were found to be coordinated to Nd^{3+} and Yb^{3+} ions. The M-O bond distance was found to be shortest for Yb and longest for uranyl ion. For hydrated metal nitrate, two water molecules are seen to be coordinated to the uranyl ion, whereas three water molecules are found to be coordinated to the Nd^{3+} and Yb^{3+} ions. The M-O (O of NO_3) length was seen to be increased due to inclusion of water molecules in the first coordination sphere. The M-O (O of NO_3) length was found to be shorter than the M-O (O of H_2O) length.The optimized geometries of UO_2^{2+}, Nd^{3+}, and Yb^{3+} ions with nitrate ion towards DH2EHA (1:2 stoichiometry) are displayed in Figure 3.3. The nitrate ion was coordinated in bidentate mode, whereas one O atom from each DH2EHA was coordinated to the metal ion. In the case of UO_2^{2+} ion, two DH2EHA molecules and two nitrate ions lead to hexa coordinated complex. The calculated values of M-O (O of DH2EHA and NO_3)bondlength was found to be in good agreement with the X-ray diffraction data.

In the case of Nd/Yb ions, two DH2EHA molecules in monodentate mode and three nitrate ions were coordinated to the metal ion in bidentate mode leading to an octa-coordinated complex. The M-O distance (amidic O) was found to be longest for uranyl and shortest for Yb ion. Further,in order to compare the structure of $UO_2(NO_3)_2.2DH2EHA$ complex with the results

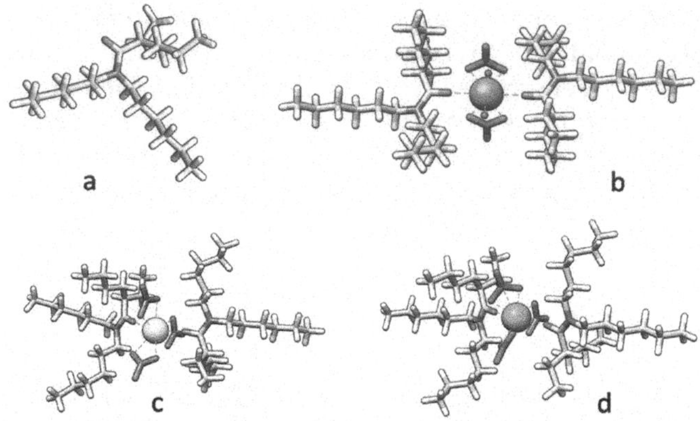

FIGURE 3.3
Optimized structure of DH2EHA and metal nitrate complexes at the B3LYP/SVP level of theory. (a) DH2EHA, (b) $UO_2(NO_3)_2$.2DH2EHA, (c) $Nd(NO_3)_3$.2DH2EHA, (d) $Yb(NO_3)_3$.2DH2EHA. (Reproduced from Kumar, V. et al., *Chemistry Select*, 2: 2348–2354, 2017.)

TABLE 3.1
Structural Parameters in Å at the BP86 and B3LYP Level of Theory Using SVP Basis Set along with the Crystallographic Data

Method	M-O (C=O)	M-O(NO_3^-)
BP86	2.390	2.515
B3LYP	2.401	2.528
Experiment	2.353	2.523

Source: Kumar, V. et al., *Chemistry Select*, 2: 2348–2354, 2017.

obtained using B3LYP, the complex was reoptimized at the BP86 level of theory (Table 3.1).

The values of M-O (O of DH2EHA and O of NO_3) bond length using both the B3LYP and BP86 were very close to each other and were also found to be in fair agreement with the values from X-ray data. In addition to the 1:2 stoichiometry, the 1:3 stoichiometry of the complexes of $M(NO_3)_3$.3DH2EHA were also optimized (Figure 3.4). Both the complexes (Nd,Yb) are nona-coordinated but the average Nd-O (C=O) distance is 2.475Å, which is slightly higher than that of average Yb –O (C=O) distance, 2.374Å.

3.2.2 Binding and Free Energy of Complexation

The biding energy (ΔE) which measures the interaction of metal ions towards a ligand is ofhigh importance for understanding the preferential selectivity among competitive metal ions and was therefore evaluated as:

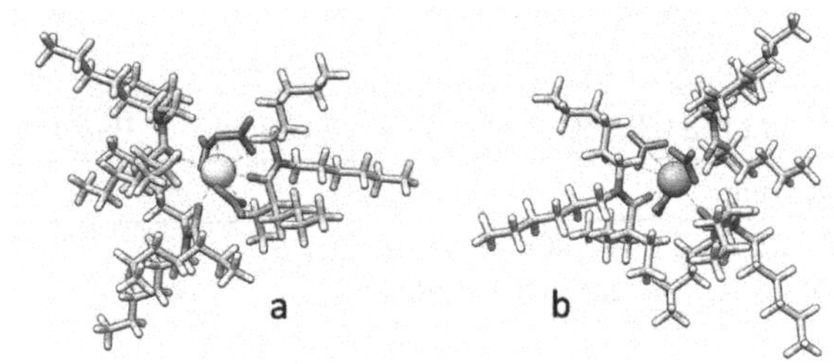

FIGURE 3.4
Optimized structure of metal nitrate complexes with DH2EHA (1:3 stoichiometry) at the B3LYP/SVP level of theory. (a) $Nd(NO_3)_3.3DH2EHA$, (b) $Yb(NO_3)_3.3DH2EHA$. (Reproduced from Kumar, V. et al., *Chemistry Select*, 2: 2348–2354, 2017.)

$$\Delta E = E_{UO2(NO3)2.2DH2EHA} - (E_{UO2(NO3)2} + 2E_{DH2EHA}) \quad (3.3)$$

$$\Delta E = E_{M(NO3)3.xDH2EHA} - (E_{M(NO3)3} + xE_{DH2EHA}) \quad (3.4)$$

Here, E represents the total electronic energy of the respective chemical species.In order to mimic the solution environment, the energy of chemical species was evaluated both in the aqueous and organic phase. The solution phase binding energy using the metal ion nitrate (**scheme-1**) was shown unable to capture the experimental binding trend. In the case of metal nitrate, the first sphere of coordination was partly occupied by the donors. Hence, the hydrated metal nitrate (**scheme-2**) was then used for the complexation. Though, the solution phase binding energy indicated the selectivity order: $UO_2^{2+} > Nd^{3+} > YB^{3+}$ as per experiment (Table 3.2), the calculated values of free energy of complexation indicateda different selectivity trend: $YB^{3+} > UO_2^{2+} > Nd^{3+}$. This mismatch was perhaps due to the use of inappropriate stoichiometry of the extracted species.

In order to demonstrate the correct selectivity trend, the calculation was performed using 1:3 M:L stoichiometry for Nd/Yb ions and the calculated results are tabulated in Tables 3.3 and 3.4 respectively. In the case of **scheme-1**, the calculated value of free energy of complexation was found to

TABLE 3.2
Extraction of Metal Ion from Simulated Sodium Diuranate (SDU)

Elements	U	Y	Yb	Er	Dy	Fe	Mn	Nd
Conc. (g/L)	30	2.08	1.076	0.174	0.022	4.0	0.05	0.5
D_M	5.4	0.05	0.001	0.002	0.0015	0.0016	0.0015	0.003
Separation factor(β)	–	108	5400	2700	3600	3300	3600	1800

Source: (Reproduced from Kumar, V. et al., *Chemistry Select*, 2: 2348–2354, 2017.)
Note: Organic phase: 1 M DH2EHA/n-dodecane. Aqueous phase: SDU solution, HNO_3=3 M.

TABLE 3.3

Calculated Value of Binding Energy and Free Energy of Extraction (kcal/mol) (1:3 stoichiometry) at the B3LYP Level of Theory Using SVP Basis Set (**scheme-1**)

Complexation Reaction	ΔE	ΔH	ΔG
$Nd(NO_3)_{3(aq)}$ + $3DH2EHA_{(org)}$ = $Nd(NO_3)_3.3DH2EHA_{(org)}$	–62.59	–63.78	–17.43
$Yb(NO_3)_{3(aq)}$ + $3DH2EHA_{(org)}$ = $Yb(NO_3)_3.3DH2EHA_{(org)}$	–61.33	–62.52	–10.58

TABLE 3.4

Calculated Value of Binding Energy and Free Energy of Extraction (kcal/mol) (1:3 stoichiometry) at the B3LYP Level of Theory Using SVP Basis Set (**scheme-2**)

Complexation Reaction	ΔE	ΔH	ΔG
$Nd(NO_3)_3(H_2O)_{3(aq)}$ + $3DH2EHA_{(org)}$ = $Nd(NO_3)_3.3DH2EHA_{(org)}$ + $3H_2O_{(aq)}$	–12.90	–12.90	1.75
$Yb(NO_3)_3(H_2O)_{3(aq)}$ + $3DH2EHA_{(org)}$ = $Yb(NO_3)_3.3DH2EHA_{(org)}$ + $3H_2O_{(aq)}$	–7.86	–7.86	10.43

be reduced for both the ions in the solution phase. The computed selectivity order follows the trend: UO_2^{2+} > Nd^{3+} > YB^{3+} as observed in the experiment. In the case of **scheme-2**, the calculated value of free energy of complexation was found to be reduced further compared to **scheme-1**. Even the calculated values of free energy of complexation for both Nd and Yb ions were found to be positive in the solution phase, indicating negligible extraction in the organic phase,which was confirmed by the very small value of distribution constant in the extraction experiment. Therefore, DFT studies not only offer the molecular level insights but also help in understanding the correct metal ion selectivity towards ligand by incorporating appropriate metal-ligand stoichiometry, which otherwise was not possible through experimental means only.

3.3 Complexation of UO_2^{2+} and Pu^{4+} towards N, N-dihexyloctanamide (DHOA)

In the preceding section, it has been established that DH2EHA selectively picks up uranyl ion over rare earths and thus can be used for the recovery of uranium from phosphoric acid-based feed solution. In a similar context, the possibility of using N,N-dialkyl amides for the reprocessing of fast breeder reactor (FBR) fuels [37], where concentration of Pu^{4+} ion is high, was explored. The N,N-dihexyloctanamide (DHOA) was proven to be the most promising candidate [54,55] for mutual separation of UO_2^{2+} and Pu^{4+} ions. The literature regarding molecular level understanding of UO_2^{2+} and Pu^{4+}

ions with DHOA is rather limited. In view of this, DFT in conjunction with the Born-Haber thermodynamical cycle [20,21,56–59] along with COSMO were used for studying competitive selectivity between UO_2^{2+} and Pu^{4+} ions using DHOA [60] by Boda et al.

The structure of DHOA and its complexes with UO_2^{2+} and Pu^{4+} ions were optimized at the BP86/SVP level of theory [44,45]. Relativistic ECP was used for U and Pu [49]. BP86 functional consists of Becke B88 exchange functional [61] and Perdew P86 correlation functional [62], which was found to be successful in estimating the molecular properties. The structures were optimized with BP86 functional and hessian calculations were performed at the same level of theory. Single point energies were calculated with the B3LYP functional [43,63] using triple zeta valence plus polarization (TZVP) basis set [64]. Quintet spin state was used for Pu^{4+} ion. The solvent effects in the energetic was inducted using COSMO [52,65] model. The default COSMO radii were used for all the elements except U (2.22Å) and Pu (2.22Å) for which default Bondi radii was used. The gas phase structures were used for the calculation of single point energy in COSMO phase. The solvation energy for metal ions is computed using implicit and explicit solvation model. The free energy of extraction (ΔG_{ext}) is the most important parameter as it can be used as yardstick for the comparative selectivity of the metal ion. The metal ion-ligand complexation reaction is modelled using the following stoichiometric reaction:

$$M^{n+}_{(aq)} + nN0^{-}_{3(aq)} + mL_{(org)} \xrightarrow{\Delta G_{ext}} \left[ML_m(N0_3)_n\right]_{(org)},$$

$$(M = UO2/Pu, n = 0,2 \text{ and } 4 \text{ and } m = 1,2 \text{ and } 3) \tag{3.5}$$

The change in Gibbs free energy of extraction, ΔG_{ext}, in Equation 3.5 was obtained by the Born-Haber thermodynamic cycle shown in Figure 3.5 (**scheme-1**), in terms of the free energy change in gas phase, $\Delta G_{(gas)}$, and the solvation free energies of the products and reactants, $\Delta\Delta G_{(sol)}$. The overall complexation reaction is manifested by ΔG_{ext} as:

$$\Delta G_{gas} = G_{ML_m(NO_3)_{n(gas)}} - \left(G_{M^{n+}(gas)} + nG_{NO^{-}_{3(gas)}} + mG_{L(gas)}\right) \tag{3.6}$$

$$\Delta\Delta G_{(sol)} = \Delta G_{sol[ML_m(NO_3)_n]} - \left(\Delta G_{sol[M^{n+}]} + n\Delta G_{sol[NO_3^-]} + m\Delta G_{sol\,[L]}\right) \tag{3.7}$$

$$\Delta G_{ext} = \Delta G_{gas} + \Delta\Delta G_{(sol)} \tag{3.8}$$

B3LYP functional predicts the thermodynamical properties quite well [66]. The interaction energy was also calculated using the hybrid PBE0 [67], which

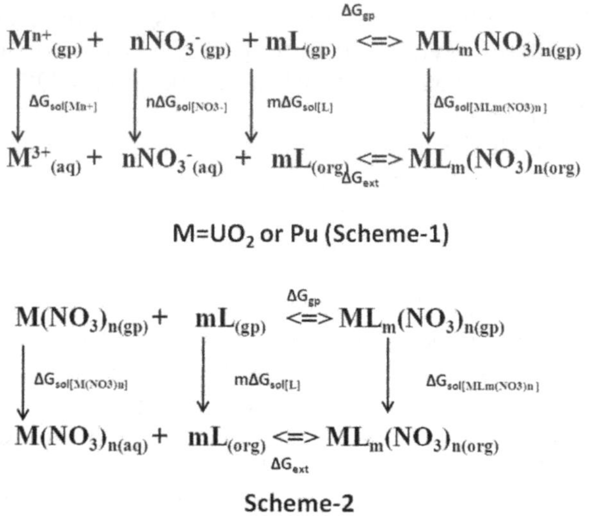

FIGURE 3.5
Thermodynamic cycle (Born-Haber) for the evaluation of free energy of extraction.

was shown to have good reliability for "f" element species [68]. MOLDEN was used for molecular visualization [69].

3.3.1 Structure and Structural Parameters of Various Chemical Species

From experimental and computational studies it has been reported that the average number of water molecules coordinated to the uranyl ion in the equatorial plane is close to 5 [70–73] which prompted to model the uranyl ion with 5 water molecules. The calculated bond length of U = O (1.769Å) with BP86/SVP level of theory is found to be very close to the experimental value of 1.766Å. The optimized structures of $UO_2(DHOA)_2(NO_3)_2$ and $Pu(DHOA)_3(NO_3)_4$ are presented in Figure 3.6.

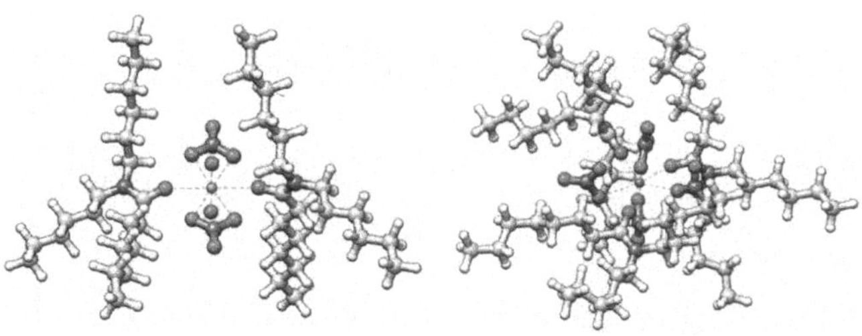

FIGURE 3.6
Optimized complexes of $UO_2(DHOA)_2(NO_3)_2$ and $Pu(DHOA)_3(NO_3)_4$. (Reproduced from Boda, A. et al., *Polyhedron*, 123: 234–242, 2017.)

In the case of $UO_2(DHOA)_2(NO_3)_2$ complex, the central U atom was found to be enclosed by eight O atoms in a hexagonal bi-pyramidal configuration as observed in the X-ray diffraction experiment [74]. Two O atoms of the uranyl ion were placed in the axial position, whereas two carbamoyl O from two DHOA ligands together with the two bidentate nitrate ions lie in the hexagonal equatorial plane as reported in the experiment [74]. In the case of $Pu(DHOA)_3(NO_3)_4$, the central Pu^{4+} ion was found to be coordinated to the three carbamoyl O of three DHOA ligands and four nitrate ions (3-bi-dentate, 1-mono dentate) leading to distorted deca coordination sphere. There are four O atoms lying in the equatorial plane (three O from three DHOA, one O from one nitrate). The calculated structural parameters for $UO_2(DHOA)_2$ $(NO_3)_2$ complex were found to be in excellent agreement with the X-ray diffraction results [74], which validates the use of present computational methods.

The metal ions are extracted from the aqueous feed solution, where it remains in a strongly hydrated form, hence, it is necessary to compute the solvation energy of the metal ions for a correct calculation of the extraction energy. The effect of solvent was modelled using cluster solvation model [75].

$$M^{m+}_{(gas)} + (H_2O)_{n(aq)} \rightarrow M^{m+}(H_2O)_{n(aq)} \tag{3.9}$$

The cluster of 5 and 9 water molecules was used for UO_2^{2+} and Pu^{4+} respectively. The minimum energy structures of hydrated $\left[UO_2(H_2O)_5\right]^{2+}$ and $\left[Pu(H_2O)_9\right]^{4+}$ clusters are displayed in Figure 3.7.

For $\left[UO_2(H_2O)_5\right]^{2+}$, the U=O bond distance (1.769Å) is elongated compared to free UO_2^{2+} ion (1.713Å). In the case of hydrated clusters of UO_2^{2+} and Pu^{4+} ions, the U-O and Pu-O bond distances are found to be longer than the values observed in its complexes with DHOA, indicating the strong complexing nature of DHOA with UO_2^{2+} and Pu^{4+} ions. The free energy of solvation for nitrate ion was taken from the earlier study [20]. The computed values of free energy of solvation for UO_2^{2+}, Pu^{4+}, and NO_3^- ions are –383.52, –1456.09, and –51.03 kcal/mol, respectively. The calculation of solvation free energy for

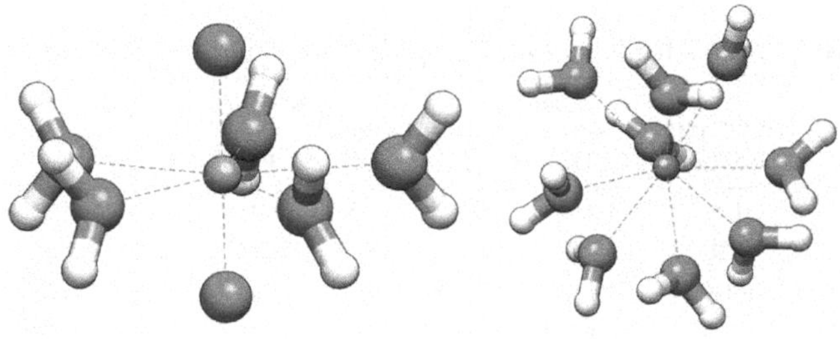

FIGURE 3.7
Optimized structures of water clusters of UO_2^{2+} and Pu^{4+} ions, i.e. (a) $\left[(UO_2(H_2O)_5\right]^{2+}$, (b) $\left[Pu((H_2O)_9\right]^{4+}$. (Reproduced from Boda, A. et al., *Polyhedron*, 123: 234–242, 2017.)

ions was corrected using standard state entropy corrections [75]. Interesting to note that though the solvation free energy of Pu^{4+} ion is much higher than that of UO_2^{2+} ion, Pu^{4+} ion is preferentially selected over the UO_2^{2+} ion. This fact can be well explained by calculating the ΔG_{ext} of the ions towards DHOA. The ΔG_{ext} for both UO_2^{2+} and Pu^{4+} ions with DHOA was evaluated using the Born-Haber thermodynamic cycle (Figure 3.5).

3.3.2 Free Energy of Extraction

The ΔG_{ext} of UO_2^{2+} and Pu^{4+} with DHOA in different organic solvent was determined using the calculated values of free energy of solvation of DHOA, $UO_2(DHOA)_2$ $(NO_3)_2$, and $Pu(DHOA)_3$ $(NO_3)_4$ in the organic solvents, namely: toluene ($\varepsilon = 2.38$), chloroform ($\varepsilon = 4.81$), octanol ($\varepsilon = 10.3$), NPOE ($\varepsilon = 23.9$), and nitrobenzene (NB) ($\varepsilon = 34.82$). The calculated value of ΔG_{sol} for DHOA and all the metal ion complexes were found to be increased with increase in the dielectric constant of the solvent. The difference in the free energy of solvation of complexes of UO_2^{2+} and Pu^{4+} with DHOA with increasing the dielectric constant of the solvent is found to be higher compared to its corresponding complexes with nitrate ion. This might be due to the screening of the charge on the metal ion by nitrate ion from the solvent which in turn reduces the free energy of solvation. Further, the ΔG_{ext} for $UO_2(DHOA)_2(NO_3)_2$ and $Pu(DHOA)_3(NO_3)_4$ in different solvents were calculated and given in Table 3.5.

The calculated values of ΔG_{ext} for both UO_2^{2+} and Pu^{4+} follows the order: NB > NPOE> octanol > chloroform > toluene > dodecane, which was corroborated by the experimental investigation (Table 3.6). The ΔG_{ext} value for UO_2^{2+}

TABLE 3.5

Calculated ΔG_{ext} of Complexes (kcal/mol)

Complex	Dodecane	Toluene	Chloroform	Octanol	NPOE	NB
$UO_2(DHOA)_2$ $(NO_3)_2$	−17.33	−17.53	−17.84	−17.99	−18.06	−18.07
$Pu(DHOA)_3$ $(NO_3)_4$	−30.76	−33.43	−38.61	−42.03	−43.98	−44.45

TABLE 3.6

Experimental D Values for U and Pu Using DHOA in Different Diluents

Solvent	Dielectric Constant	HNO_3		$HClO_4$		HNO_3		$HClO_4$	
		D_U	P_U	D_U	P_U	D_{Pu}	P_{Pu}	D_{Pu}	P_{Pu}
Dodecane	1.8	3.9 (4.3)	0.008	5	0.04	8 (7.5)	0.01	15.6	0.05
Toluene	2.38	7.9	0.008	9.8	0.05	17.2	0.01	29.8	0.05
Chloroform	4.81	13.2	0.009	16.7	0.05	25.2	0.01	38.2	0.07
Octanol	10.3	17.8	0.01	21.5	0.06	32.6	0.02	55.5	0.07
NPOE	23.9	25.1	0.01	30.7	0.08	37.3	0.02	61.2	0.08
Nitrobenzene	34.82	26.9	0.01	43.8	0.08	39.0	0.03	72.3	0.09

Source: Boda, A. et al., *Polyhedron*, 123: 234–242, 2017.

ion with nitrate ion is found to be highest in NB (–10.82 kcal/mol) and lowest in dodecane (–10.08 kcal/mol). Similar behavior is observed for Pu^{4+} ion also, and the value of ΔG_{ext} was found to be higher than that of UO_2^{2+} ion, indicating the selectivity towards Pu^{4+} ion compared to UO_2^{2+} ion with DHOA.

Ligand concentration 0.5 M DHOA, Values in the parenthesis for dodecane is from the literature, P is the partition coefficient of the metal ions, i.e. D values in the specified diluents without ligand DHOA.

Thus, the present study helped in understanding the fundamental mechanism of selectivity of Pu^{4+} ion over UO_2^{2+} ion. Furthermore, new ligand might be designed for the efficient and selective extraction of Pu^{4+} ion over UO_2^{2+} ion by tuning the electronic properties of DHOA by changing the alkyl chains.

3.4 Complexation Selectivity of UO_2^{2+} and Pu^{4+} Ion towards Tetramethyl Diglycolamide (TMDGA)

Recently, a new class of extractant, namely diglycolamide [76–82], has been developed which showedample advantages over other extractants due to ease of synthesis, milder stripping condition, and complete incinerability. Among various diglycolamide, N, N, N′, N′-tetraoctyldiglycolamide (TODGA) and N, N, N′, N′-tetra-2-ethylhexyl diglycolamide (TEHDGA) have been extensively studied for the removal of actinides and lanthanides from the nuclear waste. TODGA and TEHDGA are highly soluble in non-polar dodecane solvent and shows high distribution ratio for tri and tetravalent actinides over hexavalent actinide. Very recently, ALSEP (An-Ln separation) process has successfully demonstrated the partitioning of the minor actinides from the lanthanides using neutral TODGA and TEDGA as extractant along with an acidic extractant [83]. There are many experimental studies available for the extraction of actinides with diglycolamideextractant [76–86] but theoretical studies [87,88] are limited as to which can elucidate the type of complexation of actinides in different oxidation state through molecular level understanding and hence can assist in choosing the efficient extractants for actinides. Therefore, an attempt was made by Pahan et al. [53] to investigate the contrasting selectivity of diglycolamide ligand compared to TBP for UO_2^{2+} and Pu^{4+} ions using free energy, which helps for further modification of the extractant and suitable organic solvents for efficient extraction of actinides.

They have used tetramethyldiglycolamide (TMDGA) instead of tetraoctylderivative to keep the computational system tractable and economical. It is expected that the structural parameters will be similar for both TMDGA and TODGA as observed earlier for TODGA and N,N′-dimethyl-N, N′-diheptyl-3-oxapentanediamide (DMDHOPDA) [25]. The consideration of shorter alkyl chain molecule instead of longer alkyl chain molecule is quite familiar in computational chemistry [89–93]. It is likely that the solubility of tetraoctyl derivative of DGA in dodecane should be higher than that of tetramethyl derivative which is confirmed by the calculated partition coefficients of TODGA (logD = 11) and TMDGA (logD = –4.39) using COSMOtherm software [94]. In view of that, the ΔG_{ext} is predictable to be higher with TODGA compared to TMDGA. The structure of TMDGA and its complexes with UO_2^{2+} and Pu^{4+} ions were optimized at the BP86/SVP level of theory [45] as it was found to be quite successful in predicting the molecular geometries [95]. In addition, BP86 does not include non-local Hartree-Fock contribution, henceit is reasonably faster for predicting the geometry and vibrational frequencies.The geometries have been optimized with BP86 functional but total energies were calculated with B3LYP functional [43] using TZVP basis set with usual ECP core potential [49]. The heavier Ln/An element has filled f orbital electrons, which causes a large relativistic effect and therefore, scalar relativistic effects were incorporated in the present calculation [17,81,24,95, 96]. The solvent effect in the energetic was incorporated using COSMO [52], where default COSMO radii were used for all the elements except Pu^{4+} ion, for which the value was taken as 2.0. The solvation energy for metal ions in water is computed using explicit COSMO solvation model whichis an improved model, where the polarization charges of the solute is calculated in a continuum using scaled conducting boundary condition and has been used successfully for the free energy calculation of actinides in solution [97]. The change in Gibbs free energy of extraction, ΔG_{ext}, in Equation 3.5 can be obtained by the thermodynamic cycle (Born-Haber) as shown in Figure 3.5, in terms of $\Delta G_{(gas)}$ and $\Delta\Delta G_{(sol)}$.

3.4.1 Structure and Structural Parameters of Various Chemical Species

Three O donor atoms are lying in the same plane and are projectedin the same direction as seen from the optimized structure of TMDGA (Figure 3.8). The calculated bond distance of amide C=O was found to be 1.22Å. The optimized complexes of UO_2^{2+} and Pu^{4+} ions with TMDGA in 1:1 and 1:2 stoichiometry are also displayed in Figures 3.8 and 3.9, respectively.

In the complexes of metal ions with TMDGA, the C=O bond distance was found to be lengthened compared to free TMDGA. The metal ions were found to be coordinated via three O donor atoms of TMDGA. In 1:2 stoichiometry, two TMDGA units are coordinated to the UO_2^{2+} ion from the opposite direction in identical fashion almost in the same plane, whereas in the case of Pu^{4+} ion two TMDGA units are coordinated in a perpendicular fashion to each other. The U-O bond distance (1.752Å) of uncomplexed UO_2^{2+} was slightly lengthened (1.765Å) due to complexation with TMDGA. The M-O (amide O) bond distance for UO_2^{2+} was found to be longer (2.249Å) than for Pu^{4+} ion (2.185Å) indicating stronger interaction for Pu^{4+} ion. The M-O (amide O) distance was found to be shorter than the M-O (ethereal O) distance (2.519Å for UO_2^{2+} and 2.43Å for Pu^{4+} ion) and plays the dominant role in the coordinated interaction. The strong coordinating ability of amide

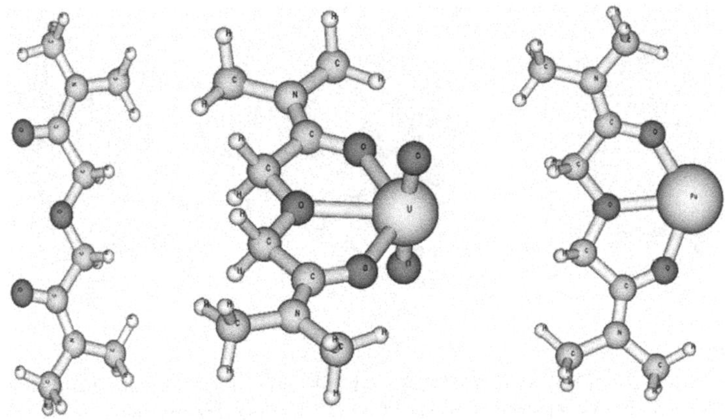

FIGURE 3.8
Optimized structures of (a) TMDGA and complexes of (b) UO_2^{2+}, and (c) Pu^{4+}, ions with TMDGA in the absence of nitrate ion (1:1) at the BP/SVP level. (Reproduced from Pahan, S. et al., *Theoretical Chemistry Accounts*, 134: 41–57, 2015.)

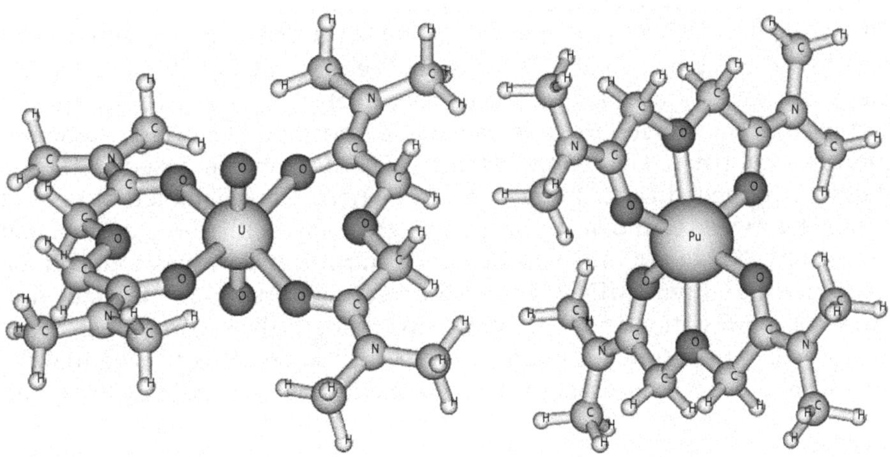

FIGURE 3.9
Optimized structures of complexes of UO_2^{2+} and Pu^{4+} ions with TMDGA in the absence of nitrate ion (1:2) at the BP/SVP level of theory. (Reproduced from Pahan, S. et al., *Theoretical Chemistry Accounts*, 134: 41–57, 2015.)

O over ethereal O is further confirmed from a large red shifting (342cm^{-1}) in the amide C=O stretching frequency compared to the red shift (163cm^{-1}) in the ethereal O frequency after complexation with the metal ions.

From solvent extraction, it was reported that UO_2^{2+} ion does not form 1:3 species with TMDGA, whereas Pu^{4+} ion does exist as 1:3 species in solution [3,98] and hence 1:3 complex of Pu^{4+} ion was further optimized, where three

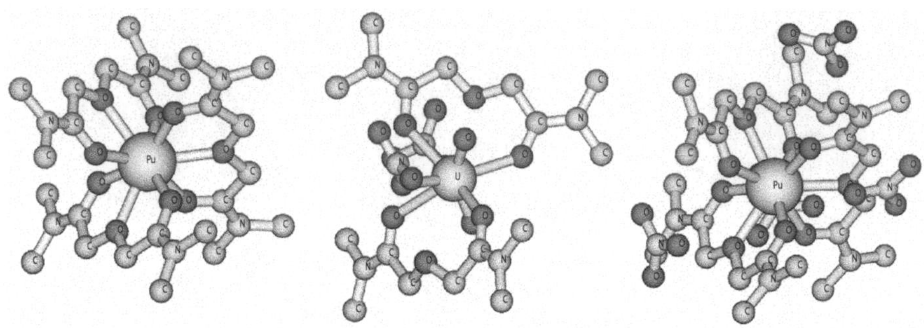

FIGURE 3.10
Optimized complexes of Pu^{4+} with TMDGA in the absence and presence of nitrate ion (1:3) at the BP/SVP level of theory (H atoms are excluded). (Reproduced from Pahan, S. et al., *Theoretical Chemistry Accounts*, 134: 41–57, 2015.)

TMDGA are coordinated to the Pu^{4+} ion in a distorted tricapped trigonal prismatic fashion (Figure 3.10).

The optimized complexes of UO_2^{2+} ion with two TMDGA and Pu^{4+} ion with three TMDGA in presence of nitrate [3,98] are also displayed in Figure 3.10. The structural parameters are listed in Table 3.7.

In metal ion- TMDGA complex, the C=O bond distance was found to be lengthened compared to free TMDGA. In the case of $UO_2(NO_3)_2(TMDGA)_2$, the central UO_2^{2+} ion was found to be coordinated to two TMDGA ligands and two nitrate ions in mono-dentate mode leading to distorted octahedral coordination, where U-O bond distance (1.752Å) of uncomplexed UO_2^{2+} was slightly lengthened (1.82Å) due to complexation. The calculated structural parameters were found to be in good agreement with the experimental results [85]. For $Pu(NO_3)_4(TMDGA)_3$, the central Pu^{4+} ion was found to be coordinated to three TMDGA ligands in nona coordinated mode in the first sphere of coordination and four nitrate ions reside in the second sphere leading to a charge neutralized species which is subsequently extracted into the organic phase. The coordination structure is described as twisted tricapped

TABLE 3.7
Structural Parameters of $UO_2(NO_3)_2(TMDGA)_2$ and $Pu(NO_3)_4(TMDGA)_3$ at the BP/SVP Level of Theory

	M-O(-C=O)			C=O	M-O (NO_3)	M-O (ethereal O)			U-O_{ax}		
Complex	**cal**	**exp**	**Δ**			**cal**	**exp**	**Δ**	**cal**	**exp**	**Δ**
$UO_2(NO_3)_2$ $(TMDGA)_2$	2.54	2.42[a]	0.12	1.24	2.41	3.06	2.61[a]	0.44	1.80	1.75[a]	0.05
$Pu(NO_3)_4$ $(TMDGA)_3$	2.38	2.29[b]	0.08	1.26		2.56	2.47[b]	0.09			

[a] Tian, G. et al., *Chemistry: A European Journal*, 15: 4172–4181, 2009.
[b] Reilly, S.D. et al., *Chemical Communication*, 48: 9732–9734, 2012.

trigonal prismatic with three C–O–C ether O atoms occupying the capping positions and the two trigonal faces formed by the C=O atom groups. The M-O (amide O) bond distance for UO_2^{2+} was found to be (2.542Å) in qualitative agreement with the experimental results of 2.421Å. Similarly, the M-O (amide O) bond distance for Pu^{4+} ion (2.381Å) was found to be in qualitative agreement with the X-ray results of 2.299Å. Further, the M-O (ethereal O) for UO_2^{2+} ion was found to be 3.06Å, which is little bit off from the experimental value of 2.614Å. For Pu^{4+} ion, M-O (ethereal O) was found to be 2.561Å in close agreement with the experimental results of 2.470Å. It is interesting to note that the M-O (amide O) bond distance in both $Pu(NO_3)_4(TMDGA)_3$ (2.38Å) and $UO_2(NO_3)_2(TMDGA)_2$ (2.54Å) were found to be longer due to the presence of nitrate ion compared to the absence of nitrate ion (2.424Å and 2.341Å for UO_2^{2+} and Pu^{4+}, respectively) as the nitrate reduces the effective positive charge on the metal ion, thus weakening the ion-dipole electrostatic interaction. The calculated structural parameters of $Pu(NO_3)_4(TMDGA)_3$ were found to be in good agreement with the experimental results [86].

3.4.2 Solvation and Free Energy of Extraction

The solvent effect in the energetic was integrated using the monomer model [99] as:

$$M_{(gas)} + mH_2O_{(aq)} \rightarrow M(H_2O)_{m(aq)} (M/m = UO22^+/5 \text{ and } Pu^{4+}/9) \quad (3.10)$$

The hydrated Pu^{4+} and UO_2^{2+} ions were used as per earlier reports [99,100]. In the case of hydrated clusters of Pu^{4+} and UO_2^{2+}, the Pu-O (2.452Å) and U-O (2.464Å) (O of H_2O) bond distances were found to be longer than the values observed in its complexes with TMDGA, indicating the strong complexing nature of TMDGA with Pu^{4+} and UO_2^{2+} ions.Thehydration free energy for nitrate ion was taken from the earlier study [20]. Further, to assess the etiquette of optimizing geometry in the gas phase and single point energy in the solvent phase, re-optimization of the $UO_2^{2+}(H_2O)_5$, $Pu^{4+}(H_2O)_9$, $UO_2(NO_3)_2$, and $Pu(NO_3)_4$ in water were done. The computed values of free energy of solvation after standard state entropy corrections for Pu^{4+} and UO_2^{2+} using monomer water solvation model are –1413.86 and –365.06 kcal/mol, respectively. The reported experimental solvation energy of Pu^{4+} ion is –1549.00kcal/mol [101]. It is worth to mention that though the solvation free energy of Pu^{4+} ion is much higher than that of UO_2^{2+} ion, Pu^{4+} ion is preferentially selected over the UO_2^{2+} ion. This preferential selectivity of Pu^{4+} ion over UO_2^{2+} ion can be well addressed by calculating ΔG_{ext} of the ions with the ligands. The ΔG_{ext} for Pu^{4+} and UO_2^{2+} was calculated using thermodynamic cycle (**scheme-1**, Figure 3.5). The free energy of solvation of TMDGA and its complexes with Pu^{4+} and UO_2^{2+} and without nitrate ion are given in Tables 3.8 and 3.9. The difference in the free energy of solvation of complexes of Pu^{4+} and UO_2^{2+} with

TABLE 3.8

Calculated Values of Free Energy (kcal/mol) Using Explicit Monomer Water Model without Nitrate Ion at the B3LYP/TZVP Level of Theory (**scheme-1** of **Figure 3.5**)

Di-Electric Constant (ε)	$\Delta G_{sol\,(L)}$	$\Delta G_{sol\,(MLn)}$ UO_2^{2+}	$\Delta G_{sol\,(MLn)}$ Pu^{4+}	$\Delta\Delta G_{sol}$ UO_2^{2+}	$\Delta\Delta G_{sol}$ Pu^{4+}	ΔG_{ext} UO_2^{2+}	ΔG_{ext} Pu^{4+}
2	−7.31	−51.23	−174.19	328.63	1261.60	15.92	131.84
5	−14.85	−93.76	−317.67	301.18	1140.76	−11.52	10.99
10	−18.37	−110.83	−374.90	291.16	1094.09	−21.55	−35.67
20	−20.42	−120.04	−405.70	286.04	1069.43	−26.67	−60.33
40	−21.53	−124.83	−421.69	283.46	1056.75	−29.25	−73.00

TABLE 3.9

Calculated Values of Free Energy (kcal/mol) Using Explicit Monomer Water Model with Nitrate Ion at the B3LYP/TZVP Level of Theory (**scheme-1, Figure 3.5**)

Di-Electric Constant (ε)	$\Delta G_{sol\,(L)}$	$\Delta G_{sol\,(ML(n)(NO3)m}$ UO_2^{2+}	$\Delta G_{sol\,(ML(n)(NO3)m}$ Pu^{4+}	$\Delta\Delta G_{sol}$ UO_2^{2+}	$\Delta\Delta G_{sol}$ Pu^{4+}	ΔG_{ext} UO_2^{2+}	ΔG_{ext} Pu^{4+}
2	−7.31	−16.40	−21.54	465.52	1618.38	0.50	−92.97
5	−14.85	−32.06	−42.21	464.98	1620.33	−0.03	−91.02
10	−18.37	−38.93	−51.20	465.12	1621.91	0.10	−89.44
20	−20.42	−42.82	−56.18	465.32	1623.07	0.30	−88.28
40	−21.53	−44.89	−58.82	465.46	1623.75	0.44	−87.60

TMDGA was found to be higher compared to its corresponding complexes with nitrate. This was due to the nitrate ions screening the charge on the metal ion from the solvent, which reduces the free energy of solvation. The solvent-metal ion interaction in the solvent phase decreases the interaction of metal ion with ligand, thereby decreasing the free energy.

From Table 3.8, it is seen that in the absence of nitrate ion, the value of $\Delta\Delta G_{sol}$ supersedesthe value of ΔG_{gas} for both UO_2^{2+} and Pu^{4+} ions, leading to a positive ΔG_{ext} in a solvent of dielectric constant 2. In the presence of nitrate ion, ΔG_{gas} supersedes the $\Delta\Delta G_{sol}$ for only Pu^{4+} ion, leading to a negative ΔG_{ext} (see Table 3.9) at dielectric constant 2, but remains positive for UO_2^{2+} ion, which clearly demonstrates the role of nitrate ion in the complexation. Further, the value of ΔG_{ext} for Pu^{4+} ion (−92.97 kcal/mol) was found to be much higher than that of UO_2^{2+} ion (0.50 kcal/mol) as reported in the solvent extraction [102] (ΔG_{Pu4+} = −11.74 kcal/mol, ΔG_{UO22+} = −6.67 kcal/mol) and the predicted distribution constants [64] (D_{Pu} = 55.7, D_U = 1.43). Recently, the preferential selectivity of Pu^{4+} ion over UO_2^{2+} ion towards DHOA was established using combined experiment and DFT studies, which were shown to be reversed with TBP, where the preferential selectivity of UO_2^{2+} ion over Pu^{4+} ion was observed [103]. Amide-based TMDGA also shows the preferential selectivity of Pu^{4+} ion over UO_2^{2+} ion in contrast to TBP. Further, the screening of solvent was performed by varying dielectric constant up to 40. From Table 3.8, it is seen that in the absence of nitrate ion, $\Delta\Delta G_{sol}$ supersedes the ΔG_{gas} for both UO_2^{2+} and Pu^{4+} ions, leading to a positive ΔG_{ext} (see Table 3.8) in solvent of dielectric constant 2 but above 5, ΔG_{ext} becomes negative. Both the solvation energy of the ligand as well as the metal ion-ligand complexes are found to be increased with increasing dielectric constant of the solvent, leading to decreased $\Delta\Delta G_{sol}$, which in turn increases the ΔG_{ext} for both Pu^{4+} and UO_2^{2+} ion with increasing dielectric constant. Whereas in presence of nitrate ion, ΔG_{ext} (Table 3.9) was found to be decreased with increased dielectric constant due to increase in $\Delta\Delta G_{sol}$ for both Pu^{4+} and UO_2^{2+} ion except at $\varepsilon = 5$, where ΔG_{ext} for UO_2^{2+} ion was found to be negative compare to positive value at $\varepsilon = 2$. Earlier, TMDGA was recommended as stripping agents [104,105] due to low partition coefficient in aqueous organic biphasic system. The low partitioning suggests that TMDGA has some solubility in non-polar solvent like dodecane, indicating few TMDGA molecules might have transferred from aqueous to dodecane phase, which seems to be reasonable. Further, it is logical to state that as the free energy of extraction with TMDGA is found to be high in low dielectric constant, it is expected to be valid for TODGA also. The present methodology for the prediction of bonding, interaction, and thermodynamics can be easily adapted to the bigger analogue like TODGA, which in fact shows high extraction in low dielectric constant solvent like dodecane for the extraction of Pu^{4+} and UO_2^{2+} ions.

In **scheme-1**, the metal and nitrate ions were considered to be isolated hydrated species. It is interesting to study when the metal and nitrate ions remain as an ion pair in highly acidic aqueous solution (>3N HNO_3) as mentioned in Equation (3.11). In the presence of nitrate ions, the hydrated metal ions prefer to coordinate with the nitrate ions, which act as bidentate ligand. Recently, the hydrated nitrate ion pair has been tested as a model for free energy calculation [89]. The stability of the hydrated metal nitrate cluster during complexation with nitrate ion can be checked using the free energy of the following complexation reaction:

$$M^{+}(H_2O)_{n(gas)} + mNO^{-}_{3(gas)} \rightarrow M(NO_3)_m(H_2O)_{r(gas)}$$
$$+ pH_2O\left(n/m/r/p = 5/2/2/3 \text{ for } UO_2^{2+} \text{ and } 9/4/2/7 \text{ for } Pu^{4+}\right) \quad (3.11)$$

The calculated value of free energy (Table 3.10) for the above reaction was found to be negative indicating the possibility of forming the hydrated nitrate ion pair species in gas phase. According to the Gibbs free energy, the metal nitrate hydrates are more stable than the metal hydrates indicating

TABLE 3.10

Calculated Thermodynamic Parameters of Reaction (kcal/mol) at the B3LYP/TZVP Level of Theory

Reaction	ΔH	ΔS*	ΔG	ΔG_{sol}
$Pu^{4+}(H_2O)_9 + 4NO_3^- = Pu(NO_3)_4(H_2O)_2 + 7H_2O$	–899.98	0.083	–924.76	–51.40
$Pu^{4+}(H_2O)_{10} + 4NO_3^- = Pu(NO_3)_4(H_2O)_2 + 8H_2O$	–877.08	0.110	–909.92	–52.57
$Pu^{4+}(H_2O)_9 + 4NO_3^- = Pu(NO_3)_4(H_2O)_2 + 6H_2O$	–907.50	0.046	–927.97	–51.19
$Pu^{4+}(H_2O)_{10} + 4NO_3^- = Pu(NO_3)_4(H_2O)_3 + 7H_2O$	–882.93	0. 072	–913.14	–52.36
$UO_2^{2+}(H_2O)_5 + 2NO_3^- = UO_2(NO_3)_2(H_2O)_2 + 3H_2O$	–298.84	0.021	–305.29	–14.17
$Pu^{4+}(H_2O)_9 + 4NO_3^- = Pu(NO_3)_4 + 9H_2O$	–874.19	0.163	–922.97	–58.05
$Pu^{4+}(H_2O)_{10} + 4NO_3^- = Pu(NO_3)_4 + 10H_2O$	–851.29	0.190	–908.14	–59.22
$UO_2^{2+}(H_2O)_5 + 2NO_3^- = UO_2(NO_3)_2 + 5H_2O$	–260.58	0.091	–287.94	–10.41

that nitrate ions have stronger coordination ability than water. Similar studies with phosphate anion was earlier reported by Dixon et al. [19]. In order to study the ion pair model (Ref. 25 by Wang et al.), the free energy of extraction was then further computed using thermodynamic cycle of **scheme-2**. The solvation of metal nitrate has been considered using explicit solvation model. The metal nitrate with first sphere of solvation was solvated as:

$$M(NO_3)_{m(gas)} + nH_2O_{(aq)} \rightarrow M(NO_3)_m(H_2O)_{n(aq)}$$
$$m/n = 2,\ 2 \text{ for } UO_2^{2+} \text{and } 3/4 \text{ for } Pu^{4+} \tag{3.12}$$

The computed value of ΔG_{ext}, for the transfer of metal ions from aqueous to dodecane phase using **scheme-2** is given in Table 3.11. The values of ΔG_{gp} and $\Delta\Delta G_{sol}$ were found to be positive, which makes ΔG_{ext} positive suggesting that the extraction is unfavourable. The extraction of hydrated metal nitrate using explicit monomer model is not workable due topositive ΔG_{ext}.

Further, the free energy for the complexation reaction using metal nitrate and hydrated metal nitrate with TMDGA ligand in 1:3 stoichiometry was also evaluated. The calculated free energy in the solution phase was found to be positive for all the metal nitrate and hydrated metal nitrate species. Hence, the results using the concept of hydrated metal nitrate was unable to capture the experimental selectivity trend, whereas present results based on thermodynamic cycle using separate hydrated metal ion and nitrate species did capture the experimental selectivity. Further, the ΔG_{ext} using Equation (3.5) for implicitly hydrated species were also calculated. The ΔG_{ext}

TABLE 3.11

Calculated Value of Thermodynamic Parameters (kcal/mol) for the Extraction of Different Hydrated Nitrate of UO_2^{2+} and Pu^{4+} Ions with TMDGA Using B3LYP/TZVP Level of Calculation at 298.15K (**scheme-2**)

M^{m+}	$\Delta G_{(gp)}$	$\Delta G_{sol(M(NO3)n}$	$\Delta G_{sol(L)}$	$\Delta G_{sol(MLn(NO3)m}$	ΔG_{ext}	$\Delta\Delta G_{(sol)}$
$UO_2(NO_3)_2(H_2O)_2$	20.29	−29.88	−7.31	−16.40	55.72	35.42
$Pu(NO_3)_4(H_2O)_3$	9.24	−16.86	−7.31	−21.54	25.61	16.36
$Pu(NO_3)_4(H_2O)_2$	9.24	−16.86	−7.31	−21.54	26.51	17.26

was found to be positive for UO_2^{2+} but negative for Pu^{4+} ion, though experimentally, the extraction of both ions were predicted to be thermodynamically favourable and thus failed to capture the experimental results. The ΔG_{ext} using thermodynamic cycle as per **scheme-1** (Table 3.9) was found to be negative for Pu^{4+} ion (−92.97 kcal/mol), whereas it was small positive for UO_2^{2+} ion (0.50 kcal/mol). The preferential extraction of Pu^{4+} over UO_2^{2+} ion was noticed in the experiment also. Further, in order to compare the extraction ability of TMDGA over TBP, the ΔG_{ext} of Pu^{4+} and UO_2^{2+} ion with TBP in dodecane was also calculated and here also the ΔG_{ext} for Pu^{4+} (−88.85 kcal/mol) was found to be higher than that of UO_2^{2+} ion (−35.59 kcal/mol). But, the ΔG_{ext} for UO_2^{2+} ion with TBP was found to be not only negative but much higher than that of with TMDGA, whereas the ΔG_{ext} for Pu^{4+} ion with TBP was found to be reduced compared to TMDGA. The difference in free energy between Pu^{4+} ion and UO_2^{2+} with TBP (53.26 kcal/mol) was found to be reduced compared to TMDGA (93.47 kcal/mol,) indicating more interaction of UO_2^{2+} with TBP over TMDGA, though reversal selectivity was not captured.

3.5 Binding of UO_2^{2+} and Pu^{4+} Ions with Ethylene Glycol Methacrylate Phosphate Anchored Graphene Oxide

Gaphene oxide (GO) has several oxygen-containing functional groups such as -COOH, -C=O, -O-, and -OH on the basal plane, edges, and defects. The presence of O bearing functional groups make GO the most attractive carbon nanomaterials as these groups act as the linkers for chemical modifications, provide hydrophilicity, and bind with heavy metal ions [106–111]. The removal of actinides such as Am(III), Th(IV), Pu(IV), Np(V), and U(VI) from water has been found to be effective by GO [112–117]. The major issue for using pristine GO is that it sorbs heavy metal ions by the electrostatic interactions that makes it non-selective, and sorption becomes highly dependent upon pH and ionic strength.A chemical modification of GO is required to make it selective for given metal ions that is important for the sequestration of target ions from the complex aqueous media. The various strategies have been used for anchoring functional groups bearing polymer chains on the GO platform for the removal of actinides and radionuclide from the aqueous streams. It has been reported that the cross-linked polymer formed by ethylene glycol methacrylate phosphate (EGMP) sorbs UO_2^{2+} and Pu^{4+} ions preferentially from a wide variety of aqueous matrices such as seawater and 3–4 mol L^{-1} HNO_3 [118–121]. Chappa et al. [122] has reported the DFT computations for the complexation of UO_2^{2+} and Pu^{4+} ions with the EGMP molecules anchored on a GO platform in the presence of nitrate ions to study the origin of selectivity of GO-EGMP towards Pu^{4+} ionto establish the experimental findings [122].

Optimization of the graphene oxide, GO (considering one epoxy group on the surface, one carboxyl group on the edge sp^2 carbon, and one hydroxyl attached with the edge sp^2 carbon, on a 5 × 5 armchair edged graphene unit cell composed of 58 carbon atomswas done at the DFTlevel using Turbomole program [45]. The equilibrium structure of ethylene glycol methacrylate phosphate - graphene oxide, GO-EGMP (phosphate unit has been covalently connected through phosphoesteric bond with hydroxyl group on the graphene sheet) [123], and UO_2^{2+} and Pu^{4+} complexes with GO-EGMP in the presence of nitrate anions were carried out at the B3LYP/SVP level of theory [61,62]. The single point energies of the studied systems were calculated with B3LYP functional using TZVP basis set. All calculation relating Pu^{4+} ions was performed using its quintet spin state. The relativistic effective core potential (ECP) was used, where 60 electrons are kept in the core of U and Pu [49]. The solvent effect in the energetic was incorporated employing COSMO model [52]. The free energy (ΔG) of complexation (Equation 3.5) of UO_2^{2+} and Pu^{4+} ions with GO-EGMP in the presence of nitrate ions arereported elsewhere [124].

3.5.1 Structure and Thermodynamics

The optimized structure of GO-EGMP is given in Figure 3.11(a). The phosphoryl (P=O) and carbonyl (C=O) groups of EGMP covalently fixed on the GO are involved in the complexation with metal ions along with nitrate as shown in Figure 3.11(b). Pu^{4+} ion formed an 8-coordinated complex, where two nitrates bonded in bidentate and the other two in monodentate mode. UO_2^{2+} ion was found to be a 7-coordinate structure where one nitrate coordinated in bidentate and another in monodentate fashion. Pu^{4+} showed the smaller M-O bond distances for both the phosphoryl and carbonyl than that of uranyl ions. This seems to indicate that Pu^{4+} ion formed a stronger coordination complex with the EGMP-GO as compared to UO_2^{2+} ions.

The free energy of complexation (ΔG), which could be correlated with the experimental D values of metal ions has also been found to be higher by

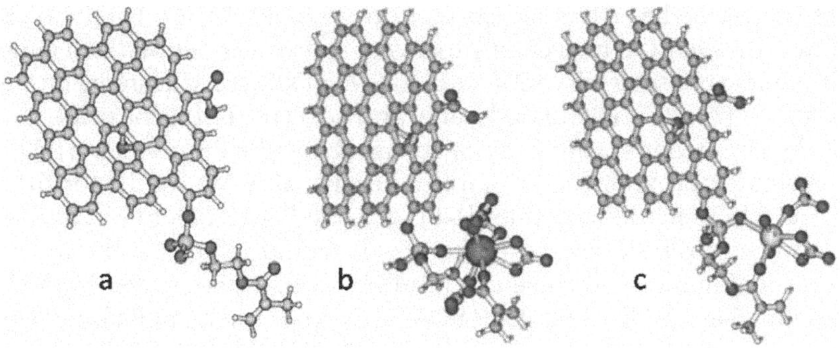

FIGURE 3.11
Optimized structure of (a) GO-EGMP, (b) complex of plutonium nitrate, and (c) complex of uranyl nitrate. C, O, N, and P atoms are shown in green, red, orange, and blue, respectively. (Reproduced from Chappa, S. et al., *The Journal of Physical Chemistry B*, 120: 2942–2950, 2016.)

–296.11 kcal mol^{-1} for Pu^{4+} ion with respect to UO_2^{2+} ion. Both for Pu^{4+} and UO_2^{2+} ions, ΔG were decreased from the enthalpy of complexation (ΔH) due to the negative contribution of entropy change.Thus, the origin of high selectivity of single EGMP on GO towards Pu^{4+} ions appears to be associated with stable 8-coordinated complex formation involving bidentate single EGMP and mono/bidentate four nitrate ions. Contrary to this, UO_2^{2+} form lower stability complex with single EGMP and nitrate ions.

3.6 Uranyl-TBP Complexes at the Aqueous-Organic Interface

The transfer of uranyl ions from aqueous to the organic phase occurs via many molecular-level events, which is very difficult to capture from the experiments alone. Earlier,molecular dynamics (MD) simulations [125–130] have been employed to investigate the complexation mechanism at the interface. Wipff et al. [128] reported that when the TBP concentration is increased to 60% or a high concentration of nitric acid is added to the aqueous phase, a disordered medium called 'third phase' is formed. However, their simulations were carried out by approximating uranyl nitrate as an ion pair instead of dissociated UO_2^{2+} and NO_3^- ions. Apart from this approximation, chloroform or $CO_2$was used as representative of organic phase instead ofdodecane. Further, the third phase formation [131–133] with increased TBP concentration or acid is a characteristic function of diluents needs to be re-investigated. They [128,129,134] ignored the complexation mechanism of uranyl and nitrate ions (by taking the associated complex of uranyl complex), which might influence the extraction mechanism and need to be further investigated. Also, the extraction of charged uranyl species into the organic phase need to be verified as it contradicts the conventional theory. Therefore, Sahu et al. has revisited the molecular process of uranyl extraction from aqueous to the organic phase using MD simulation [135,136] with different concentrations of TBP (20–50% TBP in dodecane) and acid (1M-6M nitric acid).

All the simulations were performed using LAMMPS package [137]. The united atom approximation [138] for dodecane, all atom model for the TBP [139] and TIP3P model for water [140] was used. The potential parameters for uranyl and nitrate ions are taken from literature [136]. For nitric acid, two different models were used: neutral HNO_3 and dissociated H_3O^+/NO_3^-, each with all atom OPLS model [141,142]. The pair interactions between the atoms are calculated using L-J potential [138,143]. All the non-bonded pair interactions were calculated using Lorentz-Berthelot mixing rules $\sigma_{ij} = (\sigma_i + \sigma_j)/2$ and $\varepsilon_{ij} = \sqrt{\varepsilon_i \varepsilon_j}$ with a cutoff of 12Å. During simulations, the long range electrostatic interactions are incorporated using PPPM method [144] with a precision of 10^{-5} and cutoff of 15Å. For bonds and angles, harmonic potential model [137] has been used. All systems were first relaxed by 3000 steps of energy minimization [145] followed by equilibration using NPT ensemble [137] for 30 ns with usual periodic boundary conditions. The production run was performed with NVT ensemble for 70ns. The temperature was maintained at 300 K by the Nose-Hoover thermostat. The MD trajectories and velocities were calculated using the velocity verlet algorithm [146] with a time step of 1fs, in conjunction with SHAKE constraints [147]. VMD package [148] was used to visualize the trajectories.

3.6.1 Structural Parameters

The simulations results (Figure 3.12) show that within few picoseconds, most of the uranyl ions are migrated to the interface. The time evolution show that the water molecules from the hydrated uranyl complexes are migrated to the interface. As a result, the attraction between uranyl and nitrate ions is increased and water molecules can be easily replaced by the TBPs at the interface, thus creating a favorable environment for the formation of uranyl nitrate complex.

It is seen that the TBP is surface active and sits at the organic side of the interface (Figure 3.13). It is observed that the TBP molecules form ambiphilic structure at the interface with their phosphoryl O pointing towards the water phase and hydrophobic alkyl chain to the organic phase, whereas, TBP molecules in the surroundings of uranyl ions point towards the U. Further, with the addition of acid, TBP structuring at the interface is modified. Also, the molecular orientation of TBPs is observed to be greatly influenced by the concentration of TBP in the organic phase and the acid in the aqueous phase. Higher concentration of HNO_3 intend to water pocket formation in the organic phase and are observed to be orientated in such a way that polar oxygen (O2=P) group of TBP points towards the water pool and hydrophobic alkyl chain to the organic phase, which indicates a micelle-like structure of

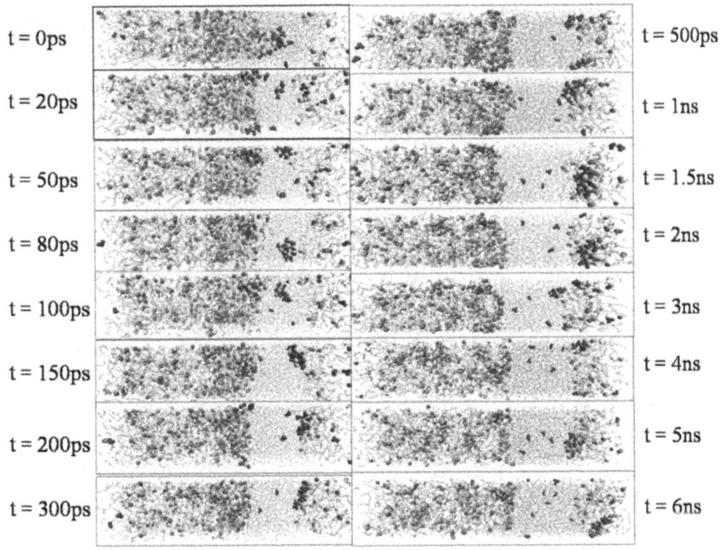

FIGURE 3.12
Time evolution of the aqueous/organic interface; snapshot. (Reproduced from Sahu, P. et al., *Physical Chemistry Chemical Physics*, 18: 23769–23784, 2016. With permission from PCCP owner societies.)

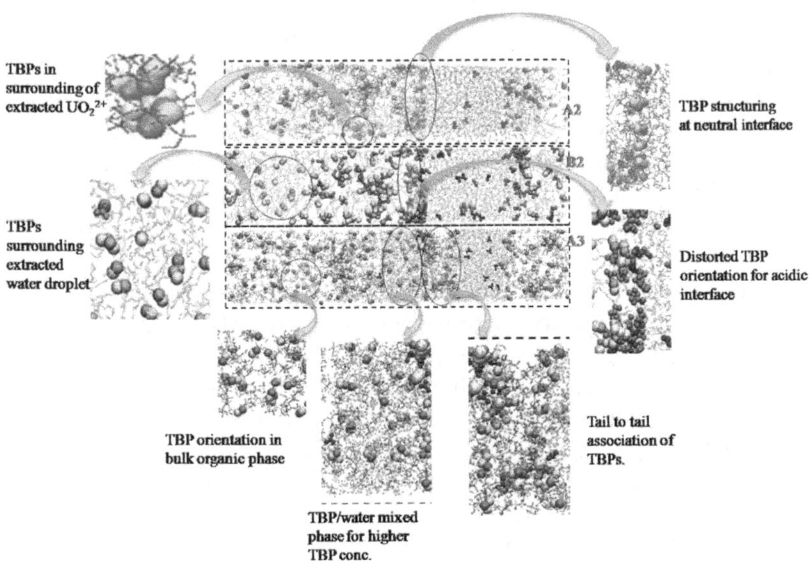

FIGURE 3.13
Water/organic interface for neutral vs acidic interface. (Reproduced from Sahu, P. et al., *Physical Chemistry Chemical Physics*, 18: 23769–23784, 2016. With permission from PCCP owner societies.)

TBPs around the water droplet in the organic phase. During MD simulations, none of the TBPs are observed to migrate towards the water phase, due to poor water solubility. Experimental studies show the water solubility of 0.4 g/lt in the TBP phase, however, when TBP is mixed with the organic it attracts less water compared to pure TBP.

The extraction of water is increased with increase in the concentration of TBP. The extracted uranyl complexes are found to be attached by 1–3 water molecules only, which indicates that the water extraction to the organic phase is favored not only by uranyl extraction but also from the TBPs exchange from the interface to the bulk organic phase. The percentage extraction of water is increased with HNO_3 from 1M to 3M, after which it remains nearly constant. The results reveal that for small concentration of H_3O^+/NO_3^-, N_{H2O} is significantly high, which arises due to the salting out effect of NO_3^-. For dissociated H_3O^+/NO_3^- ions, abundance of NO_3^- is quite high, which enhances the nitrate complexation with the uranyl at the interface and thus reduces the TBP accumulation in the surrounding of interfacial UO_2^{2+}, which leads to the formation of TBP.H_2O adducts at the interface. The TBP.H_2O complex, being neutral in nature, are transferred to the bulk organic, thus pulling the water molecules in the organic phase. However, the water extraction at the interface is reduced with further increase in H_3O^+/NO_3^- concentration. Also, with the higher concentration of TBP and HNO_3, the extracted water molecules form a cluster surrounded by TBP/dodecane phase. Nevertheless, in the case of neutral interface, the extracted water molecules form a micro emulsion with mixed TBP and water phase, known as the third phase. However, no such third phase was observed for ionic H_3O^+/NO_3^- as the extracted water molecules remain dispersed in the bulk organic phase. The extraction of HNO_3 in the organic phase is first increased and becomes nearly constant with increase in HNO_3, indicating the saturation of organic phase by acid as reported by the experiment. The extraction of H_3O^+ ions to the organic phase was facilitated by the strings, formed by H_3O^+ and NO_3^-, which are sometimes attached to the TBP or uranyl ions at the interface.

The first sphere coordination of the uranyl ions depends on the interfacial conditions and their location with respect to the interface. From the simulation trajectories, it was noticed that the uranyl ions at the interface are not separated but form loose bound clusters, in which each uranyl ion is coordinated by NO_3^-/TBP/H_2O, some of which are shared by more than two uranyl ions (see Figure 3.14). Further, electronic reorganization of uranyl-nitrate complex happens at the interface and some of the nitrates are diffused back to the aqueous phase. During this reorganization, nitrate ions are seen to be replaced by TBP, making the complex hydrophobic in nature and therefore facilitating their migration to the organic phase. Figure 3.14 shows the possible complexation of UO_2^{2+} with TBPs and NO_3^-. Results show 1:1, 1:2, 1:3, 1:4, and 1:5 adducts of UO_2^{2+}:TBP as reported earlier. The present MD

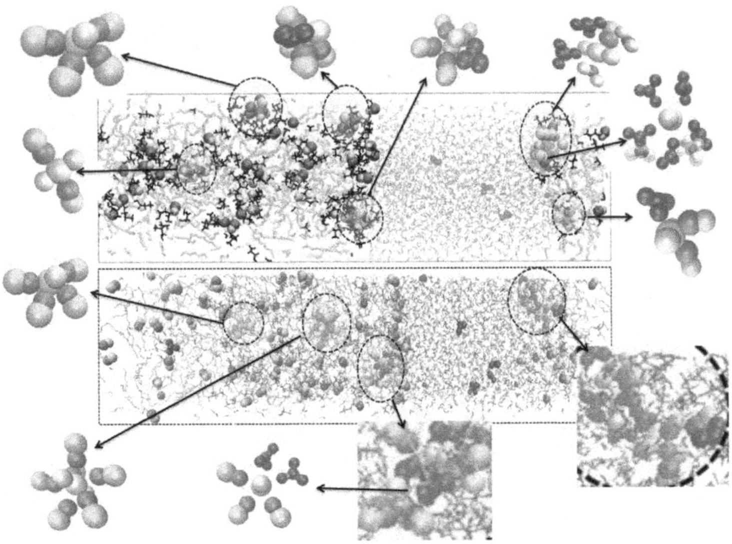

FIGURE 3.14
Enlarged view of uranyl complexation with TBPs and NO_3^- species for neutral interface. UO_2^{2+} and NO_3^- are presented by yellow and blue color respectively, whereas red and green balls respectively represent the O(O=P) and P of TBP. (Reproduced from Sahu, P. et al., *Physical Chemistry Chemical Physics*, 18: 23769–23784, 2016. With permission from PCCP owner societies.)

simulations capture the formation of one 1:6 adduct of UO_2^{2+}:TBP also. The formation of hexa-coordinated UO_2^{2+} is not unanticipated; indeed it represents the multifaceted nature of UO_2^{2+} complexation during the intermediate steps of various microscopic process, which has also been confirmed by the crystallographic data. For UO_2^{2+} ions near the organic side of the interface have higher affinity for TBP than NO_3^- or H_2O, and thus easily migrate to the bulk organic phase. The extracted uranyl complexes with TBP mostly have 1:3 and 1:4 stoichiometry, however the overall stoichiometry of UO_2^{2+}:TBP is observed to be in the range of 1 to 6.

MD simulations with equimolar composition of neutral HNO_3 and H_3O^+/NO_3^- ions were also performed for 1M, 3M, and 6M acid using 33% TBP in the organic phase. The TBPs in the organic phase form 1:1, 2:1, and 3:1 TBP: H_3O^+ complexes as shown in Figure 3.15. However, with HNO_3 only 1:1 complex of TBP.HNO_3 was formed. The uranyl ion forms complexes with following composition: $[UO2.4TBP.1NO3]^{-1}$, UO_2.2TBP.3NO_3, $[UO2.3TBP.1NO3]^{-1}$ and UO_2.5TBP. There was no direct contact of acid molecules H_3O^+ or HNO_3 with the uranyl ion. The UO_2^{2+} ions are seen to be coordinated by the one or maximum two HNO_3 molecules and only with one H_3O^+ ion (see Figure 3.15). The MD results reveal that the number of UO_2^{2+} coordinating water. ([a,b] Adapted with permission from ref. [48].

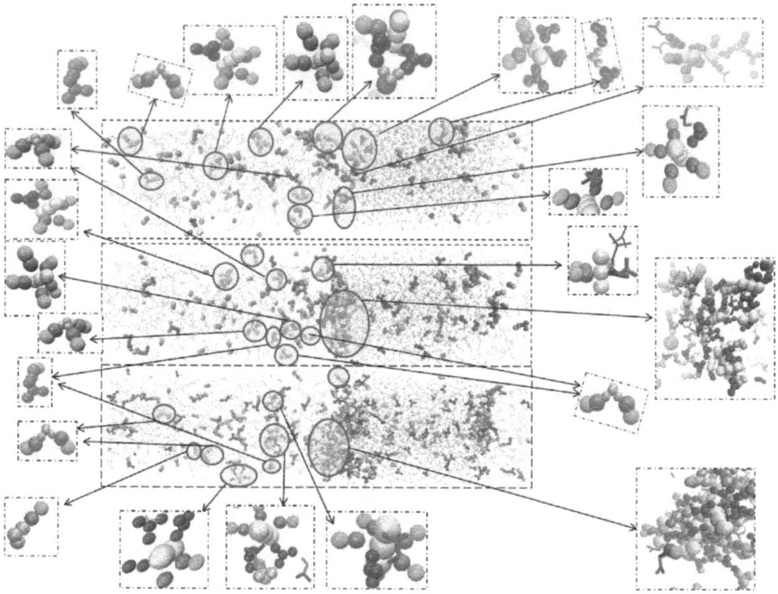

FIGURE 3.15
Snapshot for various UO_2^{2+} and TBP complexation (red and green balls represent the O(=P) and P of TBP, yellow balls UO_2^{2+} and blue balls the NO_3^-, whereas H_3O^+ and HNO_3 are represented by silver and purple color respectively. (Reproduced from Sahu, P. et al., *Physical Chemistry Chemical Physics*, 18: 23769–23784, 2016. With permission from PCCP owner societies.)

3.7 Separation of Minor Actinides using N-Donor Containing Extractants

In the preceding sections, complexation mechanism of UO_2^{2+} and Pu^{4+} ions were discussed by applying DFT and MD simulations. It is of utmost importance to separate the minor actinides from the lanthanides for effective nuclear waste reprocessing. At present, only few studies have been reported on understanding the complexation of Am^{3+} and Eu^{3+} with N-donor based CyMe4-BTBP [149,150,151] and CyMe4-BTPhen [152,153,154]. Benay et al. [149,152] performed the MD simulations to study the complexation and interfacial behavior of CyMe4-BTBP and CyMe4-BTPhen with Eu^{3+} ion in an ocatanol/water medium. Lan et al. [150,151] performed DFT studies on the complexation of Am^{3+} and Eu^{3+} ion with BTBP. Xiao et al. [154] reported DFT calculations of Am^{3+} and Eu^{3+} ion with CyMe4-BTPhen at the B3LYP level of theory for different type of complexes. Yang et al. [153] reported DFT and MP2 calculations of Am^{3+} and Eu^{3+} ions with CyMe4-BTPhen for only $\left[ML_2(NO_3)\right]$ complexes. All these studies are restricted to either BTBP or BTPhen and thus lack completeness. Therefore, detailed investigations are required to understand the selectivity of CyMe4-BTPhen over CyMe4-BTBP. To address this issue, the following complexation reaction has been considered:

$$M_{(aq)} + NO^-_{3(aq)} + 2L_{(org)} \xrightarrow{\Delta G_{ext}} \left[ML_2(NO_3)\right]^{2+}_{org} \text{ (M = } Am^{3+} \text{and } Eu^{3+}) \quad (3.13)$$

The change in Gibbs free energy, ΔG_{ext}, in Equation 3.13 obtained by the thermodynamic cycle (Figure 3.5) in conjunction with the COSMO model. The cosmo radii for Am and Eu was taken as 1.99Å and1.90Å, respectively. The dielectric constant, ε for octanol was used as 10. The optimized complexes of Am^{3+} and Eu^{3+} with CyMe4-BTPhen are presented in Figure 3.16.

The calculated Am-N1 (2.680Å) and Eu-N1 (2.681Å) distances are found to be same for $\left[ML_2(NO_3)\right]^{2+}$ with CyMe4-BTPhen, whereas Am-N2 distance (2.589Å) is shorter than Eu-N2 (2.641Å). The Eu-O bond distance is found to be also longer than Am-O distance for $\left[ML_2(NO_3)\right]^{2+}$. The optimized complexes of Am^{3+} and Eu^{3+} with CyMe4-BTBP are presented in Figure 3.17.

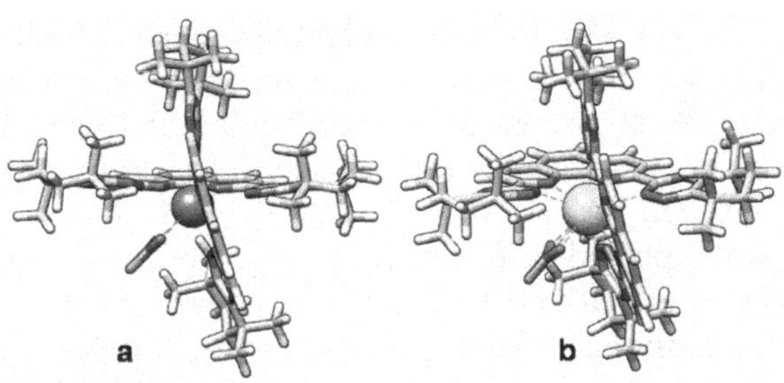

FIGURE 3.16
Optimized 1:2 complexes of (a) Am^{3+} and (b) Eu^{3+} with CyMe4-BTPhen at the BP86/TZVP level of theory.

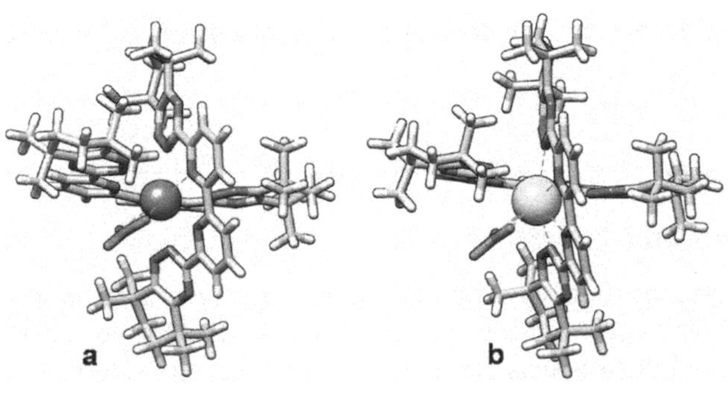

FIGURE 3.17
Optimized 1:2 complexes of (a) Am^{3+} and (b) Eu^{3+} with CyMe4-BTBP at the BP86/TZVP level of theory.

TABLE 3.12

ΔG_{ex} of Am^{3+}/Eu^{3} with CyMe4-BTPhen and CyMe4-BTBP. The Values in the Parenthesis Represent ΔG_{ex} of Am^{3+}/Eu^{3+} with CyMe4-BTBP. '#' and '$' Represent cis- and Trans-CyMe4-BTBP Respectively

Complex	ΔG_{ext} (kcal/mol)
$M^{3+} + 2L + NO_3^- \Rightarrow ML_2(NO_3)^{2+}$	–45.43/–42.25 (–52.03/–50.50)[#] (–41.05/–39.52)[$]

In the case of CyMe4-BTBP for $\left[ML_2(NO_3)\right]^{2+}$ type complex, the Am-N1 (2.602Å)distance is found to be shorter than Eu-N1 (2.681Å). Similarly, Am-N2 (2,657Å) is shorter than Eu-N2 (2.684Å). The Am-N1/Eu-N1 bond distances are found to be shorter for CyMe4-BTBP and Am-N2/Eu-N2 bond distances are shorter for the complexes with CyMe4-BTPhen. The calculated structural parameters with the available experimental data for Eu^{3+} complex are found to be in very good agreement.

The free energy of Am^{3+} and Eu^{3+} with cis, trans-CyMe4-BTBP, and CyMe4-BTPhen for $\left[ML_2(NO_3)\right]^{2+}$ complex shows that CyMe4-BTPhen ($\Delta\Delta G = \Delta G_{ex,\ Am3+}-\Delta G_{ex,\ Eu3+} = -3.18$) is more selective than CyMe4-BTBP ($\Delta\Delta G = \Delta G_{ex,\ Am3+}-\Delta G_{ex,\ Eu3+} = -1.53$ towards both cis and trans isomer) for Am^{3+} over Eu^{3+} ion. The ΔG_{ext} of 1:2 complexes were presented in Table 3.12. For CyMe4-BTBP, complexation of two conformers, namely cis and trans with Am^{3+} and Eu^{3+} ion, were studied for the $M^{3+} + 2L + NO_3^- \Rightarrow ML_2(NO_3)^{2+}$ reaction. The values of ΔG_{ext} are reported in Table 3.12.

From the table, it was found that the ΔG_{ext} value of cis-CyMe4-BTBP for Am^{3+} (–52.03) ion is found to be higher than trans-CyMe4-BTBP (–41.05). A similar trend was observed with Eu^{3+} ion, i.e. cis-CyMe4-BTBP (–50.50) has higher free energy than trans-CyMe4-BTBP (–39.52). The ΔG_{ext}values of cis-CyMe4-BTBP for both Am^{3+} (–52.03) and Eu^{3+} (–50.50) ions are found to be higher than corresponding values of CyMe4-BTPhen, i.e –45.43 and –42.25. In the case of trans-CyMe4-BTBP, the free energy for both Am^{3+} and Eu^{3+} ions are found to be smaller than corresponding values of CyMe4-BTPhen.

3.8 Conclusion

Density functional theory-based electronic structure calculations have been shown to be very useful in interpreting the coordinating environment of different radionuclide towards ligands. The interaction strength of extractant molecules with the radionuclide can be decisively predicted and thus can be used as a prescreening criteria. The preferential selectivity of radionuclide towards a particular extractant can be well predicted by determining the free energy of extraction using thermodynamic cycle, which is correlated with the experimentally determined distribution constants as confirmed by determining the free energy of UO_2^{2+} and Pu^{4+} towards DH2EHA, DHOA, TMDGA, and GO-EGMP. COSMO solvation model can be authoritatively used for the calculation of structural and thermodynamic properties in the solution phase. Even the diluents for solvent extraction can be screened using DFT-COSMO combination. It can be stated that the DFT-based computational methods have played and are still playing a key role in understanding the separation mechanism of Am^{3+}-Eu^{3+} ions from nuclear waste with varieties of extractants. In addition, high-fidelity MDsimulations were shown to capture the experimentally observed migration of uranyl nitrate from the aqueous to the organic phase. MD simulations on several systems involving various acid models reveal the effect of acid in the aqueous phase, organic phase, and at the interface. Simulations with varied TBP and acid concentration yield the trends similar to experiments and thus are helpful in planning the experiments.

Acknowledgments

Authors are grateful to Dr. Sadhana Mohan, Associate Director, ChEG and Mr. K.T. Shenoy, Head, ChED, BARC, Mumbai, for their continuous encouragement and support. Authors also acknowledge the Anupam computational facility of Computer Division for providing computer resources meant for advanced Modelling and simulation techniques.

References

1. http://www.iaea.org/pris.
2. The Nuclear Fuel Cycle: From Ore to Waste, ed. P.D. Wilson, Oxford University Press, Oxford, UK, 1996.
3. Ansari, S.A., Pathak, P., Mohapatra, P.K., Manchanda, V.K. (2012) Chemistry of Diglycolamides: Promising Extractants for Actinide Partitioning. *Chemical Review* 112: 751–1772.
4. Madic, C., Testard, F., Hudson, M.J., Liljenzin, J., Christansen, B., Ferrando, M., Facchini, A., Geist, A., Modolo, G., Espartero, G., Mendoza, J. (2004) New Solvent Extraction Processes for Minor Actinides—Final Report, *CEA-Report 6066.*
5. Nash, K.L., Madic, C., Mathur, J.N., Lacquement, J. (2006) Actinide, Separation and Technology. *The Chemistry of the Actinide and Transactinide Elements,* Springer, The Netherlands, 3rd ed., 4:2622 edited by Morss, L.R., Edelstein, N.M., Fuger, J., Katz, J.J. (2006).
6. Sasaki, Y., Sugo, Y., Suzuki, S., Tachimori, S. (2001) The novel extractants, Diglycolamides for the Extraction of Lanthanides and Actinides in HNO3-Dodecane System. *Solvent Extraction Ion Exchange* 19: 91–103.

7. Delmau, L.H., Bonnesen, P.V., Engle, N.L., Haverlock, T.J., Sloop, V., Moyer, B.A. (2006) Combined Extraction of Cesium and Strontium from Alkaline Nitrate. *Solvent Extraction Ion Exchange* 24: 197–225.
8. Dietz, M.L., Horowitz, E.P., Rhoads, S., Bartsch, R.A., Krzykawski, J. (1996) Extraction of Cesium from Acidic Nitrate Media Using Macrocyclic Polyethers: The Role of Organic Phase Water. *Solvent Extraction Ion Exchange* 14: 1–12.
9. Visser, A.E., Jensen, M.P., Laszak, I., Nash, K.L., Choppin, G.R., Rogers, R.D. (2003) Uranyl Coordination Environment in Hydrophobic Ionic Liquids: An In Situ Investigation. *Inorganic Chemistry* 42: 2197–2199.
10. Gelis, A.V., Lumetta, G.J. (2014) Actinide Lanthanide Separation Process: ALSEP. *Industrial & Engineering Chemistry* 2014, 53: 1624–1631.
11. Antonio, M.R., McAlister, D.R., Horwitz, E.P. (2015) An Europium (III) Diglycolamide Complex: Insights into the Coordination Chemistry of Lanthanides in Solvent Extraction. *Dalton Transactions* 44: 515–521.
12. Kaltsoyannis, K. (2013). Does Covalency Increase or Decrease across the Actinide Series? Implications for Minor Actinide Partitioning. *Inorganic Chemistry* 52: 3407–3413.
13. Benay, G., Schurhammer, R., Desaphy, J., Wipff, G. (2011) Substituent Effects on BTP's Basicity and Complexation Properties with Ln III Lanthanide Ions. *New Journal of Chemistry* 35: 184–189.
14. Benay, G., Schurhammer, R., Wipff, G. (2010) BTP-Based Ligands and Their Complexes with Eu3+ at "Oil"/Water Interfaces. A Molecular Dynamics Study. *Physical Chemistry Chemical Physics* 12: 11089–11102.
15. Bhattacharyya, A., Ghanty, T., Mohapatra, P.K., Manchanda, V.K. (2011) Selective Americium(III) Complexation by Dithiophosphinates: A Density Functional Theoretical Validation for Covalent Interactions Responsible for Unusual Separation Behavior from Trivalent Lanthanides. *Inorganic Chemistry* 50: 3913–3921.
16. Kriz, J., Dybal, J., Makrlik, E., Vanura, P., Moyer, B.A. (2011) Interaction of Cesium Ions with Calix[4]arene-bis(t-octylbenzo-18-crown-6): NMR and Theoretical Study. *Journal of Physical Chemistry B* 115: 7578–7587.
17. Keith, J.M., Batista, R. (2012) Theoretical Examination of the Thermodynamic Factors in the Selective Extraction of Am3+ from Eu3+ by Dithiophosphinic Acids. *Inorganic Chemistry* 51: 13–15.
18. Cao, X., Heidelberg, D., Clupka, J., Dolg, M. (2010) First-Principles Study of the Separation of AmIII/CmIII from EuIII with Cyanex301. *Inorganic Chemistry* 49: 10307–10315.
19. Jackson, V.E., Gutowski, K.E., Dixon, D.A. (2013) Density Functional Theory Study of the Complexation of the Uranyl Dication with Anionic Phosphate Ligands with and without Water Molecules. *Journal of Physical Chemistry A* 117: 8938–8957.
20. Boda, A., Ali, Sk.M. (2012) Density Functional Theoretical Investigation of Remarkably High Selectivity of the Cs+ Ion over the Na+ Ion toward Macrocyclic Hybrid Calix-Bis-Crown Ether. *Journal of Physical Chemistry A* 116: 8615–8623.
21. Ali, Sk. M., Joshi, J.M., Singha Deb, A.K., Boda, A., Shenoy, K.T., Ghosh, S.K. (2014) Dual Mode of Extraction for Cs+ and Na+ Ions with Dicyclohexano-18-crown-6 and Bis(2-propyloxy)calix[4]crown-6 in Ionic Liquids: Density Functional Theoretical Investigation. *RSC Advances* 4: 22911–22925.
22. Ali, Sk.M. (2014) Design and Screening of Suitable Ligand/Diluents Systems for Removal of Sr2+ Ion from Nuclear Waste: Density Functional Theoretical Modelling. *Computational & Theoretical Chemistry* 1034: 38–52.

23. Ali, Sk.M. (2014) Thermodynamical Criteria of the Higher Selectivity of Zirconium Oxycations over Hafnium Oxycations towards Organophosphorus Ligands: Density Functional Theoretical Investigation. *European Journal of Inorganic Chemistry* 2014: 1533–1545.
24. Roy, L.E., Bridges, N.J., Martin, L.R. (2013) Theoretical Insights into Covalency Driven f Element Separations. *Dalton Transactions* 42: 2636–2642.
25. Wang, C.J., Lan, J.H., Wu, Q.Y., Zhao, Y.L., Wang, X.K., Chai, Z.F., Shi, W.Q. (2014) Theoretical Insights into the Separation of Am (III) over Eu (III) with PhenBHPPA. *Dalton Transactions* 42: 8713–8720.
26. Lumetta, G.J., Rapko, B.M., Garza, P.A., Hay, B.P., Gilbertson, R.D., Weakley, T.J., Hutchison, J.E. (2002) Deliberate Design of Ligand Architecture Yields Dramatic Enhancement of Metal Ion Affinity. *Journal of American Chemical Society* 124: 5644–5645.
27. Sood, D.D., Patil, S.K. (1996) Chemistry of Nuclear Fuel Reprocessing: Current Status. *Journal of Radioanalytical and Nuclear Chemistry* 203: 547–573.
28. Mathur, J.N., Murali, M.S., Nash, K.L. (2001) Actinide Partitioning—A Review. *Solvent Extraction Ion Exchange* 19: 357–390.
29. Nash K.L., Madic C., Mathur J.N., Lacquement J. (2008) Actinide Separation Science and Technology. In: Morss L.R., Edelstein N.M., Fuger J. (eds) The Chemistry of the Actinide and Transactinide Elements. Springer, Dordrecht.
30. Manchanda, V.K., Ruikar, P.B., Sriram, S., Nagar, M.S., Pathak, P.N., Gupta, K.K., Singh, R.K., Chitnis, R.R., Dhami, P.S., Ramanujam, A. (2001) Distribution Behavior of U (VI), Pu (IV), Am (III), and Zr (IV) with N, N-dihexylOctanamide under Uranium-Loading Conditions. *Nuclear Technology* 134, 231–240.
31. Foreman, M.R.S., Hudson, M.J., Drew, M.G.B., Hill, C., Madic, C. (2006) Complexes Formed between the Quadridentate, Heterocyclic Molecules 6,6-bis-(5,6-dialkyl-1,2,4-triazin-3-yl)-2,2-bipyridine (BTBP) and Lanthanides(III): Implications for the Partitioning of Actinides(III) and Lanthanides(III). *Dalton Transactions* 6: 1645–1653.
32. Hudson, M.J., Boucher, C.E., Braekers, D., Desreux, J.F., Drew, M.G.B., Foreman, M.R.S., Harwood, L.M., Hill, C., Madic, C., Marken, F., Youngs, T.G.A. (2006) New bis (triazinyl) Pyridines for Selective Extraction of Americium (III). *New Journal of Chemistry* 30: 1171–1183.
33. Magill, J., Berthou, V., Haas, D.,Galy, J.,Schenkel, R., Wiese, H.W., Heusener, G., Tommasi, J., Youinou, G. (2003) Impact Limits of Partitioning and Transmutation Scenarios on the Radiotoxicity of Actinides in Radioactive Waste. *Journal of Nuclear Engineering* 42: 263–277.
34. Afsar, A., Laventine, D.M., Harwood, L.M., Hudson, M.J., Geist, A. (2013) Utilizing Electronic Effects in the Modulation of BTPhen Ligands with Respect to the Partitioning of Minor Actinides from Lanthanides. *Chemical Communication* 49: 8534–8536.
35. Hudson, M.J., Harwood, L.M., Laventine, D.M., Lewis, F.W. (2013) Use of Soft Heterocyclic N-Donor Ligands to Separate Actinides and Lanthanides. *Inorganic Chemistry* 52: 3414–3428.
36. Katz, J.J., Morss, L.R., Edelstein, N.M., Fuger, J. (2008) *The Chemistry of the Actinide and Transactinide Elements*. Springer, Dordrecht.
37. Siddall, T.H. (1960) Effects of Structure of N, N-Disubstituted Amides on Their Extraction of Actinide and Zirconium Nitrates and of Nitric Acid. *Journal of Physical Chemistry* 64: 1863–1866.

38. Siddall, T.H. (1961) USAEC Report DP – 541, E. I. Du Pont de Nemours and Co. Alken, SC.
39. Manchanda, V.K., Pathak, P.N. (2004) Amides and Diamides as Promising Extractants in the Back End of the Nuclear Fuel Cycle: An Overview. *Separation and Purification Technology* 35: 85–103.
40. Gupta, K.K., Manchanda, V.K., Subramanian, M.S., Singh, R.K. (2000) N,N-Dihexyl Hexanamide: A Promising Extractant for Nuclear Fuel Reprocessing. *Separation Science and Technology* 35: 1603–1617.
41. Pathak, P.N., Prabhu, D.R., Kanekar, A.S., Manchanda, V.K. (2009) Recent R&D Studies Related to Coprocessing of Spent Nuclear Fuel Using N,N-Dihexyloctanamide. *Separation Science and Technology* 44: 3650–3663.
42. Kumar, V., Mondal, S., Mallavarapu, A., Ali, Sk. M., Singh, D.K. (2017) High Complexation Selectivity of U (VI) over Rare Earths by N, N-Dihexyl-2-ethylhexanamide (DH2EHA): Experimental and Theoretical Evidence. *Chemistry Select* 2: 2348–2354.
43. Becke, A.D. (1993) Density-Functional Thermochemistry. III. The Role of Exact Exchange. *Journal of Chemical Physics* 98: 5648–5652; Lee, C., Wang, W., Parr, R.G. (1988) Development of the Colle-Salvetti Correlation-Energy Formula into a Functional of the Electron Density. *Physical Review B* 37: 785–789.
44. Schaefer, A., Horn, H., Ahlrichs, R. (1992) Fully Optimized Contracted Gaussian Basis Sets for Atoms Li to Kr. *Journal of Chemical Physics* 97: 2751–2777.
45. Ahlrichs, R., Bar, M.,Haser, M., Horn, H., Kolmel, C. (1989) Electronic Structure Calculations on Workstation Computers: The Program System Turbomole. *Chemical Physics Letter* 162: 165–169; TURBOMOLE V6.0 2009, a development of University of Karlsruhe and Forschungszentrum Karlsruhe GmbH, 1989–2007, TURBOMOLE GmbH.
46. Xiao, C.L., Wu, Q.Y., Wang, C.Z., Zhao, Y.L., Chai, Z.F., Shi, W.Q. (2014) Quantum Chemistry Study of Uranium (VI), Neptunium (V), and Plutonium (IV, VI) Complexes with Preorganized Tetradentate Phenanthrolineamide Ligands. *Inorganic Chemistry* 53: 10846–10853.
47. Ali, Sk.M. (2017) Enhanced Free Energy of Extraction of Eu 3+ and Am 3+ Ions towards Diglycolamide Appended Calix[4]arene: Insights from DFT-D3 and COSMO-RS Solvation Models. *Dalton Transactions* 46: 10886–10898.
48. Wu, H., Wu, Q.Y., Wang, C.Z., Lan, Z.H., Liu, Z.R., Chai, Z.F., Shi, W.Q. (2016) New Insights into the Selectivity of Four 1,10-phenanthroline-Derived Ligands toward the Separation of Trivalent Actinides and Lanthanides: A DFT-Based Comparison Study. *Dalton Transactions* 45: 8107–8117.
49. Dolg, M., Stoll, H., Preuss, H. (1989) Energy-Adjusted ab initio Pseudopotentials for the Rare Earth Elements. *Journal of Chemical Physics* 90: 1730–1734; Kuchle, W., Dolg, M., Stoll, H., Preuss, H. (1994) Energy-Adjusted Pseudopotentials for the Actinides. Parameter Sets and Test Calculations for Thorium and Thorium Monoxide. *Journalof Chemical Physics* 100: 7535–7542; Cao, X., Dolg, M., (2004) Segmented Contraction Scheme for Small-Core Actinide Pseudopotential Basis Sets. *Journal of Molecular Structure (THEOCHEM)* 673: 203–209; Cao, X., Dolg, M., (2002) Segmented Contraction Scheme for Small-Core Lanthanide Pseudopotential Basis Sets. *Journal of Molecular Structure (THEOCHEM)* 581: 139–147.
50. Shamov, G.A., Schreckenbach, G., Vo, T.N. (2007) A Comparative Relativistic DFT and ab initio Study on the Structure and Thermodynamics of the Oxofluorides of Uranium (IV), (V) and (VI). *Chemistry: A European Journal* 13: 4932–4947.

51. Sengupta, A., Sk., J. Boda, A., Ali, Sk.M. (2016) An Amide Functionalized Task Specific Carbon Nanotube for the Sorption of Tetra and HexaValent Actinides: Experimental and Theoretical Insight. *RSC Advances* 6: 39553–39562.
52. Klamt, A. (1995) Conductor-Like Screening Model for Real Solvents: A New Approach to the Quantitative Calculation of Solvation Phenomena. *Journal of Physical Chemistry* 99: 2224–2235.
53. Pahan, S., Boda, A., Ali, Sk.M. (2015) Density Functional Theoretical Analysis of Structure, Bonding, Interaction and Thermodynamic Selectivity of Hexavalent Uranium (UO2 2+) and Tetravalent Plutonium (Pu4+) Ion Complexes of Tetramethyl Diglycolamide (TMDGA). *Theoretical Chemistry Accounts* 134: 41–57.
54. Gasparini, G.M., Grossi, G. (1980) Application of N, N-dialkyl Aliphatic Amides in the Separation of Some Actinides. *Separation Science and Technology* 15: 825–844.
55. Gupta, K.K., Manchanda, V.K., Sriram, S., Thomas, G., Kulkarni, P.G., Singh, R.K. (2000) Third Phase Formation in the Extraction of Uranyl Nitrate by N, N-dialkylAliphatic Amides. *Solvent Extraction Ion Exchange* 18: 421–439.
56. Boda, A., Joshi, J.M., Ali, Sk. M., Shenoy, K.T., Ghosh, S.K. (2013) Density Functional Theoretical Study on the Preferential Selectivity of Macrocyclic Dicyclohexano-18-crown-6 for Sr+2 Ion over Th+4 Ion during Extraction from an Aqueous Phase to Organic Phases with Different Dielectric Constants. *Journal of Molecular Modeling* 19: 5277–5291.
57. Boda, A., Ali, Sk.M., Rao, H., Ghosh, S.K. (2011) DFT Modeling on the Suitable Crown Ether Architecture for Complexation with Cs+ and Sr2+ Metal Ions. *Journal of Molecular Modeling* 17: 1091–1108.
58. Boda, A., Ali, Sk.M., Shenoy, K.T., Ghosh, S.K. (2013) Density Functional Theoretical Modeling of Selective Ligand for the Separation of Zr and Hf Metal Oxycations (ZrO2+ and HfO2+). *Separation Science and Technology* 48: 2397–2409.
59. Singha Deb, A.K., Ali, Sk.M., Shenoy, K.T. (2015) Unanticipated Favoured Adsorption Affinity of Th(IV) Ions towards Bidentate Carboxylate Functionalized Carbon Nanotubes (CNT–COOH) over Tridentate Diglycolamic Acid Functionalized CNT: Density Functional Theoretical Investigation. *RSC Advances* 5: 80076–80088.
60. Boda, A., Singha Deb, A.K., Sengupta, A., Ali, Sk.M., Shenoy, K.T. (2017) Elucidation of Complexation of Tetra and Hexavalent Actinides towards an Amide Ligand in Polar and Non-Polar Diluents: Combined Experimental and Theoretical Approach. *Polyhedron* 123: 234–242.
61. Becke, A.D. (1988) Density-Functional Exchange-Energy Approximation with Correct Asymptotic Behavior. *Physical Review A* 38: 3098–3100.
62. Perdew, J.P. (1986) Density-Functional Approximation for the Correlation Energy of the Inhomogeneous Electron Gas. *Physical Review B* 33: 8822–8824.
63. Becke, A.D. (1993) A New Mixing of Hartree–Fock and Local Density-Functional Theories. *Journal of Chemical Physics* 98: 1372–1377.
64. Neese, F. (2009) Prediction of Molecular Properties and Molecular Spectroscopy with Density Functional Theory: From Fundamental Theory to Exchange-Coupling. *Coordination Chemistry Review* 253: 526–563.
65. Klamt, A., Schuurmann, G. (1993) COSMO: A New Approach to Dielectric Screening in Solvents with Explicit Expressions for the Screening Energy and Its Gradient. *Journal of Chemical Society Perkin Transactions* 2: 799–805.

66. Shamov, G.A., Schreckenbach, G. (2005) Density Functional Studies of Actinyl Aquo Complexes Studied Using Small-Core Effective Core Potentials and a Scalar Four-Component Relativistic Method. *Journal of Physical Chemistry* A 109: 10961–10974.
67. Adamo, C., Barone, V. (1999) Toward Reliable Density Functional Methods without Adjustable Parameters: The PBE0 Model. *Journal of Chemical Physics* 110: 6158–6170.
68. Vetere, V., Adamo, C., Maldivi, P. (2000) Performance of the parameter Free'PBE0 Functional for the Modeling of Molecular Properties of Heavy Metals. *Chemical Physics Letter* 325: 99–105.
69. Schaftenaar, G., Noordik, J.H. (2000) Molden: A Pre- and Post-Processing Program for Molecular and Electronic Structures. *Journal of Computer-Aided Molecular Design* 14: 123–134.
70. Neuefeind, J., Soderholm, L., Skanthakumar, S. (2004) Experimental Coordination Environment of Uranyl (VI) in Aqueous Solution. *Journal of Physical Chemistry A* 108: 2733–2739.
71. Allen, P.G., Bucher, J.J., Shuh, D.K., Edelstein, N.M., Reich, T. (1997) Investigation of Aquo and Chloro Complexes of UO22+, NpO2+, Np4+, and Pu3+ by X-ray Absorption Fine Structure Spectroscopy. *Inorganic Chemistry* 36: 4676–4683.
72. Semon, L., Boehme, C., Billard, I., Hennig, C., Lutzenkirchen, K., Reich, T., Rossberg, A., Rossini, I. Wipff, G. (2001) Do Perchlorate and Triflate Anions Bind to the Uranyl Cation in an Acidic Aqueous Medium? A Combined EXAFS and Quantum Mechanical Investigation. *Chemistry Physical Chemistry* 2: 591–598.
73. Gutowski, K.E., Dixon, D.A. (2006) Predicting the Energy of the Water Exchange Reaction and Free Energy of Solvation for the Uranyl Ion in Aqueous Solution. *The Journal of Physical Chemistry A* 110: 8840–8856.
74. Kannan, S., Deb, S.B., Gamare, J.S., Drew, M.G.B. (2008) Coordination and Separation Studies of the Uranyl Ion with Iso-butyramide Based Ligands: Synthesis and Structures of [UO2(NO3)2(iC3H7CON{iC4H9}2)2] and [UO2(C6 H5COCHCOC6H5)2(iC3H7CON{iC3H7}2)]. *Polyhedron* 27: 2557–2562.
75. Bryantsev, V.S., Diallo, M.S., Goddard III, W.A. (2008) Calculation of Solvation Free Energies of Charged Solutes Using Mixed Cluster/Continuum Models. *The Journal of Physical Chemistry B* 112: 9709–9719.
76. Panja, S., Mohapatra, P.K., Tripathi, S.C., Manchnada, V.K. (2009) Studies on Uranium (VI) Pertraction across a N, N, N N-tetraoctyldiglycolamide (TODGA) Supported Liquid Membrane. *Journal of Membrane Science* 337: 274–281.
77. Sasaki, Y., Sugo, Y., Suzuki, S., Kimura, T. (2005) A Method for the Determination of Extraction Capacity and Its Application to N, N, N, N-tetraalkylderivatives of Diglycolamide-monoamide/n-dodecane Media. *Analytica Chimica Acta* 543: 31–37.
78. Nave, S., Modolo, G., Madic, C., Testard, F. (2008) Aggregation Properties of N,N,N,N-Tetraoctyl-3-oxapentanediamide (TODGA) in n-Dodecane. *Solvent Extraction Ion Exchange* 22: 527–551.
79. Modolo, G., Asp, H., Schreinemachers, C., Vijgen, V. (2007) Development of a TODGA Based Process for Partitioning of Actinides from a PUREX Raffinate Part I: Batch Extraction Optimization Studies and Stability Tests. *Solvent Extraction Ion Exchange* 25: 703–721.

80. Modolo, G., Asp, H., Vijgen, V., Malmbeck, R., Magnusson, D., Sorel, C. (2008) Demonstration of a TODGA-Based Continuous Counter-Current Extraction Process for the Partitioning of Actinides from a Simulated PUREX Raffinate, Part II: Centrifugal Contactor Runs.*Solvent Extraction Ion Exchange* 26: 62–76.
81. Ansari, S.A., Pathak, P.N., Hussain, M., Parmar, A.K., Prasad, A.K., Manchanda, V.K. (2006) Extraction of Actinides Using N, N, N, N-tetraoctyldiglycolamide (TODGA): A Thermodynamic Study. *Radiochimica Acta* 94: 307–312.
82. Brown, J., McLachlan, F., Sarsfield, M., Taylor, R., Modolo, G., Wilden, A. (2012) Plutonium Loading of Prospective Grouped Actinide Extraction (GANEX) Solvent Systems Based on Diglycolamide Extractants. *Solvent Extraction Ion Exchange* 30: 127–141.
83. Mowafy, E.A.,Aly, H.F. (2007)Synthesis of Some N, N, N, N-Tetraalkyl-3-Oxa-Pentane-1, 5-diamide and Their Applications in Solvent Extraction. *Solvent Extraction Ion Exchange* 25: 205–224.
84. Kannan, S., Moody, M.A., Barnes, C.L., Duval, P.V. (2008) Lanthanum (III) and Uranyl (VI) Diglycolamide Complexes: Synthetic Precursors and Structural Studies Involving Nitrate Complexation. *Inorganic Chemistry*47: 4691–4695.
85. Tian, G., Rao, L., Teat, S.G., Liu, G. (2009) Quest for Environmentally Benign Ligands for Actinide Separations: Thermodynamic, Spectroscopic, and Structural Characterization of UVI Complexes with Oxa-diamide and Related Ligands. *Chemistry: A European Journal* 15: 4172–4181.
86. Reilly, S.D., Gaunt, A.J., Scott, B.L., Modolo, G., Iqbal, M., Verboom, W., Sarsfield, M.J. (2012) Plutonium (IV) Complexation by Diglycolamide Ligands—Coordination Chemistry Insight into TODGA-Based Actinide Separations. *Chemical Communication* 48: 9732–9734.
87. Gong, Y., Hu, H.S., Rao, L., Li, J., Gibson, J.K. (2013) Experimental and Theoretical Studies on the Fragmentation of Gas-Phase Uranyl–, Neptunyl–, and Plutonyl–Diglycolamide Complexes. *The Journal of Physical Chemistry B* 117: 10544–10550.
88. Gong, Y., Hu, H.S., Tian, G., Rao, L., Li, J., Gibson, J.K. (2013) A Tetrapositive Metal Ion in the Gas Phase: Thorium (IV) Coordinated by Neutral Tridentate Ligands. *Angew Chemistry* 52: 6885–6888.
89. Ansari, S.A., Mohapatra, P.K., Ali, Sk.M., Sengupta, A., Bhattacharyya, A., Verboom, W. (2016) Understanding the Complexation of Eu 3+ with Three Diglycolamide-Functionalized Calix[4]arenes: Spectroscopic and DFT Studies. *Dalton Transactions* 45: 5425–5429.
90. Hirata, M., Guilbaud, P., Doblera, M., Tachimori, S. (2003) Molecular Dynamics Simulations for the Complexation of Ln 3+ and UO 2 2+ Ions with Tridentate Ligand Diglycolamide (DGA). *Physical Chemistry Chemical Physics* 5: 691–695.
91. Ye, X., Cui, S., de Almeida, V.F., Hay, B.P., Khomami, B. (2010) Uranyl Nitrate Complex Extraction into TBP/Dodecane Organic Solutions: A Molecular Dynamics Study. *Physical Chemistry Chemical Physics* 12: 15406–15409.
92. Ali, Sk. M., Pahan, S., Bhattacharyya, A., Mohapatra, P.K. (2016) Complexation Thermodynamics of Diglycolamide with f-Elements: Solvent Extraction and Density Functional Theory Analysis. *Physical Chemistry Chemical Physics* 18: 9816–9828.
93. Gopi Krishna, G., Reddy, R.S., Raghunath, P., Bhanuprakash, K., Lakshmi, K., Choudary, B.M. (2004) A Computational Study of Ligand Interactions with Hafnium and Zirconium Metal Complexes in the Liquid–Liquid Extraction Process. *The Journal of Physical Chemistry B* 108: 6112–6120.

94. Eckert, F., Klamt, A. (2008) COSMOtherm, version C2.1, Release 01.08; COSMOlogic GmbH & Co, KG, Leverkusen, Germany.
95. Justin, J.P., Sundararajan, M., Vincent, M.A., Hiller, I.H. (2009) The Geometric Structures, Vibrational Frequencies and Redox Properties of the Actinyl Coordination Complexes ([AnO 2 (L) n] m; An= U, Pu, Np; L= H 2 O, Cl–, CO32–, CH3CO2–, OH–) in Aqueous Solution, Studied by Density Functional Theory Methods. *Dalton Transactions* 2009: 5902–5909.
96. Ali, Sk. M., Boda, A., Singha Deb, A.K., Sahu, P., Shenoy, K.T. (2017) Computational Chemistry Assisted Design and Screening of Ligand-Solvent Systems for Metal Ion Separation.Frontiers in Computational Chemistry, Bentham Science Publishers, Sharjah, UAE.
97. Schreekenbach, G., Shamov, G.A. (2010) TheoreticalActinide Molecular Science. *Accounts of Chemical Research* 43: 19–29.
98. Zhu, Z.X.,Sasaki, Y., Suzuki, H., Suzuki, S., Kimura, T. (2004) Cumulative Study on Solvent Extraction of Elements by N, N, N′, N′-tetraoctyl-3-oxapentanediamide (TODGA) from Nitric Acid into n-dodecane. *Analytica Chimica Acta* 527: 163–168.
99. Zhan, C.G., Dixon, D.A. (2001) Absolute Hydration Free Energy of the Proton from First-Principles Electronic Structure Calculations. *The Journal of Physical Chemistry A* 105: 11534–11540.
100. Ciupka, J., Cao-Dolg, X., Wiebke, J., Dolg, M. (2010) Computational Study of Lanthanide (III) Hydration. *Physical Chemistry Chemical Physics* 12: 13215–13223.
101. Marcus, Y. (1991) Thermodynamics of Solvation of Ions. Part 5.—Gibbs Free Energy of Hydration at 298.15 K. *Journal of Chemical Society Faraday Transactions* 87: 2995–2999.
102. Ansari, S.A., Pathak, P.N., Hussain, M., Parmar, A.K., Prasad, A.K., Manchanda, V.K. (2006) Extraction of Actinides Using N, N, N′, N′-tetraoctylDiglycolamide (TODGA): A Thermodynamic Study. *Radiochimica Acta* 94: 307–312.
103. Verma, P.K., Kumari, N., Pathak, P.N., Sadhu, B., Sundararajan, M., Aswal, V.K., Mohapatra, P.K. (2014) Investigations on Preferential Pu(IV) Extraction over U(VI) by N,N-Dihexyloctanamide versus Tri-n-butyl Phosphate: Evidence through Small Angle Neutron Scattering and DFT Studies. *The Journal of Physical Chemistry A* 118: 3996–4004.
104. Wei, M., He, Q., Feng, X., Chen, J. (2012) Physical Properties of N,N,N,N-tetramethyl Diglycolamide and Thermodynamic Studies of Its Complexation with Zirconium, Lanthanides and Actinides. *Journal of Radioanalytical and Nuclear Chemistry* 293: 689–697.
105. Meng, W., Xiaogui, F., Jing, C. (2013) Studies on Stripping of Actinides, and Some Fission and Corrosion Products from Loaded TRPO by N,N,N′,N′-Tetramethyl Diglycolamide. *Separation Science Technology* 48: 741–748.
106. Zhao, G., Li, J.X., Ren, X.M., Chen, C.L., Wang, X.K. (2011) Few-Layered Graphene Oxide Nanosheets as Superior Sorbents for Heavy Metal Ion Pollution Management. *Environment Science & Technology* 45: 10454–10462.
107. Mauter, M.S., Elimelech, M. (2008) Environmental Applications of Carbon-Based Nanomaterials. *Environment Science & Technology* 42: 5843–5859.
108. Carpio, I.E.M., Mangadlao, J.D., Nguyen, H.N., Advincula, R.C., Rodrigues, D.F. (2014) Graphene Oxide Functionalized with Ethylenediamine TriaceticAcid for Heavy Metal Adsorption and Anti-Microbial Applications. *Carbon* 77: 289–301.

109. Smith, S.C., Rodrigues, D.F. (2015) Carbon-Based Nanomaterials for Removal of Chemical and Biological Contaminants from Water: A Review of Mechanisms and Applications. *Carbon* 91: 122–143.
110. Eigler, S., Hirsch, A. (2014) Chemistry with Graphene and Graphene Oxide—Challenges for Synthetic Chemists. *Angew Chemistry International Edition* 53: 7720–7738.
111. Thomas, H.R., Marsden, A.J., Walker, M., Wilson, N.R., Rourke, J.P. (2014) Sulfur-Functionalized Graphene Oxide by Epoxide Ring-Opening. *Angew Chemistry International Edition* 53: 7613–7618.
112. Romanchuk, A.Y., Slesarev, A.S., Kalmykov, S.N., Kosynkinz, D.V., Tour, J.M. (2013) Graphene Oxide for Effective Radionuclide Removal. *Physical ChemistryChemical Physics* 15: 2321–2327.
113. Zhao, G.X., Wen, T., Yang, X., Yang, S.B., Liao, J.L., Hu, J., Shao, D.D., Wang, X.K. (2012) Preconcentration of U (VI) Ions on Few-Layered Graphene Oxide Nanosheets from Aqueous Solutions. *Dalton Transactions* 41: 6182–6188.
114. Sun, Y.B., Wang, Q., Chen, C.L., Tan, X.L., Wang, X.K. (2012) Interaction between Eu(III) and Graphene Oxide Nanosheets Investigated by Batch and Extended X-ray Absorption Fine Structure Spectroscopy and by Modeling Techniques. *Environmental Science and Technology* 46: 6020–6027.
115. Li, Z., Chen, F., Yuan, L., Liu, Y., Zhao, Y., Chai, Z., Shi, W. (2012) Uranium (VI) Adsorption on Graphene Oxide Nanosheets from Aqueous Solutions. *Chemical Engineering Journal* 210: 539–546.
116. Zhao, G.X., Ren, X.M., Gao, X., Tan, X.L., Li, J.X., Chen, C.L., Huang, Y., Wang, X.K. (2011) Removal of Pb (II) Ions from Aqueous Solutions on Few-Layered Graphene Oxide Nanosheets. *Dalton Transactions* 40: 10945–10952.
117. Song, W., Wang, X., Wang, Q., Shao, D., Wang, X. (2015) Plasma-Induced Grafting of Polyacrylamide on Graphene Oxide Nanosheets for Simultaneous Removal of Radionuclides. *Physical Chemistry Chemical Physics* 17:398–406.
118. Das, S., Pandey, A.K., Athawale, A.K., Manchanda, V.K. (2009) Adsorptive Preconcentration of Uranium in Hydrogels from Seawater and Aqueous Solutions. *Industrial & Engineering Chemistry Research* 48: 6789–6796.
119. Das, S., Pandey, A.K., Athawale, A.K., Subramanian, M., Seshagiri, T.K., Khanna, P.K., Manchanda, V.K. (2011) Silver Nanoparticles Embedded Polymer Sorbent for Preconcentration of Uranium from Bio-Aggressive Aqueous Media. *Journal of Hazardous Materials* 186: 2051–2059.
120. Das, S., Pandey, A.K., Athawale, A.K., Natarajan, V., Manchanda, V.K. (2012) Uranium Preconcentration from Seawater Using Phosphate Functionalized Poly (propylene) Fibrous Membrane. *Desalination & Water Treatment* 38: 114–120.
121. Vasudevan, T., Das, S., Sodaye, S., Pandey, A.K., Reddy, A.V.R. (2009) Pore-functionalized Polymer Membranes for Preconcentration of Heavy Metal Ions. *Talanta* 78: 171–177.
122. Chappa, S., Singha Deb, A.K., Ali, Sk.M., Debnath, A.K., Aswal, D.K., Pandey, A.K. (2016) Change in the Affinity of Ethylene Glycol Methacrylate Phosphate Monomer and Its Polymer Anchored on a Graphene Oxide Platform toward Uranium(VI) and Plutonium(IV) Ions. *The Journal of Physical Chemistry B* 120: 2942–2950.
123. Guo, Y., Bao, C., Song, L., Yuan, B., Hu, Y. (2011) In Situ Polymerization of Graphene, Graphite Oxide, and Functionalized Graphite Oxide into Epoxy Resin and Comparison Study of On-the-Flame Behavior. *Industrial & Engineering Chemistry Research*50: 7772–7783.

124. Singha Deb, A.K., Ali, Sk.M., Shenoy, K.T., Ghosh, S.K. (2015) Adsorption of Eu3+ and Am3+ ion towards Hard Donor-Based Diglycolamic Acid-Functionalised Carbon Nanotubes: Density Functional Theory Guided Experimental Verification. *Molecular Simulation* 41: 490–503.
125. Beudaert, P., Lamare, V., Dozol, J.F., Troxler, L., Wipff, G. (1998) Theoretical Studies on Tri-n-butyl Phosphate: MD Simulations in Vacuo, in Water, in Chloroform and at a Water/Chloroform Interface. *Solvent Extraction and Ion Exchange* 16: 597–618.
126. Ye, X., Cui, S., de Almeida, V., Khomami, B. (2009) Interfacial Complex Formation in Uranyl Extraction by Tributyl Phosphate in Dodecane Diluent: A Molecular Dynamics Study. *The Journal of Physical Chemistry B* 113: 9852–9862.
127. Ye, X., Smith, R.B., Cui, S., de Almeida, V., Khomami, B. (2010) Influence of Nitric Acid on Uranyl Nitrate Association in Aqueous Solutions: A Molecular Dynamics Simulation Study. *Solvent Extraction and Ion Exchange* 28: 1–18.
128. Baaden, M., Burgard, M., Wipff, G. (2001) TBP at the Water-oil Interface: The Effect of TBP Concentration and Water Acidity Investigated by Molecular Dynamics Simulations. *The Journal of Physical Chemistry B* 105: 11131–11141.
129. Baaden, M., Schurhammer, R., Wipff, G. (2002) Molecular Dynamics Study of the Uranyl Extraction by Tri-n-butylphosphate (TBP): Demixing of Water/"Oil"/TBP Solutions with a Comparison of Supercritical CO2 and Chloroform. *The Journal of Physical Chemistry B* 106: 434–441.
130. Jayasinghe, M., Beck, T.L. (2009) Molecular Dynamics Simulations of the Structure and Thermodynamics of Carrier-Assisted Uranyl Ion Extraction. *The Journal of Physical Chemistry B*113: 11662–11671.
131. Chiarizia, R., Jensen, M.P., Borkowski, M., Ferraro, J.R., Thiyagarajan, P., Littrell, K.C. (2003) Third Phase Formation Revisited: The U(VI), HNO3–TBP, n-Dodecane System. *Solvent Extraction and Ion Exchange* 21: 1–27.
132. Plaue, J., Gelis, A., Czerwinski, K., Thiyagarajan, P., Chiarizia, R. (2006) Small-Angle Neutron Scattering Study of Plutonium Third Phase Formation in 30% TBP/HNO3/Alkane Diluent Systems. *Solvent Extraction and Ion Exchange* 24: 283–298.
133. Deepika, P., Sabharwala, K.N., Srinivasan, T.G., Rao, P.R.V. (2010) Studies on the Use of N,N,N′,N′-Tetra(2-ethylhexyl) Diglycolamide (TEHDGA) for Actinide Partitioning. I: Investigation on Third-Phase Formation and Extraction Behavior. *Solvent Extraction and Ion Exchange* 28: 184–201.
134. Baaden, M., Berny, F., Wipff, G. (2001) The Chloroform/TBP/Aqueous Nitric Acid Interfacial System: A Molecular Dynamics Investigation. *Journal of Molecular Liquids* 90: 1–9.
135. Sahu, P., Ali, Sk.M., Shenoy, K.T. (2016) Passage of TBP–Uranyl Complexes from Aqueous–Organic Interface to the Organic Phase: Insights from Molecular Dynamics Simulation. *Physical Chemistry Chemical Physics* 18: 23769–23784.
136. Sahu, P., Ali, Sk.M., Shenoy, K.T. (2017) TBP Assisted Uranyl Extraction in Water-Dodecane Biphasic System: Insights from Molecular Dynamics Simulation. *Chemical Product and Process Modeling* 12: DOI: 10.1515/cppm-2016-0024.
137. Plimpton, S. (1995) Fast Parallel Algorithms for Short-Range Molecular Dynamics. *Journal of Computational Physics* 117: 1–19.
138. Cornell, W.D., Cieplak, P., Bayly, C.I., Gould, I.R., Merz, K.M., Ferguson, D.M., Spellmeyer, D.C., Fox, T., Caldwell, J.W., Kollman, P.A. (1995) A Second Generation Force Field for the Simulation of Proteins, Nucleic Acids, and Organic Molecules. *Journal of the American Chemical Society* 117: 5179–5197.

139. Cui, S., de Almeida, V., Hay, B.P., Ye, X., Khomami. (2012) Molecular Dynamics Simulation of Tri-n-butyl-Phosphate Liquid: A Force Field Comparative Study. *The Journal of Physical Chemistry B* 116: 305–313.
140. Jorgensen, W.L., Chandrasekhar, J., Madura, J.D., Impey, R.W., Klein, M.L. (1983) Comparison of Simple Potential Functions for Simulating Liquid Water. *Journal of ChemicalPhysics* 79: 926–935.
141. Gutle, C., Demaison, J., Rudolph, H.D. (2009) AnharmonicForce Field and Equilibrium Structure of Nitric Acid. *Journal of Molecular Spectroscopy* 254: 99–107.
142. Siu, S.W.I., Pluhackova, K., Backmann, R.A. (2012) Optimization of the OPLS-AA Force Field for Long Hydrocarbons. *Journal of Chemical Theory & Computation* 8: 1459–1470.
143. Spieser, S. A. H.,Leeflang, B.R., Kroon-Batenburg, L.M.J., Kroon, J. (2000) A Force Field for Phosphoric Acid: Comparison of Simulated with Experimental Data in the Solid and Liquid State. *The Journal of Physical Chemistry A* 104: 7333–7338.
144. Luty, B.A., Gunsteren, W.V. (1996) Calculating Electrostatic Interactions Using the Particle-Particle Particle-Mesh Method with Nonperiodic Long-Range Interactions. *The Journal of Physical Chemistry* 100: 2581–2587.
145. Trangenstein, J.A. (1987) Customized Minimization Techniques for Phase Equilibrium Computations in Reservoir Simulation. *Chemical Engineering Science* 42: 2847–2863.
146. Frenkel, D., Smit, B. (2002) Understanding Molecular Simulation. Academic Press, Sydney.
147. Ryckaert, J.P., Ciccotti, G., Berendsen, H.J.C. (1977) Numerical Integration of the Cartesian Equations of Motion of a System with Constraints: Molecular Dynamics of n-alkanes. *Journal of Computational Physics* 23: 327–341.
148. Humphrey, W., Dalke, A., Schulten, K. (1996) VMD: Visual Molecular Dynamics. *Journal of Molecular Graphics* 14: 33–38.
149. Benay, G., Schurhammer, R., Wipff, G. (2011) Basicity, Complexation Ability and Interfacial Behavior of BTBPs: A Simulation Study. *Physical Chemistry Chemical Physics* 13: 2922–2934.
150. Lan, J.H., Shi, W.Q., Yuan, L.Y., Feng, Y.X., Zhao, Y.L., Chai, Z.F. (2011) Thermodynamic Study on the Complexation of Am (III) and Eu (III) with Tetradentate Nitrogen Ligands: A Probe of Complex Species and Reactions in Aqueous Solution. *The Journal of Physical Chemistry A* 116: 504–511.
151. Lan, J.H., Shi, W.Q., Yuan, L.Y., Zhao, Y.L., Li, Z., Chai, Z.F. (2011) Trivalent Actinide and Lanthanide Separations by Tetradentate Nitrogen Ligands: A Quantum Chemistry Study. *Inorganic Chemistry* 50: 9230–9237.
152. Benay, G., Wipff, G. (2013) Oil-Soluble and Water-Soluble BTPhens and Their Europium Complexes in Octanol/Water Solutions: Interface Crossing Studied by MD and PMF Simulations. *The Journal of Physical Chemistry B* 117: 1110–1122.
153. Yang, Y., Hu, S., Fang, Y., Wei, H., Wang, D., Yang, L., Zhang, H., Luo, S. (2015) Density Functional Theory Study of the Eu (III) and Am (III) Complexes with Two 1, 10-Phenanthroline-Type Ligands. *Polyhedron* 95: 86–90.
154. Xiao, C.L., Wang, C.Z., Lan, Z.H., Yuan, L.Y., Zhao, Y.L., Chai, Z.F., Shi, W.Q. (2014) Selective Separation of Am (III) from Eu (III) by 2, 9-Bis (dialkyl-1, 2, 4-triazin-3-yl)-1, 10-phenanthrolines: A Relativistic Quantum Chemistry Study. *Radiochimica Acta* 102: 875–886.

4

Aqueous Separation in Metal-Organic Frameworks: From Experiments to Simulations

Krishna M. Gupta and Jianwen Jiang

CONTENTS

4.1 Introduction

Metal-Organic Frameworks (MOFs) are a novel class of nanoporous materials [1]. They can be produced by judicious selection of inorganic and organic building blocks, and further modified via post-synthetic. Compared with other porous materials (e.g. zeolites), the degree of diversity and multiplicity in MOFs is substantially more extensive [2]. By varying metal coordination environments and multi-topic organic linkers, thousands of different MOFs have been synthesized [3,4]. Due to the extremely tunable structure, high porosity (up to 90%) and large surface area (beyond 6000 m^2/g), MOFs have drawn significant attention in diversified fields such as gas storage and separation [5,6], liquid separation [7], chemical sensing [8], and catalysis [9,10].

Liquid separation in MOFs is relatively new and it can be further categorized into aqueous and organic separation based on the solvent used. In this chapter, our focus is on aqueous separation with water as the solvent. In such a condition, the underlying behavior of MOFs upon exposure to water is crucial [11]. Thus, first, we discuss water stability of MOFs, then outline experiments on the applications of MOFs in aqueous separation including removal of cations, anions, and organics. With rapidly growing computational resources, molecular simulation has become an indispensable tool and played an increasingly important role in materials science and engineering. Simulation at an atomistic/molecular level can provide microscopic insight that otherwise is experimentally intractable, and thus elucidates underlying physics from the bottom-up. In addition, simulation can secure the fundamental interpretation of experimental observation and guide the rational selection and design of novel materials [12]. Therefore, we finally showcase simulation studies examining MOFs for aqueous separation such as water desalination, biofuel purification, and other aqueous mixtures (e.g. water/dimethyl sulfoxide, water/amino acids, and water/glucose).

4.2 Water Stability

Most of the earlier reported MOFs are sensitive to water due to the breaking of metal-ligand bonds. For example, a prototype MOF (MOF-5) was observed to undergo structural degradation in humid air [13]. As water or moisture is commonly present in most industrial processes, the instability of MOFs in water is considered a bottleneck for practical applications.

The complete study of water stability of MOFs includes in humid vapor along with the aqueous phase. Low et al. examined a series of MOFs by varying organic linker, pore structure, metal node, and coordination under exposure to 1 mol% steam for a few hours. Figure 4.1 illustrates the steam stability map of MOFs. Among the MOFs under study, ZIF-8, MOF-74, and MIL-110/-101 were found to be the most stable; the metal-ligand bond strength was identified to be a key criterion governing steam stability [14]. To evaluate water-induced distortion of MOF-5, molecular simulations were conducted by Greathouse and Allendorf. It was postulated that the relatively weak Zn-O interaction in MOF-5 allows for possible attack by water molecules, and consequently, the lattice structure degrades even at a low water content. Nevertheless, the structure completely collapses at a high water concentration if the ligand oxygen atoms coordinated to Zn are replaced by oxygen atoms of water [15]. A systematic investigation of water vapor adsorption and subsequent structural analysis was conducted by Schoenecker et al. for MOFs, including open metal sites (HKUST-1 and Mg-MOF-74), amine-functional groups (UiO-66-NH_2 and DMOF-1), carboxylate coordination (UMCM-1, HKUST-1, and Mg-MOF-74) and nitrogen coordination (DMOF-1). Due to weakly coordinated zinc-carboxylate, the crystal structure

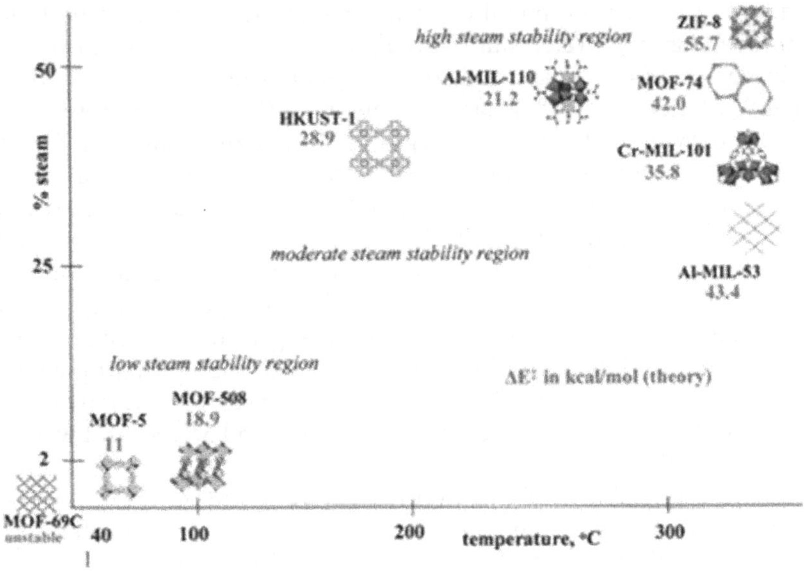

FIGURE 4.1
Steam stability map of MOFs. For a given MOF, the position represents its maximum structural stability by XRD measurement. The number (in kcal/mol) is the activation energy of ligand displacement by a water molecule estimated by molecular modeling. (Reprinted from Low, J. J. et al., *Journal of the American Chemical Society*, 131: 15834–15842, 2009. With permission.)

of UMCM-1 was observed to completely degrade in water vapor. A complete loss of crystallinity under 90% relative humidity occurred in DMOF-1. With higher stability of Zr clusters, however, both the parent and functionalized UiO-66 showed good structural stability [16]. The stability of MOFs in aqueous solutions instead of water vapor was examined by Matzger and coworkers on a timescale from hours to months using powder X-ray diffraction. Metal clusters seemed to be important in determining the stability, as zinc acetate containing MOFs (MOF-5 and MOF-177) were found less stable than those containing copper paddle-wheel, whereas MIL-100 with trinuclear chromium clusters was highly stable [17].

Generally, metal clusters with higher oxidation states are less susceptible to water, resulting in rigid structures, even prolonged exposure to moist environment. To synthesize water-stable MOFs, Fe^{3+}, Cr^{3+}, and Zr^{4+} clusters were recommended with commonly used carboxylate linkers. For example, Férey and coworkers reported Fe- and Cr-based MIL-100 and MIL-101 with good stability in water and other solvents [18]. MOFs containing Zr_6O_8 were found exceptional hydro-stability with decomposition temperature above 500 °C, and they were resistant to most chemicals and sustained crystallinity even after exposure to 10 tons/cm^2 of external pressure. These MOFs belong to the well-known UiO and PCN families [19,20].

In addition to varying metal cluster, organic linker can also be tailored to improve stability. Water stable MOFs were synthesized from nitrogen-containing ligands such as imidazolates, pyrazolates, triazolates, and tetrazolates [21]. Zeolitic-imidazolate frameworks (ZIFs) are the representative of this type. With zeolite-like topologies, ZIFs usually consist of Zn^{2+}/Co^{2+} together with imidazolate, and possess high chemical and thermal stability [22,23]. Alternatively, pyrazolate was used as a linker along with different metal clusters including Ni^{2+}, Cu^{2+}, Zn^{2+}, and Co^{2+} to synthesize microporous MOFs, which exhibited great hydrothermal stability [24]. Another pathway to enhance water stability of MOFs is through functionalization (e.g. introduce hydrophobic pores), thus preventing or reducing water-framework interaction. For example, fluorinated MOFs were revealed highly hydrophobic and there was no detectable water adsorption even at near 100% relative humidity [25]. The absence of attractive interaction between FMOF-1 and water was also confirmed [26].

In brief, water stable MOFs should be able to maintain crystallinity and porosity even after water intrusion into the frameworks. Generally, the breaking of metal-ligand bonds occurs due to hydrolysis reaction, leading to the decomposition of MOFs. To avoid this detrimental reaction, strong coordination bonds and significant steric hindrance to water are required [27]. There are several factors governing water stability such as metal-ligand bond strength, coordination number and oxidation state of metal center, basicity and functionality of linker, as well as framework dimensionality; these are discussed in details elsewhere [28]. Among these, linker basicity is the most important for the structural stability of MOFs in aqueous media. Metal clusters with higher oxidation states are likely to produce more stable MOFs towards exposure to water. As discussed above, various attempts such as metal exchange, ligand modification, and framework functionalization can be adopted to improve water stability of MOFs.

4.3 Experimental Studies

Aqueous separation is generally targeted to remove hazardous compounds such as toxic cations, anions, and organics for water treatment. Due to the exceptional water stability, UiO, MIL, and ZIF families of MOFs are mostly explored for this type of application.

4.3.1 Cations

With increasing world population and economic development, a large amount of toxic metal ions (e.g. Cd^{2+} and Hg^{2+}), as well as radioactive ions (e.g. U^{4+} and Ba^{2+}), have been introduced into water. These cations can accumulate in living organisms and lead to dysfunction in nervous, circulatory, and immune systems [29]. It is thus crucial to remove them from water and minimize health and environmental risks.

Two isostructural mesoporous MOFs named PCN-100 and PCN-101 with cavities up to 2.73 nm were designed using $Zn_4O(CO_2)_6$ as a secondary building unit and two extended ligands containing amino functional groups TATAB (4,4′,4″-s-triazine-1,3,5-triyltri-p-aminobenzoate) and BTATB (4,4′,4″-(benzene-1,3,5-triyltris(azanediyl))tribenzoate). It was observed that PCN-100 could capture Cd^{2+} and Hg^{2+} ions by forming complex with TATAB, a coordination mode similar to that in aminopyridinato complex [30]. Mon et al. reported a robust and water-stable bioMOF featuring hexagonal channels decorated with methionine residues, which was capable of selectively capturing CH_3Hg^+ and Hg^{2+} from water by decreasing the concentration from 10 ppm to the much safer values of 27 and 6 ppb, respectively [31]. To remove Cd^{2+} from aqueous solution, Wang et al. synthesized a sulfonic acid functionalized MOF [$Cu_3(BTC)_2$–SO_3H] and demonstrated a high Cd^{2+} uptake capacity of 88.7 mg/g, surpassing the benchmark adsorbents. As shown in Figure 4.2, the strong adsorption was because of the chelation between Cd^{2+} ion and –SO_3H group. Moreover, the MOF adsorbed with Cd^{2+} could be readily regenerated and recycled without a significant loss of uptake capacity [32]. UiO-66 and its derivatives have also been tested. For instance, the removal of trace amount of Hg^{2+} along with other toxic ions such as Bi^{3+}, Zn^{2+}, Pb^{2+}, and Cd^{2+} were investigated using UiO-66 [33]. A remarkable 99.9% removal of Hg^{2+} from 10 ppm to below 0.01 ppm was achieved by incorporating thiol groups on the terephthalate (BDC) ligand in UiO-66 [34]. Furthermore, UiO-66(Zr)–2COOH was reported to exhibit a high selectivity (up to about 27) for the selective removal of Cu^{2+} over Ni^{2+} from aqueous solution. The reason was attributed to the unique chelation effect of two carboxyl groups on the adjacent organic ligand as well as the Jahn–Teller effect [35]. Meng et al. reported a water stable, pillar-layer MOF formulated as $[Zn(trz)(H_2betc)_{0.5}]\cdot$DMF. This MOF exhibited exceptional stability with uncoordinated carboxyl groups and showed a significant response to the removal of transition metal ions. Particularly, it could effectively and selectively adsorb Cu^{2+} ions and is

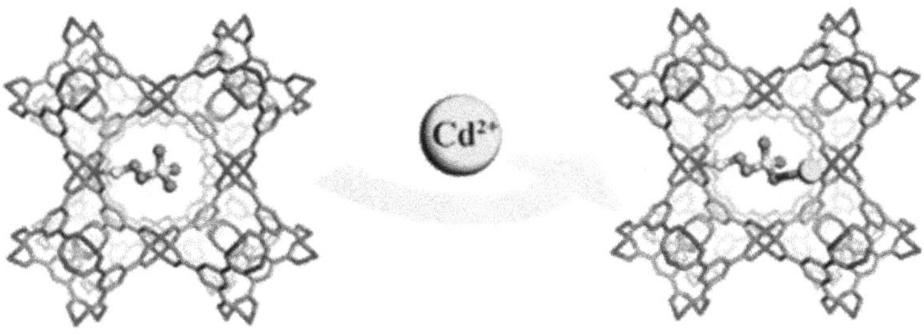

FIGURE 4.2
Cd^{2+} adsorption in $Cu_3(BTC)_2$–SO_3H. (Reprinted from Wang, Y. et al., *Journal of Materials Chemistry*, 3: 15292–15298, 2015. With permission.)

potentially useful for chromatographic separation of Cu^{2+}/Co^{2+} ions [36]. In addition, two series of MOFs were prepared and tested for Pb^{2+} adsorption, the first type with identical ligand and different metal ions (Zn, Zr, Cr, and Fe) named MOF-5, UiO-66, MIL-101(Cr), and MIL-53(Fe), while the second type with identical metal (Zn) and different ligands (methyl-2-imidazolate, BDC, and BDC-NH_2) named ZIF-8, MOF-5, and IRMOF-3. Among these, ZIF-8 was observed to exhibit the highest adsorption capacity [37].

Naturally, radioactive ions may exist in water and they are harmful to human being [38]. HKUST-1 was examined for the removal of uranium (UO_2^{2+}, $UO_2(OH)^+$, and $UO_2(OH)^{2+}$) in the concentration range of 10–800 mg/L and temperature range of 298–318 K. The highest uranium uptake was observed at 800 mg/L and 318 K [39]. Carboni et al. prepared three UiO-68 topologies using amino-TPDC or TPDC bridging ligands containing orthogonal phosphorylurea groups. The stable and porous phosphoryl-derived MOFs were shown highly efficient in adsorbing uranyl ions (UO_2^{2+}) with a saturation capacity up to 217 mg/g. As shown in Figure 4.3, the efficient adsorption is promoted by the cooperative binding between uranyl and phosphoryl groups in a tetrahedral pocket, as well as the suitable inter-atom distances [40]. Uranium (UO_2^{2+}) separation was also studied in MOF-76, which exhibited high sensitivity and adsorption capacity at pH = 3 [41]. Moreover, UO_2^{2+} adsorption was evaluated in UiO-66 and its amine derivative, demonstrating desirable selectivity towards UO_2^{2+} against a range of competing ions [42]. A Zn-MOF containing azo-functionalized pores was synthesized by solvothermal process. The removal of UO_2^{2+} by this Zn-MOF was found in terms of monolayer chemisorption with a large capacity of 312.32 mg/g [43]. For other radioactive contaminants like Cs^+ and Sr^{2+}, an ultra-stable MOF namely MIL-101-SO_3H was explored by Aguila et al. [44]. MIL-101-SO_3H and MOF-808-SO_4 were further examined for the removal of radioactive Ba^{2+} and showed excellent performance; particularly, a high Ba^{2+} uptake capacity of 131.1 mg/g was determined in MOF-808-SO_4, which surpasses most reported values in the literature [45].

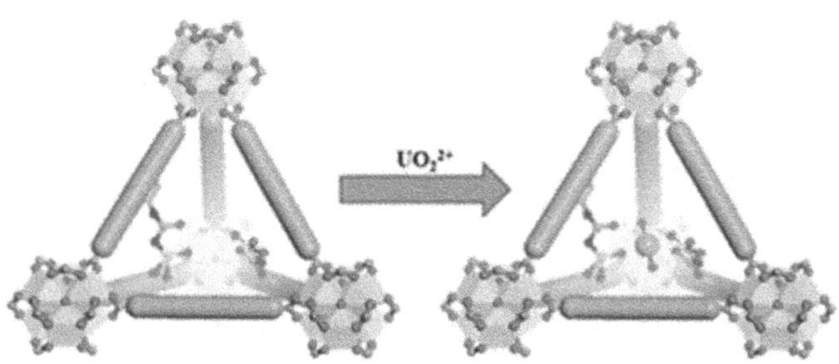

FIGURE 4.3
Schematic depicting the uranyl-binding pocket. UO_2^{2+} is coordinated with the phosphoryl oxygen in a monodentate fashion. Distances between oxygen atoms range from 4.5 to 4.8 Å, accommodating the U–O bond lengths appropriate for π-π binding. (Reprinted from Carboni, M. et al., *Chemical Science*, 4: 2396–2402, 2013. With permission.)

While it is indispensable to remove toxic and radioactive ions, recovery of precious metal ions from aqueous solutions is also desirable. Yun and coworkers synthesized Zr-MOFs and tested the adsorption of Pd^{2+}, Pt^{4+}, and Au^{3+} from strongly acidic solutions. It should be noted that these ions are present in anionic form as $PdCl_4^{2-}$, $PtCl_6^{2-}$, and $AuCl_4^-$, respectively. High adsorption capacities were found and attributed to the inner-sphere complexation between the anions and incompletely coordinated Zr atoms. Moreover, extremely high recovery for Au^{3+} was observed in UiO-66-NH_2, due to additional electrostatic attraction between the ions and protonated amines in UiO-66-NH_2 [46].

4.3.2 Anions

In aqueous solutions, many anions (e.g. F^-, $Cr_2O_7^{2-}$, PO_4^{2-}, are AsO_4^{3-}) are hazardous and need to be removed. Wong et al. designed a MOF comprised of terbium metal linked with mucic acid, which was able to adsorb F^-, Cl^-, Br^-, I^-, CN^-, and CO_3^{2-} from water. This was attributed to strong hydrogen-bonding (H-bonding) between the anions and –OH groups on mucic acid; however, PO_4^{2-} and SO_4^{2-} could not be adsorbed because of their large size [47]. Defluoridation was tested in 11 water stable MOFs including MIL-53(Fe, Cr, Al), MIL-68(Al), CAU-1, UiO-66(Zr, Hf), and ZIFs-7, -8, -9. UiO-66(Zr) exhibited an adsorption capacity of 41.36 mg/g, higher than most of the reported adsorbents. The study suggested that increasing the number of –OH groups in MOFs is an efficient strategy to improve defluoridation performance [48]. A typical aluminium-based MOF, namely MIL-96, was also examined for defluoridation of drinking water. Compared with activated alumina or nano-alumina, MIL-96 showed superior performance [49]. To efficiently trap $Cr_2O_7^{2-}$, two water stable MOFs, FIR-53 and FIR-54, were synthesized by Fu et al. with uptake capacities of 74.2 and 103 mg/g, respectively [50]. In another study, a cationic robust Zr-MOF (ZJU-101) was reported. It possessed not only a high capacity of 245 mg/g to remove $Cr_2O_7^{2-}$, but also selective capture of other ions in a very short period of time [51]. Based on a dual capture strategy, furthermore, a water stable cationic MOF was designed using Ni^{2+} metal node and a neutral ligand, which was found to simultaneously adsorb $Cr_2O_7^{2-}$ and MnO_4^- [52]. Alternatively, Aboutorabi et al. prepared a MOF (TMU-30) based on isonicotinate N-oxide with sufficient water stability to remove $HCrO_4^-$ and CrO_4^{2-}. Within 10 min, a maximum capacity (145 mg/g) was reached following the pseudo-second-order kinetics. The reason for spontaneous adsorption was explained in terms of the electrostatic interactions between these ions and N-oxide groups [53].

Arsenic and selenium based anions are toxic and carcinogenic, thus it is important to remove them from aqueous environments. Wang et al. examined UiO-66 to remove aquatic arsenate existing as H_3AsO_4, $H_2AsO_4^-$, or $HAsO_4^{2-}$ depending on solution pH. A remarkable uptake capacity of 303 mg/g was achieved, which outperforms most of the current adsorbents. Figure 4.4 illustrates the proposed adsorption mechanism of H_3AsO_4 in UiO-66. Moreover, it was found that UiO-66 could function well across a broad

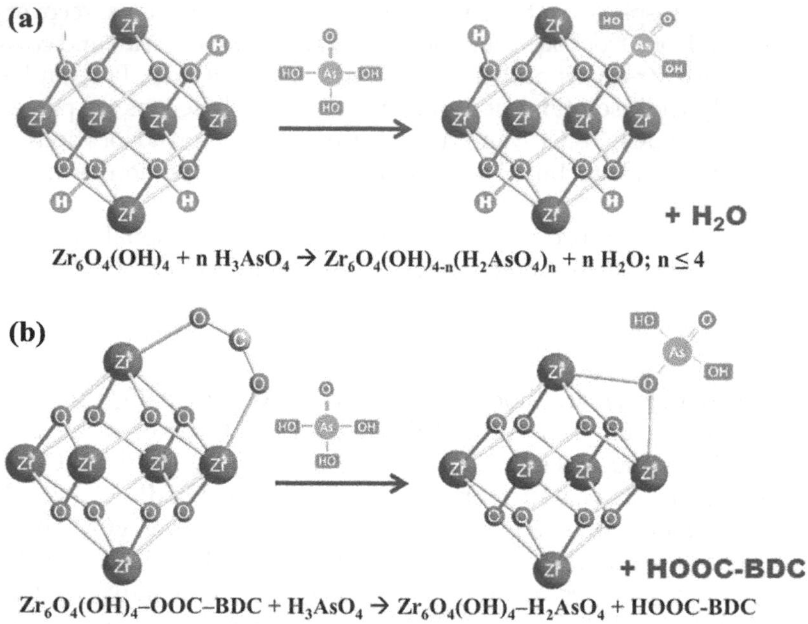

FIGURE 4.4
Adsorption mechanism of H_3AsO_4 in UiO-66 at (a) hydroxyl group and (b) BDC ligand. In (b), the H atoms are omitted for clarity. (Reprinted from Wang, C. et al., *Scientific Reports*, 5: 16613, 2015. With permission.)

pH range of 1 to 10 [54]. UiO-66 and analogues were further used to remove arsenate AsO_4^{3-} and arsenite AsO_3^{3-} from ground water (pH 6–8.5). The missing linker sites in $Zr_6(O)_4(OH)_4$ nodes were identified to be excellent binders for AsO_4^{3-}, while the thiolated linkers could selectively coordinate AsO_3^{3-} with a dual-capture strategy in reversible manner [55]. An amino-functionalized iron-based MOF, namely NH_2-MIL-88(Fe), was synthesized. Featured with unique fluorescence enhancement and high specific surface area as well as excellent stability, NH_2-MIL-88(Fe) demonstrated high sensitive detection and efficient removal of AsO_4^{3-} with a fast response time (<1 min), broad linear range (0.1–50 mM), and high sensitivity (detection limit of as low as 4.2 ppb). The study indicated that the biofunctional MOF might be an excellent candidate for AsO_4^{3-} determination and remediation from contaminated water [56]. Recently, multifuntioanl MOF hollow tubes were fabricated using a templated freeze-drying protocol. Particularly, ZIF-8@ sodium alginate (SA) tube exhibited 96.8% efficiency for AsO_4^{3-} extraction and could be recycled for three times without loss of structural integrity and porosity. Furthermore, both AsO_4^{3-} and methyl orange (MO) were extracted from a mixed solution with high efficiency (95.3% AsO_4^{3-} and 97.1% MO) [57]. On the other hand, Howarth et al. tested a series of water-stable Zr-MOFs including NU-1000, UiO-66 and its derivatives, as well as UiO-67, for their capability to adsorb and remove selenate SeO_4^{2-} and selenite SeO_3^{2-} from aqueous solutions. The MOF aperture and the number of node-based adsorption sites were pointed out to be important in the performance. Both anions were shown to bind with zirconium node in a bridging fashion. NU-1000 was shown to have the highest adsorption capacity and fastest uptake rate for both anions [58].

4.3.3 Organics

In our daily life and industrial processes, a wide variety of organic compounds (e.g. aromatics, dyes, pharmaceuticals, and drugs) are produced and introduced into water. Several review articles have summarized the studies to remove organics through MOFs from aqueous solutions [59,60]. Bai et al. reported one of the first studies to investigate the adsorption of aromatics by a two-dimensional MOF, which was formed between hexanuclear copper clusters and triazine-containing ligands exhibiting luminescent properties. It was observed that adsorption of toluene, nitrobenzene, aniline, and p/m/o-xylene from 0.5% aqueous solutions effectively quenched the luminescence due to π-π interaction and H-bonding [61]. The potential of MIL-101 was exploited for benzene adsorption in vapor as well as aqueous phase under 1000 ppm concentration. Attributed to the combination of a very large surface area and a small particle size, MIL-101 showed a higher adsorption capacity (80 wt% at saturation) than activated carbon (45 wt% at saturation) [62]. Xie et al. investigated a series of MOFs for nitrobenzene (NB) removal from water. In the presence of μ_2-OH groups in Al–O–Al, two aluminium-based MOFs, CAU-1 and MIL-68(Al), were found to outperform previously reported porous materials [63]. The adsorption of 5-hydroxymethylfurfural (HMF) from aqueous solution was examined in ZIF-8, ZIF-90, and ZIF-93. The equilibrium uptake was determined to increase as ZIF-93 < ZIF-90 < ZIF-8, following the trend of framework hydrophobicity [64]. ZIF-8 was also identified as an effective adsorbent for the adsorption of benzotriazole and tetracycline from aqueous solution [65].

Adsorption of dyes has been mostly examined in water stable MILs due to their large pore size and unique structure characteristics [66–68]. For instance, two Cr-MOFs (MIL-101 and MIL-53) were evaluated for the removal of methyl orange (MO) from wastewater. Compared with MIL-53, MIL-101 exhibited a higher adsorption capacity and a faster rate kinetic because of larger porosity and pore size. The performance of MIL-101 was further improved by grafting ethylenediamine and protonated ethylenediamine, even though the porosity and pore size were slightly reduced upon grafting [66]. Highly porous MIL-100(Fe, Cr) were tested for the removal of anionic MO and cationic methylene blue (MB). MIL-100(Fe) exhibited high adsorption uptake for both MO and MB, while MIL-100(Cr) could selectively adsorb MB from a MO/MB mixture [69]. Recently, a new microporous negatively charged MOF was reported to possess a remarkable capability to selectively adsorb and separate MB through an ion-exchange process. The size and charge of dyes were revealed to greatly affect the separation efficacy [70].

MOFs have also been examined for removal of pharmaceuticals, drugs and personal care products from contaminated aqueous solutions. Highly water stable Cr-MILs are promising candidates. MIL-101 was found to have good adsorption performance for furosemide and sulfasalazine [17]. Compared with MIL-100 and activated carbon, MIL-101 turned out a superior adsorbent for the removal of a nonsteroidal anti-inflammatory drug (naproxen) and a bioactive metabolite (clofibric acid). The high adsorption capacity was attributed to the electrostatic interaction between the anionic naproxen and positively charged MIL-101 [71]. Following this, Hu et al. used MIL-101 to separate naproxen and its metabolites from urine samples. A higher uptake was seen in MIL-101 than multi-walled carbon nanotube and C_{18} bonded silica [72]. Other MOFs were also explored for this application. ZIF-8 was examined for the adsorption of phthalic acid (H_2-PA) and diethyl phthalate (DEP) from aqueous solutions. The adsorption capacity was much higher than in commercial activated carbon and most reported adsorbents, due to the favorable electrostatic interaction between the negatively charged PA and positively charged ZIF-8 [73]. UiO-66 was examined for the adsorptive removal of a hazardous herbicide/pesticide, namely methylchlorophenoxypropionic acid (MCPP), from water. In the presence of electrostatic and π–π interactions, a very fast adsorption rate and a high adsorption capacity especially at low MCPP concentrations were observed [74]. Zhu et al. investigated the adsorption of organophosphorus pesticides, glyphosate (GP), and glufosinate (GF) in UiO-67. Because of strong affinity towards phosphoric groups and adequate pore size in UiO-67, the adsorption capacities were determined up to 3.18 mmol/g for GP and 1.98 mmol/g for GF, which are much higher than in many other adsorbents. The adsorption isotherms were fitted using the Langmuir and Freundlich models. As shown in Figure 4.5, the fitting by the Langmuir model appears better. This was attributed to the homogeneous distribution of active adsorption sites on the surface of UiO-67, as the Langmuir model assumes that the surface of adsorbent is homogeneous. Moreover, the adsorption capacities GP and GF were found to increase with increasing UiO-67 content [75]. Six water stable MOFs including MIL-101(Cr), MIL-100(Cr, Fe), MIL-96(Al), MIL-53(Cr), and UiO-66 were evaluated for the removal of hydrochloride from dilute aqueous solution. Among all, UiO-66 exhibited excellent stability towards hydrochloride and the highest adsorption capacity [76]. Recently, the adsorption behavior of two personal care products, namely anticonvulsant carbamazepine (CBZ) and antibiotic tetracycline hydrochloride (TC) on UiO-66 was examined. The maximum adsorption capacities at 25 °C were measured to be 37.2 and 23.1 mg/mg for of CBZ and TC, respectively. Based on X-ray diffraction, Fourier transform infrared spectra, scanning electron microscopy, and X-ray photoelectron (XPS) spectra, the surface interactions between UiO-66 and the two pollutants were quantified [77].

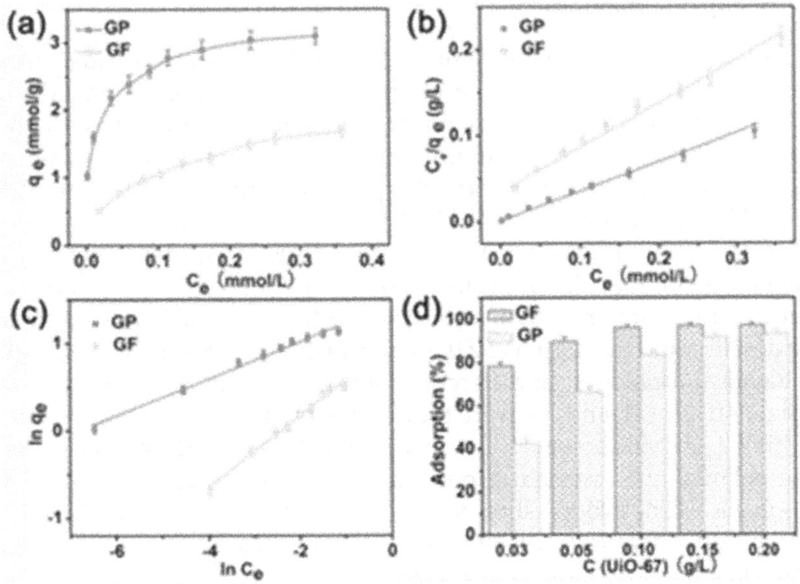

FIGURE 4.5
(a) Adsorption isotherms of GP and GF in UiO-67. (b) Langmuir model and (c) Freundlich model (25 °C, pH = 4, $C_{adsorbent}$ = 0.03 g/L). (d) Effect of UiO-67 content (25 °C, pH = 4, $C_{initial}$ = 0.05 mmol/L). (Reprinted from Zhu, X. et al., *ACS Applied Material Interfaces*, 7: 223–231, 2015. With permission.)

Overall, water stable UiO, MIL and ZIF families of MOFs have been extensively examined for aqueous separation. The pore size, π-π, and electrostatic interaction are identified as the key governing factors. With diversified pores, topologies, and structures, MOFs can adsorb a wide variety of molecules ranging from small ions to large organics. Compared with conventional zeolites and activated carbons, MOFs exhibit higher capacity and efficiency for separation. To further improve the performance, different metal clusters and organic ligands can be used to design and synthesize new functional MOFs.

4.4 Simulation Studies

To date, most simulations in MOFs have been largely focused on gas storage and separation, particularly the storage of low-carbon footprint energy carriers (e.g. H_2) and the separation of CO_2-containing gas mixtures for carbon capture [78–83]. Nevertheless, a series of simulation studies have been reported, primarily by our group, using MOFs for aqueous separation including water desalination, biofuel purification, and other aqueous mixtures.

4.4.1 Water Desalination

Seawater accounts for over 95% of water on the Earth and thus can supply abundant fresh water after economical desalination. Using molecular dynamics (MD) simulation, we demonstrated that ZIF-8 could act as a reverse osmosis (RO) membrane to desalinate seawater. Under an external pressure, water molecules were observed to permeate through ZIF-8 membrane, whereas Na^+ and Cl^- ions were blocked due to the molecular sieving of small apertures in ZIF-8. Because of the surface interaction and geometrical confinement, water molecules in the membrane experienced less H-bonding and longer lifetime compared with bulk water [84]. Subsequently, we simulated water desalination through five ZIF membranes (ZIF-25, -71, -93, -96, and -97) with identical topology but different functional groups. With a larger aperture size, ZIF-25, -71, and -96 were found to exhibit a much higher water flux than ZIF-93 and -97; however, the flux in ZIF-25, -71, and -96 was governed by the polarity of functional group rather than aperture size. ZIF-25 with hydrophobic $-CH_3$ groups showed the highest flux among ZIF-25, -71, and -96, despite the smallest aperture size. Water molecules were observed to undergo fast flushing motion in ZIF-25, but frequent jumping in ZIF-96 and particularly in ZIF-97. This study revealed the key factors (aperture size and polarity of functional groups) governing water permeation [85].

Furthermore, seawater pervaporation (PV) through ZIF-8, -93, -95, -97, and -100 was simulated. Salt rejection in the five ZIFs was predicted to be 100%. With the largest aperture, ZIF-100 possesses the highest water permeability of 5×10^{-4} kg·m/(m²·h·bar), which is higher than commercial RO membranes, as well as zeolite and graphene oxide membranes (Figure 4.6). A to-and-fro motion was seen for water in ZIF-100. In ZIF-8, -93, -95, and -97 with similar aperture size, water flux was found to depend on framework hydrophobicity. The dynamic and structural properties of water are useful to design new ZIF membranes for water desalination [86].

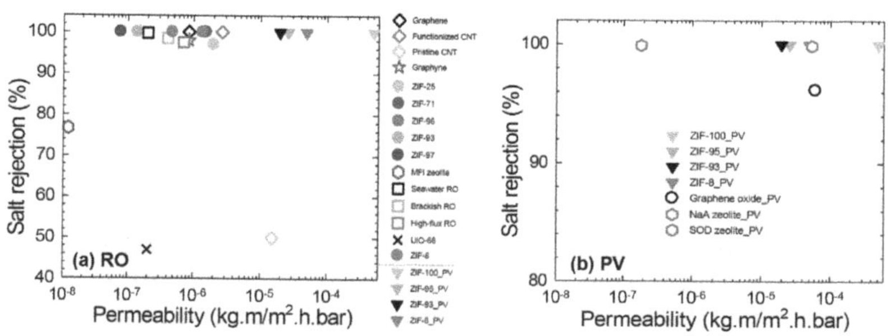

FIGURE 4.6
Salt rejection versus permeability for ZIF membranes along with other reported membranes. (Reprinted from Gupta, K. M. et al., *ACS Applied Material Interfaces*, 8: 13392–13399, 2016. With permission.)

4.4.2 Biofuel Purification

There has been considerable interest in the use of environmentally benign, carbon neutral biofuel. Nevertheless, as-produced biofuel contains a large amount of water and it is important to separate water/alcohol mixtures. The separation was estimated to account for 60–80% of total product cost [87], thus high-efficiency biofuel purification is indispensable. Combining MD and Monte Carlo simulations, we predicted the separation of ethanol/water liquid mixtures in Na-*rho*-ZMOF and $Zn_4O(bdc)(bpz)_2$ at pervaporation (PV, 50 °C) and vapor permeation (VP, 100 °C) conditions. In hydrophilic Na-*rho*-ZMOF, water was found to be preferentially adsorbed over ethanol due to its strong interaction with nonframework Na^+ ions and ionic framework. In contrast, ethanol was adsorbed more in hydrophobic $Zn_4O(bdc)(bpz)_2$ as attributed to the favorable interaction with methyl groups. At both PV and VP conditions, the permselectivities in the two MOFs were primarily determined by adsorption selectivity. As shown in Figure 4.7, the maximum permselectivity in Na-*rho*-ZMOF is about 12 at VP condition, thus Na-*rho*-ZMOF is preferable for biofuel dehydration. On the other hand, $Zn_4O(bdc)(bpz)_2$ exhibits the maximum permselectivity of 75 at PV condition and it is promising for biofuel recovery [88].

In another study, we simulated the adsorptive separation of ethanol/water mixtures in ZIF-8, -25, -71, -90, -96, and -97 with different functional groups. The selectivity of ethanol/water was found to drop with increasing ethanol composition (X_E), and largely determined by framework hydrophobicity as well as cage size. Among the six ZIFs, ZIF-8 exhibited the highest selectivity. This simulation study provides microscopic insight into the adsorption of ethanol and water in various ZIFs, and reveals the significant role of functional groups in governing biofuel purification [89]. Furthermore, five ZIFs (ZIF-68, -69, -78, -79, and -81) with isoreticular GME topology but different organic linkers were examined for the separation of ethanol/water mixtures. Among the five ZIFs, ZIF-79 with hydrophobic $-CH_3$ groups shows the highest adsorptive selectivity. Figure 4.8 illustrates the density distributions of

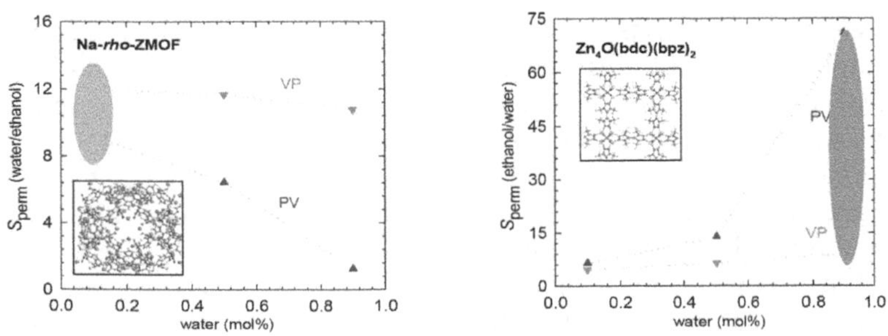

FIGURE 4.7
Permselectivities for water/ethanol mixtures in Na-*rho*-ZMOF and $Zn_4O(bdc)(bpz)_2$. (Reprinted from Nalaparaju, A. et al., *Energy & Environmental Science*, 4: 2107–2116, 2011. With permission.)

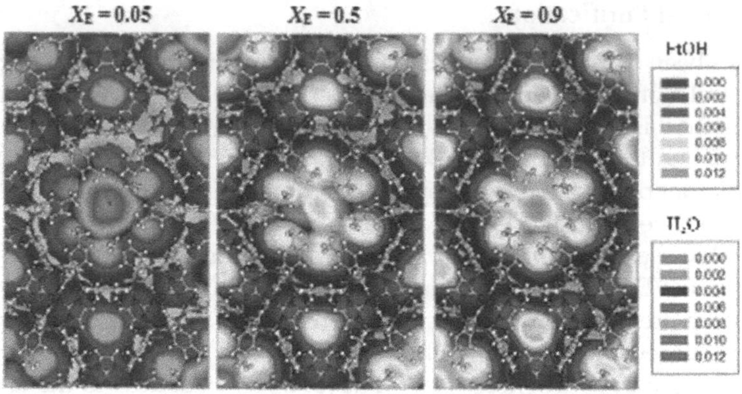

FIGURE 4.8
Density contours of ethanol/water mixtures in ZIF-79. (Reprinted from Zhang, K. et al., *American Institute of Chemical Engineers Journal*, 61: 2763–2775, 2015. With permission.)

ethanol and water in ZIF-79. At X_E = 0.05, ethanol is adsorbed in the small pores and the six corners of the large pores (near $-CH_3$ groups). With increasing X_E to 0.5, ethanol is more populated in both the small and large pores, and water in the center of the large pores is partially replaced by ethanol. The replacement of water by ethanol is almost complete at $X_E = 0.9$ as a consequence of competitive adsorption. The microscopic insights provided therein would facilitate the development of new ZIFs for biofuel purification [90].

4.4.3 Aqueous Mixtures

In addition to above discussed water desalination and biofuel purification, the separation of other aqueous mixtures was also investigated. For example, Nalaparaju and Jiang conducted a simulation study to examine the recovery of dimethyl sulfoxide (DMSO) from aqueous solution in $Zn_4O(bdc)(bpz)_2$, $Zn(bdc)(ted)_{0.5}$, and ZIF-71 [91]. In chemical industries, DMSO is often mixed with water to form biphase medium, which is used to dissolve pharmaceuticals and drugs [92], and formulate polymers, dyes, and electronics [93]. Due to a hydrophobic nature, the three MOFs were found to highly selectively adsorb DMSO from DMSO/H_2O mixtures. As shown in Figure 4.9, the selectivity initially increases with increasing composition of DMSO (X_{DMSO}), then decreases. The initial increase is due to the enhanced interaction of DMSO with multiple adsorption sites in the framework. However, the interaction strength drops when most adsorption sites are occupied; consequently, a decrease in selectivity is seen at high X_{DMSO}. The highest selectivity is up to 1700 in ZIF-71; nevertheless, the recovery capacity of DMSO in ZIF-71 is significantly lower than in $Zn_4O(bdc)(bpz)_2$ and $Zn(bdc)(ted)_{0.5}$. Therefore, $Zn_4O(bdc)(bpz)_2$ and $Zn(bdc)(ted)_{0.5}$ might be practically better among the three MOFs for DMSO recovery [91].

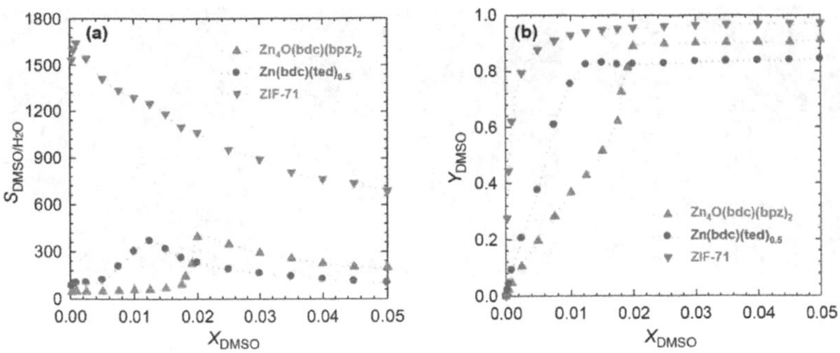

FIGURE 4.9
(a) Adsorption selectivity (b) *XY* diagram of DMSO/H_2O liquid mixtures in $Zn_4O(bdc)(bpz)_2$, $Zn(bdc)(ted)_{0.5}$ and ZIF-71. (Reprinted from Nalaparaju, A., Jiang, J. W., *Langmuir*, 28: 15305–15312, 2012. With permission.)

Hu, Chen, and Jiang examined the capability of MIL-101 for the separation of a mixture of amino acids from aqueous solution. The mixture consisted of arginine (Arg), phenylalanine (Phe), and tryptophan (Trp), differing in molecular size, weight, and charge. Among the three amino acids, Arg is the most hydrophilic and experiences the strongest interaction with water, but the weakest interaction with MIL-101; thus Arg was observed to transport the fastest in MIL-101 (Figure 4.10). Moreover, Arg was predicted to form the largest number of H-bonds with water and possess the largest hydrophilic solvent-accessible surface area. On the contrary, Trp showed the weakest interaction with water and the closest to MIL-101. By synergizing energetic and structural analysis, the separation mechanism of the three amino acids was attributed to the cooperative solute-solvent and solute-framework interactions. From this study, MIL-101 appears an interesting material for the separation of biologically important molecules [94].

We also reported a simulation study on glucose recovery from aqueous solution by adsorption in MIL-101, MIL-101-NH_2, and MIL-101-CH_3 [95]. Glucose is a useful precursor for biofuels, pharmaceuticals, and chemicals [96]. As shown in Figure 4.11, the F atom of MIL-101 was identified to be

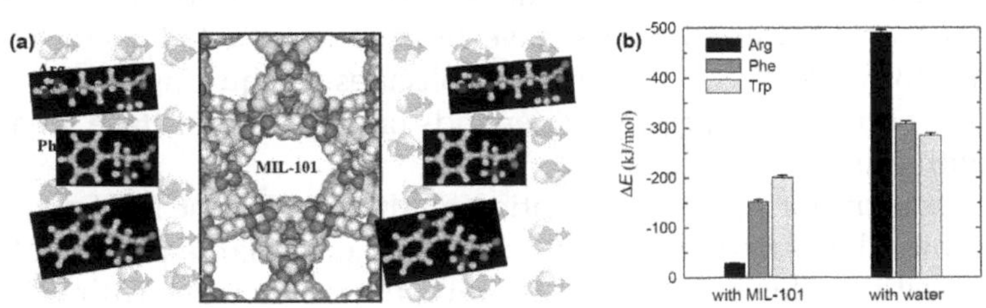

FIGURE 4.10
(a) Separation of amino acids (Arg, Phe, and Trp) from aqueous solution in MIL-101 (b) Interaction energies of amino acids with MIL-101 and water. (Reprinted from Hu, Z. Q. et al., *Langmuir*, 29: 1650–1656, 2013. With permission.)

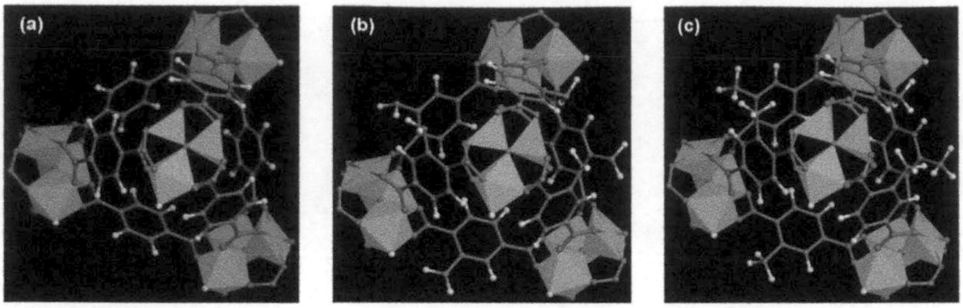

FIGURE 4.11
Supertetrahedra in (a) MIL-101 (b) MIL-101-NH_2 (c) MIL-101-CH_3. Cr_3O. (Reprinted from Gupta, K. M. et al., *Scientific Reports*, 5: 12821, 2015. With permission.)

the most favorable adsorption site. Among three MIL-101s, MIL-101 exhibited the highest adsorption capacity and recovery efficacy. Upon functionalization by -NH_2 or -CH_3 group, the steric hindrance in MIL-101 increased; consequently, the interactions between glucose and framework became less attractive, thus reducing the capacity and mobility of glucose. Moreover, in the presence of ionic liquid (1-ethyl-3-methyl-imidazolium acetate) as an impurity or adding anti-solvent (ethanol or acetone), a similar adverse effect was observed. The simulation study provides useful structural and dynamic properties of glucose in MIL-101 and it suggests that MIL-101 might be potentially useful for glucose recovery [95].

4.5 Summary

Both experimental and simulation studies for aqueous separation in MOFs are briefly summarized. The experiments are discussed on the removal of cations, anions, and organics from aqueous solutions, whereas the simulations are focused on water desalination, biofuel purification, and the separation of DMSO, amino acids, and glucose from water. With readily tunable pores and functionalities, MOFs have been demonstrated to possess the capability to remove/recover various molecules in aqueous solutions. The separation performance primarily depends on pore size, surface area, interaction with framework, etc.

A pre-requisite for aqueous separation in MOFs is that the MOFs should be water stable. This is particularly important because a large portion of MOFs are chemically unstable in water. The stability is largely determined by several factors, such as the bond strength and coordination number of metal-ligand, the oxidation state of metal cluster, the basicity and functionality of ligand, as well as framework topology and dimensionality. Currently, UiO, MIL, and ZIF families are mostly explored for aqueous separation because of their high water stability. Nevertheless, the water stability of other MOFs can be improved by tailoring metal clusters and organic ligands.

With ever-increasing demands for clean water, benign liquid fuel and high-purity drugs, it is foreseeable that more extensive studies on aqueous separation will be conducted. At this moment, experiments are dominant in this field. As computational resources rapidly evolve, however, we expect that there will be more simulation endeavors; consequently, wealthy atomic-resolution and time-resolved insights can be provided to assist the rational selection and design of new MOFs for high-performance aqueous separation.

Acknowledgments

We gratefully acknowledge the National University of Singapore, the Ministry of Education of Singapore, and the Singapore National Research Foundation for support.

References

1. Yaghi, O. M., Li, G., Li, H. (1995) Selective Binding and Removal of Guests in a Microporous Metal–Organic Framework. *Nature* 378: 703–706.
2. Furukawa, H., Cordova, K. E., O'Keefe, M., Yaghi, O. M. (2013) The Chemistry and Applications of Metal–Organic Frameworks. *Science* 341: 1230444.
3. Stock, N., Biswas, S. (2012) Synthesis of Metal-Organic Frameworks: Routes to Various MOF Topologies, Morphologies and Composites. *Chemical Reviews* 112: 933–969.
4. Cohen, S. M. (2012) Postsynthetic Methods for the Functionalization of Metal–Organic Frameworks. *Chemical Reviews* 112: 970–1000.
5. He, Y., Zhou, W., Qian, G., Chen, B. (2014) Methane Storage in Metal-Organic Frameworks. *Chemical Society Reviews* 43: 5657–5678.
6. Liu, J., Thallapally, P. K., McGrail, B. P., Brown, D. R., Liu, J. (2012) Progress in Adsorption-Based CO_2 Capture by Metal–Organic Frameworks. *Chemical Society Reviews* 41: 2308–2322.
7. Van de Voorde, B., Bueken, B., Denayer, J., De Vos, D. (2014) Adsorptive Separation on Metal–Organic Frameworks in the Liquid Phase. *Chemical Society Reviews* 43: 5766–5788.
8. Hu, Z., Deibert, B. J., Li, J. (2014) Luminescent Metal–Organic Frameworks for Chemical Sensing and Explosive Detection. *Chemical Society Reviews* 43: 5815–5840.
9. Dhakshinamoorthy, A., Garcia, H. (2014) Metal-Organic Frameworks as Solid Catalysts for the Synthesis of Nitrogen-Containing Heterocycles. *Chemical Society Reviews* 43: 5750–5765.

10. Yoon, M., Srirambalaji, R., Kim, K. (2012) Homochiral Metal-Organic Frameworks for Asymmetric Heterogeneous Catalysis. *Chemical Reviews* 112: 1196–1231.
11. Wang, C., Liu, X., Keser Demir, N., Chen, J. P., Li, K. (2016) Applications of Water Stable Metal-Organic Frameworks. *Chemical Society Reviews* 45: 5107–5134.
12. Hill, J. R., Subramanian, L., Maiti, A. (2005) Molecular Modeling Techniques in Material Sciences. Taylor & Francis: London.
13. Kaye, S. S., Dailly, A., Yaghi, O. M., Long, J. R. (2007) Impact of Preparation and Handling on the Hydrogen Storage Properties of Zn_4O(1,4-Benzenedicarboxylate)$_3$ (MOF-5). *Journal of the American Chemical Society* 129: 14176–14177.
14. Low, J. J., Benin, A. I., Jakubczak, P., Abrahamian, J. F., Faheem, S. A., Willis, R. R. (2009) Virtual High Throughput Screening Confirmed Experimentally: Porous Coordination Polymer Hydration. *Journal of the American Chemical Society* 131: 15834–15842.
15. Greathouse, J. A., Allendorf, M. D. (2006) The Interaction of Water with MOF-5 Simulated by Molecular Dynamics. *Journal of the American Chemical Society* 128: 10678–10679.
16. Schoenecker, P. M., Carson, C. G., Jasuja, H., Flemming, C. J. J., Walton, K. S. (2012) Effect of Water Adsorption on Retention of Structure and Surface Area of Metal–Organic Frameworks. *Industrial & Engineering Chemistry Research* 51: 6513–6519.
17. Cychosz, K. A., Matzger, A. J. (2010) Water Stability of Microporous Coordination Polymers and the Adsorption of Pharmaceuticals from Water. *Langmuir* 26: 17198–17202.
18. Férey, G., Mellot-Draznieks, C., Serre, C., Millange, F., Dutour, J., Surblé, S., Margiolaki, I. A. (2005) Chromium Terephthalate-Based Solid with Unusually Large Pore Volumes and Surface Area. *Science* 309: 2040–2042.
19. Cavka, J. H., Jakobsen, S., Olsbye, U., Guillou, N., Lamberti, C., Bordiga, S., Lillerud, K. P. A New Zirconium Inorganic Building Brick Forming Metal–Organic Frameworks with Exceptional Stability. *Journal of the American Chemical Society* 130: 13850–13851.
20. Bai, Y., Dou, Y., Xie, L.-H., Rutledge, W., Li, J.-R., Zhou, H.-C. (2016) Zr-Based Metal–Organic Frameworks: Design, Synthesis, Structure, and Applications. *Chemical Society Reviews* 45: 2327–2367.
21. Zhang, J.-P., Zhang, Y.-B., Lin, J.-B., Chen, X.-M. (2012) Metal Azolate Frameworks: From Crystal Engineering to Functional Materials. *Chemical Reviews* 112: 1001–1033.
22. Banerjee, R., Phan, A., Wang, B., Knobler, C., Furukawa, H., O'Keeffe, M., Yaghi, O. M. (2008) High-Throughput Synthesis of Zeolitic Imidazolate Frameworks and Application to CO_2 Capture. *Science* 319: 939–943.
23. Huang, X.-C., Lin, Y.-Y., Zhang, J.-P., Chen, X.-M. (2006) Ligand-Directed Strategy for Zeolite-Type Metal–Organic Frameworks: Zinc Imidazolates with Unusual Zeolitic Topologies. *Angewandte Chemie International Edition* 45: 1557–1559.
24. Colombo, V., Galli, S., Choi, H. J., Han, G. D., Maspero, A., Palmisano, G., Masciocchi, N., Long, J. R. (2011) High Thermal and Chemical Stability in Pyrazolate-Bridged Metal–Organic Frameworks with Exposed Metal Sites. *Chemical Science* 2: 1311–1319.

25. Yang, C., Kaipa, U., Mather, Q. Z., Wang, X., Nesterov, V., Venero, A. F., Omary, M. A. (2011) Fluorous Metal–Organic Frameworks with Superior Adsorption and Hydrophobic Properties toward Oil Spill Cleanup and Hydrocarbon Storage. *Journal of the American Chemical Society* 133: 18094–18097.
26. Nijem, N., Canepa, P., Kaipa, U., Tan, K., Roodenko, K., Tekarli, S., Halbert, J., Oswald, I. W. H., Arvapally, R. K., Yang, C., Thonhauser, T., Omary, M. A., Chabal, Y. J. (2013) Water Cluster Confinement and Methane Adsorption in the Hydrophobic Cavities of a Fluorinated Metal–Organic Framework. *Journal of the American Chemical Society* 135: 12615–12626.
27. Burtch, N. C., Jasuja, H., Walton, K. S. (2014) Water Stability and Adsorption in Metal–Organic Frameworks. *Chemical Reviews* 114: 10575–10612.
28. Qadir, N. U., Said, S. A. M., Bahaidarah, H. M. (2015) Structural Stability of Metal Organic Frameworks in Aqueous Media—Controlling Factors and Methods to Improve Hydrostability and Hydrothermal Cyclic Stability. *Microporous Mesoporous Materials* 201: 61–90.
29. Howarth, A. J., Liu, Y., Hupp, J. T., Farha, O. K. (2015) Metal–Organic Frameworks for Applications in Remediation of Oxyanion/Cation-Contaminated Water. *CrystEngComm* 17: 7245–7253.
30. Fang, Q. R., Yuan, D. Q., Sculley, J., Li, J. R., Han, Z. B., Zhou, H. C. (2010) Functional Mesoporous Metal–Organic Frameworks for the Capture of Heavy Metal Ions and Size-Selective Catalysis. *Inorganic Chemistry* 49: 11637–11642.
31. Mon, M., Lloret, F., Ferrando-Soria, J., Martí-Gastaldo, C., Armentano, D., Pardo, E. (2016) Selective and Efficient Removal of Mercury from Aqueous Media with the Highly Flexible Arms of a BioMOF. *Angewandte Chemie International Edition* 55: 11167–11172.
32. Wang, Y., Ye, G., Chen, H., Hu, X., Niu, Z., Ma, S. (2015) Functionalized Metal–Organic Framework as a New Platform for Efficient and Selective Removal of Cd^{2+} from Aqueous Solution. *Journal of Materials Chemistry* 3: 15292–15298.
33. Shahat, A., Hassan, H. M. A., Azzazy, H. M. E. (2013) Optical Metal–Organic Framework Sensor for Selective Discrimination of Some Toxic Metal Ions in Water. *Analytica Chimica Acta* 793: 90–98.
34. Yee, K.-K., Reimer, N., Liu, J., Cheng, S.-Y., Yiu, S.-M., Weber, J., Stock, N., Xu, Z. (2013) Effective Mercury Sorption by Thiol-Laced Metal–Organic Frameworks: In Strong Acid and the Vapor Phase. *Journal of the American Chemical Society* 135: 7795–7798.
35. Zhang, Y., Zhao, X., Huang, H., Li, Z., Liu, D., Zhong, C. (2015) Selective Removal of Transition Metal Ions from Aqueous Solution by Metal–Organic Frameworks. *RSC Advances* 5: 72107–72112.
36. Meng, X., Zhong, R.-L., Song, X.-Z., Song, S.-Y., Hao, Z.-M., Zhu, M., Zhao, S.-N., Zhang, H.-J. (2014) A Stable, Pillar-Layer Metal–Organic Framework Containing Uncoordinated Carboxyl Groups for Separation of Transition Metal Ions. *Chemical Communications* 50: 6406–6408.
37. Wang, L., Zhao, X., Zhang, J., Xiong, Z. (2017) Selective Adsorption of Pb^{2+} over the Zinc-Based MOFs in Aqueous Solution-Kinetics, Isotherms, and the Ion Exchange Mechanism. *Environmental Science and Pollution Research* 24: 14198–14206.
38. Xiao, C., Silver, M. A., Wang, S. (2017) Metal–Organic Frameworks for Radionuclide Sequestration from Aqueous Solution: A Brief Overview and Outlook. *Dalton Transactions* 46: 16381–16386.

39. Feng, Y., Jiang, H., Li, S., Wang, J., Jing, X., Wang, Y., Chen, M. (2013) Metal–Organic Frameworks HKUST-1 for Liquid-Phase Adsorption of Uranium. *Colloids Surfaces A* 431: 87–92.
40. Carboni, M., Abney, C. W., Liu, S., Lin, W. (2013) Highly Porous and Stable Metal–Organic Frameworks for Uranium Extraction. *Chemical Science* 4: 2396–2402.
41. Yang, W., Bai, Z.-Q., Shi, W.-Q., Yuan, L.-Y., Tian, T., Chai, Z.-F., Wang, H., Sun, Z.-M. (2013) MOF-76: From a Luminescent Probe to Highly Efficient U(VI) Sorption Material. *Chemical Communications* 49: 10415–10417.
42. Luo, B. C., Yuan, L. Y., Chai, Z. F., Shi, W. Q., Tang, Q. (2016) U(VI) Capture from Aqueous Solution by Highly Porous and Stable MOFs: UiO-66 and Its Amine Derivative. *Journal of Radioanalytical and Nuclear Chemistry* 307: 269–276.
43. Liu, S., Luo, M., Li, J., Luo, F., Ke, L., Ma, J. Adsorption Equilibrium and Kinetics of Uranium onto Porous Azo-Metal–Organic Frameworks. *Journal of Radioanalytical and Nuclear Chemistry* 310: 353–362.
44. Aguila, B., Banerjee, D., Nie, Z., Shin, Y., Ma, S., Thallapally, P. K. (2016) Selective Removal of Cesium and Strontium Using Porous Frameworks from High Level Nuclear Waste. *Chemical Communications* 52: 5940–5942.
45. Peng, Y., Huang, H., Liu, D., Zhong, C. (2016) Radioactive Barium Ion Trap Based on Metal–Organic Framework for Efficient and Irreversible Removal of Barium from Nuclear Wastewater. *ACS Applied Material Interfaces* 8: 8527–8535.
46. Lin, S., Kumar Reddy, D. H., Bediako, J. K., Song, M.-H., Wei, W., Kim, J.-A., Yun, Y.-S. (2017) Effective Adsorption of Pd(II), Pt(IV) and Au(III) by Zr-Based Metal–Organic Frameworks from Strongly Acidic Solutions. *Journal of Materials Chemistry A* 5: 13557–13564.
47. Wong, K. L., Law, G. L., Yang, Y. Y., Wong, W. T. A. (2006) Highly Porous Luminescent Terbium–Organic Framework for Reversible Anion Sensing. *Advanced Materials* 18: 1051–1054.
48. Zhao, X., Liu, D., Huang, H., Zhang, W., Yang, Q., Zhong, C. (2014) The Stability and Defluoridation Performance of MOFs in Fluoride Solutions. *Microporous Mesoporous Materials* 185: 72–78.
49. Zhang, N., Yang, X., Yu, X., Jia, Y., Wang, J., Kong, L., Jin, Z., Sun, B., Luo, T., Liu, J. (2014) Al-1,3,5-Benzenetricarboxylic Metal–Organic Frameworks: A Promising Adsorbent for Defluoridation of Water with pH Insensitivity and Low Aluminum Residual. *Chemical Engineering Journal* 252: 220–229.
50. Fu, H.-R., Xu, Z.-X., Zhang, J. (2015) Water-Stable Metal–Organic Frameworks for Fast and High Dichromate Trapping Via Single-Crystal-to-Single-Crystal Ion Exchange. *Chemistry of Materials* 27: 205–210.
51. Zhang, Q., Yu, J., Cai, J., Zhang, L., Cui, Y., Yang, Y., Chen, B., Qian, G. A (2015) Porous Zr-Cluster-Based Cationic Metal–Organic Framework for Highly Efficient $Cr_2O_7^{2-}$ Removal from Water. *Chemical Communications* 51: 14732–14734.
52. Desai, A. V., Manna, B., Karmakar, A., Sahu, A., Ghosh, S. K. (2016) A Water-Stable Cationic Metal–Organic Framework as a Dual Adsorbent of Oxoanion Pollutants. *Angewandte Chemie International Edition* 55: 7811–7815.
53. Aboutorabi, L., Morsali, A., Tahmasebi, E., Büyükgüngor, O. (2016) Metal–Organic Framework Based on Isonicotinate N-Oxide for Fast and Highly Efficient Aqueous Phase Cr(VI) Adsorption. *Inorganic Chemistry* 55: 5507–5513.

54. Wang, C., Liu, X., Chen, J. P., Li, K. (2015) Superior Removal of Arsenic from Water with Zirconium Metal-Organic Framework UiO-66. *Scientific Reports* 5: 16613.
55. Audu, C. O., Nguyen, H. G. T., Chang, C.-Y., Katz, M. J., Mao, L., Farha, O. K., Hupp, J. T., Nguyen, S. T. (2016) The Dual Capture of As(V) and As(III) by UiO-66 and Analogues. *Chemical Science* 7: 6492–6498.
56. Xie, D., Ma, Y., Gu, Y., Zhou, H., Zhang, H., Wang, G., Zhang, Y., Zhao, H. (2017) Bifunctional NH_2-MIL-88(Fe) Metal–Organic Framework Nanooctahedra for Highly Sensitive Detection and Efficient Removal of Arsenate in Aqueous Media. *Journal of Materials Chemistry A* 5: 23794–23804.
57. Chen, Y., Chen, F., Zhang, S., Cai, Y., Cao, S., Li, S., Zhao, W., Yuan, S., Feng, X., Cao, A., Ma, X., Wang, B. (2017) Facile Fabrication of Multifunctional Metal–Organic Framework Hollow Tubes to Trap Pollutants. *Journal of the American Chemical Society* 139: 16482–16485.
58. Howarth, A. J., Katz, M. J., Wang, T. C., Platero-Prats, A. E., Chapman, K. W., Hupp, J. T., Farha, O. K. (2015) High Efficiency Adsorption and Removal of Selenate and Selenite from Water Using Metal–Organic Frameworks. *Journal of the American Chemical Society* 137: 7488–7494.
59. Hasan, Z., Jhung, S. H. (2015) Removal of Hazardous Organics from Water Using Metal–Organic Frameworks: Plausible Mechanisms for Selective Adsorptions. *Journal of Hazardous Materials* 283: 329–339.
60. Dias, E. M., Petit, C. (2015) Towards the Use of Metal–Organic Frameworks for Water Reuse: A Review of the Recent Advances in the Field of Organic Pollutants Removal and Degradation and the Next Steps in the Field. *Journal of Materials Chemistry A* 3: 22484–22506.
61. Bai, Y., He, G., Zhao, Y., Duan, C., Dang, D., Meng, Q. (2006) Porous Material for Absorption and Luminescent Detection of Aromatic Molecules in Water. *Chemical Communications* 0: 1530–1532.
62. Jhung, S. H., Lee, J. H., Yoon, J. W., Serre, C., Férey, G., Chang, J. S. (2007) Microwave Synthesis of Chromium Terephthalate MIL-101 and Its Benzene Sorption Ability. *Advanced Materials* 19: 121–124.
63. Xie, L., Liu, D., Huang, H., Yang, Q., Zhong, C. (2014) Efficient Capture of Nitrobenzene from Waste Water Using Metal–Organic Frameworks. *Chemical Engineering Journal* 246: 142–149.
64. Jin, H., Li, Y., Liu, X., Ban, Y., Peng, Y., Jiao, W., Yang, W. (2015) Recovery of HMF from Aqueous Solution by Zeolitic Imidazolate Frameworks. *Chemical Engineering Science* 124: 170–178.
65. Jiang, J.-Q., Yang, C.-X., Yan, X.-P. (2013) Zeolitic Imidazolate Framework-8 for Fast Adsorption and Removal of Benzotriazoles from Aqueous Solution. *ACS Applied Material Interfaces* 5: 9837–9842.
66. Haque, E., Lee, J. E., Jang, I. T., Hwang, Y. K., Chang, J. S., Jegal, J., Jhung, S. H. (2010) Adsorptive Removal of Methyl Orange from Aqueous Solution with Metal–Organic Frameworks, Porous Chromium-Benzenedicarboxylates. *Journal of Hazardous Materials* 181: 535–542.
67. Maes, M., Schouteden, S., Alaerts, L., Depla, D., De Vos, D. E. (2011) Extracting Organic Contaminants from Water Using the Metal-Organic Framework Cr(OH)·{OOC-C_6H_4-COO}. *Physical Chemistry Chemical Physics* 13: 5587–5589.

68. Huo, S. H., Yan, X. P. (2012) Metal-Organic Framework MIL-100(Fe) for the Adsorption of Malachite Green from Aqueous Solution. *Journal of Materials Chemistry* 22: 7449–7455.
69. Tong, M., Liu, D., Yang, Q., Devautour-Vinot, S., Maurin, G., Zhong, C. (2013) Influence of Framework Metal Ions on the Dye Capture Behavior of MIL-100 (Fe, Cr) MOF Type Solids. *Journal of Materials Chemistry A* 1: 8534–8537.
70. He, Y. C., Yang, J., Kan, W. Q., Zhang, H. M., Liu, Y. Y., Ma, J. F. (2015) A New Microporous Anionic Metal-Organic Framework as a Platform for Highly Selective Adsorption and Separation of Organic Dyes. *Journal of Materials Chemistry A* 3: 1675–1681.
71. Hasan, Z., Jeon, J., Jhung, S. H. (2012) Adsorptive Removal of Naproxen and Clofibric Acid from Water Using Metal–Organic Frameworks. *Journal of Hazardous Materials* 209-210: 151–157.
72. Hu, Y., Song, C., Liao, J., Huang, Z., Li, G. (2013) Water Stable Metal–Organic Framework Packed Microcolumn for Online Sorptive Extraction and Direct Analysis of Naproxen and Its Metabolite from Urine Sample. *Journal of Chromatography A* 1294: 17–24.
73. Khan, N. A., Jung, B. K., Hasan, Z., Jhung, S. H. (2015) Adsorption and Removal of Phthalic Acid and Diethyl Phthalate from Water with Zeolitic Imidazolate and Metal–Organic Frameworks. *Journal of Hazardous Materials* 282: 194–200.
74. Seo, Y. S., Khan, N. A., Jhung, S. H. (2015) Adsorptive Removal of Methylchlorophenoxypropionic Acid from Water with a Metal–Organic Framework. *Chemical Engineering Journal* 270: 22–27.
75. Zhu, X., Li, B., Yang, J., Li, Y., Zhao, W., Shi, J., Gu, J. (2015) Effective Adsorption and Enhanced Removal of Organophosphorus Pesticides from Aqueous Solution by Zr-Based MOFs of UiO-67. *ACS Applied Material Interfaces* 7: 223–231.
76. Lan, X., Zhang, H., Bai, P., Guo, X. (2016) Investigation of Metal Organic Frameworks for the Adsorptive Removal of Hydrochloride from Dilute Aqueous Solution. *Microporous Mesoporous Material* 231: 40–46.
77. Chen, C., Chen, D., Xie, S., Quan, H., Luo, X., Guo, L. (2017) Adsorption Behaviors of Organic Micropollutants on UiO-66: Analysis of Surface Interactions. *ACS Applied Materials Interfaces* 9: 41043–41054.
78. Duren, T., Bae, Y. S., Snurr, R. Q. (2009) Using Molecular Simulation to Characterise Metal-Organic Frameworks for Adsorption Applications. *Chemical Society Reviews* 38: 1203–1212.
79. Liu, D. H., Zhong, C. L. (2010) Understanding Gas Separation in Metal-Organic Frameworks. *Journal of Materials Chemistry* 20: 10308–10318.
80. Jiang, J. W., Babarao, R., Hu, Z. Q. (2011) Molecular Simulations for Energy, Environmental and Pharmaceutical Applications of Nanoporous Materials: From Zeolites, Metal-Organic Frameworks to Protein Crystals. *Chemical Society Reviews* 40: 3599–3612.
81. Jiang, J. W. (2012) Metal-Organic Frameworks for CO_2 Capture: What Are Learned from Molecular Simulations. In *Coordination Polymers and Metal Organic Frameworks*, Ortiz, O. L.; Ramírez, L. D., Eds. Nova Science Publishers.
82. Jiang, J. W. (2012) Recent Development of in Silico Molecular Modeling for Gas and Liquid Separations in Metal–Organic Frameworks. *Current Opinion in Chemical Engineering* 1: 138–144.

83. Getman, R. B., Bae, Y. S., Wilmer, C. E., Snurr, R. Q. (2012) Review and Analysis of Molecular Simulations of Methane, Hydrogen and Acetylene Storage in Metal-Organic Frameworks. *Chemical Reviews* 112: 703–723.
84. Hu, Z. Q., Chen, Y. F., Jiang, J. W. (2011) Zeolitic Imidazolate Framework-8 as a Reverse Osmosis Membrane for Water Desalination: Insight from Molecular Simulation. *Journal of Chemical Physics* 134: 134705.
85. Gupta, K. M., Zhang, K., Jiang, J. W. (2015) Water Desalination through Zeolitic Imidazolate Framework Membranes: Significant Role of Functional Groups. *Langmuir* 31: 13230–13237.
86. Gupta, K. M., Qiao, Z. W., Zhang, K., Jiang, J. W. (2016) Seawater Pervaporation through Zeolitic Imidazolate Framework Membranes: Atomistic Simulation Study. *ACS Applied Material Interfaces*, 8: 13392–13399.
87. Ragauskas, A. J., Williams, C. K., Davison, B. H., Britovsek, G., Cairney, J., Eckert, C. A., Frederick, W. J., Hallett, J. P., Leak, D. J., Liotta, C. L., Mielenz, J. R., Murphy, R., Templer, R., Tschaplinski, T. (2006) The Path Forward for Biofuels and Biomaterials. *Science* 311: 484–488.
88. Nalaparaju, A., Zhao, X. S., Jiang, J. W. (2011) Biofuel Purification by Pervaporation and Vapor Permeation in Metal–Organic Frameworks: A Computational Study. *Energy & Environmental Science* 4: 2107–2116.
89. Zhang, K., Nalaparaju, A., Chen, Y. F., Jiang, J. W. (2014) Biofuel Purification in Zeolitic Imidazolate Frameworks: The Significant Role of Functional Groups. *Physical Chemistry Chemical Physics* 16: 9643–9655.
90. Zhang, K., Gupta, K. M., Chen, Y. F., Jiang, J. W. (2015) Biofuel Purification in Gme Zeolitic–Imidazolate Frameworks: From Ab Initio Calculations to Molecular Simulations. *American Institute of Chemical Engineers Journal* 61: 2763–2775.
91. Nalaparaju, A., Jiang, J. W. (2012) Recovery of Dimethyl Sulfoxide from Aqueous Solutions by Highly Selective Adsorption in Hydrophobic Metal–Organic Frameworks. *Langmuir* 28: 15305–15312.
92. Bhagwatwar, H. P., Phadungpojna, S., Chow, D. S., Andersson, B. S. (1996) Formulation and Stability of Busulfan for Intravenous Administration in High-Dose Chemotherapy. *Cancer Chemotherapy and Pharmacology* 37: 401–408.
93. Park, S. J., Yoon, T. I., Bae, J. H., Seo, H. J., Park, H. J. (2001) Biological Treatment of Wastewater Containing Dimethyl Sulphoxide from the Semi-Conductor Industry. *Process Biochemistry* 36: 579–589.
94. Hu, Z. Q., Chen, Y. F., Jiang, J. W. (2013) Liquid Chromatographic Separation in Metal-Organic Framework MIL-101: A Molecular Simulation Study. *Langmuir* 29: 1650–1656.
95. Gupta, K. M., Zhang, K., Jiang, J. W. (2015) Glucose Recovery from Aqueous Solutions by Adsorption in Metal–Organic Framework MIL-101: A Molecular Simulation Study. *Scientific Reports* 5: 12821.
96. Lange, J. P., Van der Heide, E., Van Buijtenen, J., Price, R. (2012) Furfural: A Promising Platform for Lignocellulosic Biofuels. *ChemSusChem* 5: 150–166.

5

Coating of Nanoparticles in Aqueous Solutions: Insights from Molecular Dynamics Simulations

Zuzana Benková and M. Natália D. S. Cordeiro

CONTENTS

5.1 Introduction

Nanoparticles (NPs) with the size range 1–100 nm between bulk materials and atomic or molecular structures possess many intriguing properties. The properties of NPs strongly depend on their size, shape, chemical composition, and structure, and are easily modulated. NPs have found wide application in various fields of life sciences, such as separation technologies, optics, electronics, histological studies, diagnostics, therapeutics, drug delivery systems, and bioimaging. Several types of NPs are employed in technological and medical applications based on (a) inorganic materials such as noble metals (Au and Ag), magnetic particles (Fe_2O_3, Fe_3O_4, TiO_2), carbon allotropes (graphene, fullerene, carbon nanotubes, nanorods, and nanohorns), silica, quantum dots, (b) organic materials such as micelles, dendrimers, liposomes, and (c) hybrid nanoparticles. The shape of nanoparticles might vary from three-dimensional (nanospheres) through two-dimensional (graphene sheets) to one-dimensional (nanorods or carbon nanotubes). The biological application of inorganic NPs requires covering of NPs by a monolayer of organic ligands in order to improve NPs biocompatibility, inhibit NPs aggregation, prevent protein adsorption, and induce favorable interactions with biological membranes. At the same time, grafting of an organic ligand monolayer enables the tuning and control of physicochemical properties of the nanoparticle surfaces by the structure and chemistry of the grafted ligands.

In this chapter, if not stated otherwise, the attention is focused mainly on atomistic computer simulations of hybrid NPs where an inorganic spherical core is coated with organic polymer chains in an aqueous environment. The grafted chains may be of synthetic origin—i.e. polyethylene oxide, polymethylmethacrylate and its derivatives, poly(*N*-alkylacrylamide), alkanethiols and their functionalized derivatives, or of natural origin—nucleic acids, amino acids, glutathione. The polymer chains are tethered to nanoparticle surfaces irreversibly by covalent bonds or reversibly by adsorption of polymers ensuing from non-bonded physicochemical interactions. In hybrid NPs, the additional variables controlling the NPs behavior are the chemistry, molar mass, and the coverage density of grafted polymers. The monolayers are often charged, which plays a significant role in penetration of NPs through biological membranes or in aggregation/dispersion processes of NPs. Of special relevance are stimuli-responsive grafted monolayers that provide unique properties of NPs. These layers respond to the changes of external conditions by adopting different conformations, which considerably modify their properties. The monolayers of mixed polymer composition are also often used for their specific properties. Different arrangements of the two polymer species are possible, namely, the random or striped mixing and the arrangement typical for Janus NPs, where the polymer monolayer continuously coating one hemisphere of NP is chemically distinct from the polymer monolayer continuously coating the other hemisphere. The NPs covered with mixed monolayers partly mimic the biological molecules and may be utilized in the development of smart machines. It has been shown that the striped morphology of NPs enables nondisruptive penetration of NPs into cells whereas analogous NPs with randomly mixed monolayer do not.

Nowadays, nanotechnology impacts a large range of fields including computer science, materials technology, engineering, and medicine. Understanding the processes involved at the nanoscale level often requires interdisciplinary studies, and computer simulations constitute an important contribution to this area. Computational techniques in the field of molecular simulations allow investigation of nanoscale-sized systems and provide insight (often at the atomistic level) into properties which are not accessible to experimental techniques. At the same time, molecular simulations help to explain and clarify experimental findings. Recently, a large number of publications reporting computer simulations of NPs in biological systems have been recorded. Different computational methods exist considering the system at different scale levels ranging from the atomistic approach (all-atom or united-atom models) through coarse-grained to mesoscale approaches. Currently, the most widely employed atomistic methods are molecular dynamics dissipative particle dynamics, and Monte Carlo.

5.2 Setting Up and Representation of Hybrid NPs in Water for Molecular Simulations

In atomistic simulations, the core of hybrid nanoparticle is usually discretized into atoms and is modeled either by a crystallographic model or by a continuous model. For instance, in the crystallographic model of a gold core, the polyhedral gold nanocluster may be assumed comprising multiple, locally flat (111), and (100) facets [1–3]. In the continuous model, the gold nanoclusters are of spherical geometry. The former model is more appropriate for larger nanoclusters with diameters >3 nm. Nearly spherical polyhedral geometry of AuNPs might be achieved when 144 gold atoms are organized in a rhombicosidodecahedron (~2 nm) way, as shown in Figure 5.1 [4–7]. Rigidity of the gold core can be preserved assuming a number of virtual constant bonds and constraints between Au atoms within the core. In order to reduce the computational time and to enable simulations of larger systems, the core of NPs might be represented in a coarse-grained way with a reduced number of core interaction sites while keeping the atomistic representation of the grafted polymer chains [8]. The correct chemistry of the coarse-grained representation of the core is preserved by setting the interparticle potentials to be the integration of the all pairwise potentials involving at least one core atom.

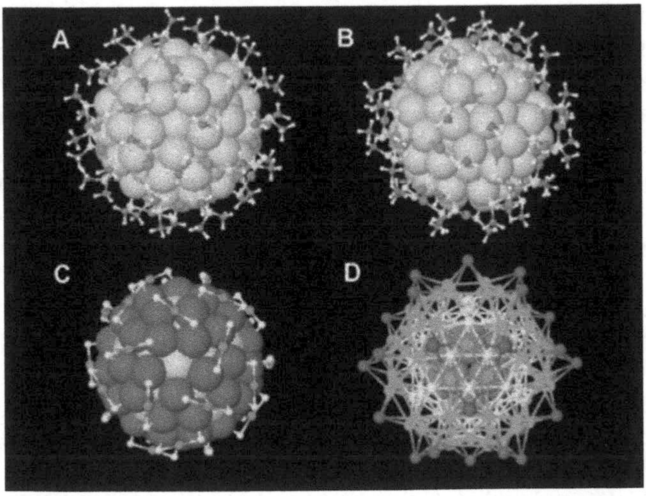

FIGURE 5.1
Equilibrated structure of $Au_{144}(SR)_{60}$ hybrid nanoparticle viewed down a 5-fold (a) and a 3-fold (b) symmetry axis. Chiral arrangement of the RS-Au-SR units covering the 60-atom surface of the Au_{114} core (c), all 144 gold atoms, with different atom shells discerned by different shades of gray (d). (Reprinted with permission from Lopez-Acevedo, O. et al., *Journal of Physical Chemistry C*, 113: 5035–5038, 2009. Copyright 2009 American Chemical Society.)

The polymer chains can be grafted onto the NPs covalently or noncovalently. In the case of AuNPs, the alkanethiols are tethered to the surface of AuNPs through Au-S linkage, usually in all-trans conformation creating self-assembled monolayers (SAMs). One should realize, however, that the SAMs on curved surface differ from SAM on planar surfaces since with the increasing distance from the surface more space becomes available for monomers. This concentration gradient increasing toward the surface of the core increases with increasing surface curvature. Interesting, for instance, is the grafting of polymer chains carried out in ref. 5 where a pair of two terminally functionalized alkanethiols mutually connected by a sulfur atom is grafted to the Au core through Au-S bond (Figure 5.1). Thus, one sulfur atom is coordinated with two Au atoms. In this way, one Au atom serves as a bridge between the two alkanethiols and the other one belongs to the peripheral layer of AuNPs. Altogether, 30 such pairs can be grafted to the core built up of 144 Au atoms. In this way, the sulfur atoms are inserted directly into the outer layer of the atoms of the gold core. In experimental investigations, the polymer chains are often grafted onto a NP through some linkers. For the sake of simplicity, such linkers are usually omitted in the atomistic simulation studies. However, if the chemical composition of grafted polymers does not allow direct chemical bond with the core surface, the linker is necessary also in the simulation studies. This is, for instance, the case of polystyrene chains grafted onto the silica core [9]. The frequently occurring combination of silica surface with poly(ethylene oxide) (PEO) chains is an example when PEO chains may be grafted directly to the silica support through an oxygen atom [10]. The noncovalent adsorption of a protecting monolayer on a nanoparticle is based on van der Waals, electrostatic, or some specific interactions between the NPs and the polymers constituting the monolayer. This reversible adsorption depends on the polarity of solvent, which is often responsible for the concurrent interactions with the component of the hybrid NPs. Strong adsorption between a NP's core and grafted polymer chains might be based on π-π stacking as has been used when pyrene-polyethylene comprised of 45 polyethylene units terminated with a pyrene unit has been anchored to the carbon single wall nanoparticle (SWNP) [11]. However, the spontaneous adsorption of pyrene moiety on the external wall of SWNT is observed in polar organic solvent (*N,N*-dimethylacetamide) while in aqueous solution, the pyrene appeared as an inefficient anchor. Since the aromatic character of carbon nanotubes (CNTs) decreases with decreasing diameter of CNT (enhanced surface curvature), CNTs with larger diameter display stronger π-π stacking.

The atomistic molecular dynamics simulations of hybrid NPs are time-demanding since a relatively large system (simulation cell) needs to be considered and sufficiently long calculations must be performed to get reliable statistics. One should realize that the number of water molecules correlates with the simulation cell. Moreover, in the studies of aggregation properties of NPs, at least two NPs are included into the simulation cell and the evaluation of attractive/repulsive forces as functions of the distance between NPs requires equilibration and production calculations carried out for a range of different distances between NPs. The simulations of a completely coarse-grained system of NPs shifts the limit of the system size and time scale to larger values but it disallows to study specific interactions, which are often significant in an aqueous medium [12].

The metal core of the hybrid NPs composed of noble metal atoms is well represented by one force field [13] compatible with commonly adopted force fields such as AMBER [14], OPLS-AA [15], CHARMM [16], CVFF [17], COMPASS [18], and PCFF [19]. The most commonly employed force fields for the representation of the remaining components in the systems of hybrid NPs in an aqueous environment are AMBER, CHARMM, and OPLS-AA with the 12–6 form of the Lennard-Jones potential. These force fields are also compatible with the force fields generated for the cores composed of many other atoms. As regards to the water molecules, the TIP3P [20] or SPC [20] force fields are usually employed to model those because of their compatibility with the above mentioned force fields.

5.3 Gold Hybrid Nanoparticles

The gold nanoparticles (AuNPs) are important in research and technology due to their salient features such as stability at ambient conditions, tunable optical properties, catalytic properties, biocompatibility, biochemical, and biomedical activities. AuNPs coated with self-assembled monolayers (SAM) of alkanethiols belong to the most widely studied systems using molecular simulations. In addition, alkanethiol AuNPs can be easily prepared. Molecular dynamics simulations of AuNPs in aqueous solutions are mostly aimed at their aggregation or dispersion properties. The effect of different functional groups terminating the alkanethiols tethered to the AuNPs on static, dynamic and thermodynamic properties are often investigated. It has been shown that the terminal groups significantly affect the aggregation of hybrid AuNPs. The hydrophobic termination of grafted chains leads to the aggregation of AuNPs in water. Although this aggregation is entropically disfavored, the enthalpic component, ensuing from the van der Waals interactions between the hydrophobic grafted chains, prevails in water as well as in dry conditions [21]. The contribution due to the van der Waals interactions increases with the number of hydrophobic contacts, which in turn increases with increasing polymer concentration (coverage density). While in the case of hydrophobic terminal methyl group, this trend is preserved also in the presence of water, the situation is different in the case of hydrophilic terminal hydroxyl group. In the absence of water, the aggregates formed by the AuNPs functionalized with S-$(CH_2)_{12}$-OH are even larger than the aggregates formed by the AuNPs functionalized with S-$(CH_2)_{12}$-CH_3 due to the additional hydrogen-bonding interactions between the hydrophilic chains. However, in the presence of water, the aggregation of the hydroxyl terminated AuNPs is weakened due to the competing hydrogen bonding between water molecules and terminal hydroxyl groups. This is well evidenced in Figure 5.2 where at the coverage densities 100%, AuNPs functionalized with S-$(CH_2)_{12}$-OH chains lose the propensity to aggregate while AuNPs functionalized with S-$(CH_2)_{12}$-CH_3 aggregate.

In order to comprehensively understand the behavior and functions of hybrid NPs in aqueous solvents, it is inevitable to investigate also the structural and dynamic properties of interfacial water molecules in the vicinity of NPs. Such a molecular dynamics study has been performed for AuNPs coated with four differently terminated alkane-1-thiols (for brevity, alkanethiol refers to alkane-1-thiol throughout this chapter) with the methyl (CH_3), carboxyl (COOH), primary amine (NH_2), and hydroxyl (OH) terminal substitutents [22]. The clean AuNPs has been used as a standard. The reduced density profiles of the water oxygen and hydrogen atoms suggest that a two-shelled water structure occurs close to the clean and nonpolar AuNPs. This water structure is diminished in the proximity of polar AuNPs. If AuNPs are substituted with polar chains the water molecules penetrate to AuNPs

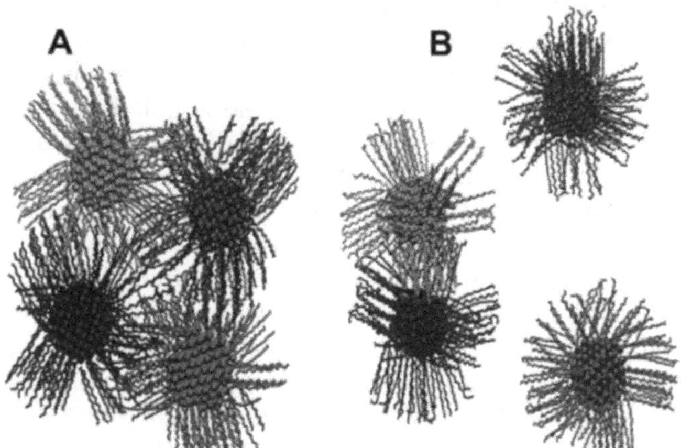

FIGURE 5.2
Final configurations of aggregates of AuNPs coated with SAM of S-$(CH_2)_{12}$-CH_3 (a) and S-$(CH_2)_{12}$-OH (b) at the coverage density of 100% and at the temperature of 300 K. Water molecules are not shown for the sake of clarity. (Reprinted from *Computational Materials Science*, 86, Meena Devi, J., Aggregation of Thiol Coated Gold Nanoparticles: A Simulation Study on the Effect of Polymer Coverage Density and Solvent, 174–179, Copyright 2014, with permission from Elsevier.)

core more deeply. On the other side, the nonpolar chains tend to repel the water molecules from the surface of AuNPs. The tail functional groups of the modified AuNPs significantly affect the orientation of water molecules. The analysis of the bivariate distributions of the preferential orientation of water molecules reveals the preferred orientations of the interfacial water molecules as illustrated in Figure 5.3. Obviously, the ligands with polar terminal substituents substantially alter the orientation of the water molecules close to the AuNP when compared with the clean AuNP or AuNP coated with the monolayer of ligands with hydrophobic terminal substituents. The formation of the hydrogen bonds between polar substituents and water molecules attenuates the formation of the hydrogen bonds between water molecules close to the AuNP. The two-shelled structure of water near the AuNP is

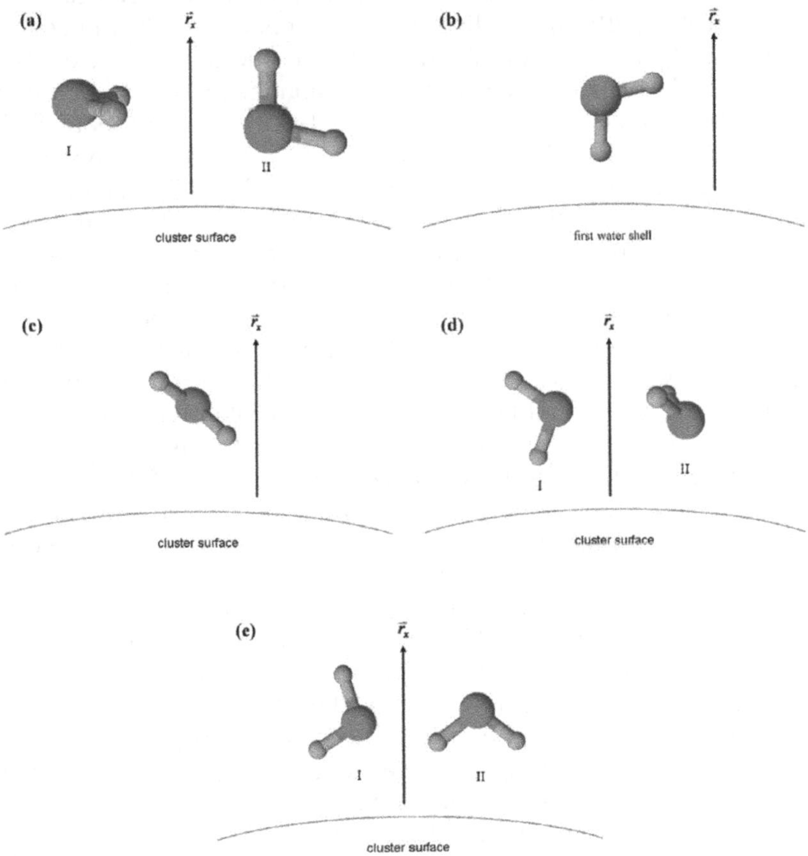

FIGURE 5.3
Preferential orientation of interfacial water molecules near clean AuNP and AuNP coated with methyl terminated alkanethiols in the first (a) and second hydration sphere (b), near AuNP coated with carboxyl (COOH) (c), primary amine (NH_2) (d), and hydroxyl (OH) (e) terminal substituents. (Reprinted with permission from Yang, A. C., Weng, C. I., *Journal of Physical Chemistry C*, 114: 8679–8709, 2010. Copyright 2010 American Chemical Society.)

capable of the increased residence time and the reduced diffusion coefficient of water molecules in the vicinity of the AuNP.

The AuNPs are supposed to function at physiological conditions and that is why understanding their function requires the knowledge of interactions of AuNPs with the ions occurring in a physiological medium. From this point of view, the charged NPs possess interesting properties. It has been well established that the negatively charged NPs can translocate through biological membranes without intrusion while the positively charged NPs intrude the biological membrane since they leave a large hole in the lipid bilayer during the translocation process. The atomistic molecular dynamics simulations appear to be a suitable tool to explore the interactions between charged ligands in the monolayer of NPs and the surrounding medium and to characterize the associated processes. Such a study has been carried out for a pair of $Au_{144}(SR)_{60}$ in a saline where the gold atoms have been organized into a rhombicosidodecahedron [7] and the outer shell of AuNPs has been complexed with 60 terminally functionalized ligands, 11-sulfanyl-undecane-1-sulfonate $S(CH_2)_{11}SO_3^-$, 5-sulfanyl-pentane-1-sulfonate, $S(CH_2)_5 SO_3^-$, 5-sulfanyl-pentane-1-ammonium, $S(CH_2)_{11}NH_3^+$, 4-sulfanylbenzoate, $SPhCOO^-$, and protonated 4-sulfanylbenzamide, $SPhCONH_3^+$. These charged substituents were neutralized with counter ions, Na^+/Cl^-. The concentration of the NaCl in saline was 150 mM. The potential of mean force was estimated to determine whether the repulsive or attractive interactions dominated within the pair of coated AuNPs. It turns out that the attraction and thus the aggregation between identically charged NPs may be induced by bridging these NPs with a counter ion. However, the effectivity of this bridging is influenced by several factors. The AuNPs coated with long and flexible chain, i.e., $S(CH_2)_{11}SO_3^-$, appear to be neutralized by Na^+ only modestly. Since the attraction between these AuNPs predominantly follows from the hydrophobic interactions of long alkane fragments when the chains are collectively bent, the aggregation may happen only at very high concentrations of AuNPs in saline. High affinity between AuNPs is observed when the AuNPs are substituted with short $(CH_2)_5SO_3^-$ ligands due to the bridge formed between two AuNPs by Na^+. The Na^+ counter ions act as bridges even more efficiently when AuNPs are substituted with more rigid and thus less fluctuating $SPhCOO^-$ ligands (Figure 5.4). The Cl^- counter ions also participate in salt bridging between AuNPs coated with short positively charged ligands but the mediated bridges are weaker in strength and fewer in number. The smaller F^- counter ions are supposed to form more and stronger bridges between positively charged hybrid AuNPs.

The aggregation of charged NPs in aqueous solutions depends not only on the size and concentration but also on the shape of counter ions. This has been demonstrated using molecular dynamics simulations of $Au_{144}(SPhCOO^-)_{60}$ and $Au_{144}(SPhCOO^-)_{30}(SPhCOOH)_{30}$ nanoparticles in different aqueous solutions containing Na^+, K^+, TMA^+ (tetramethylammonium cation), or TRS^+

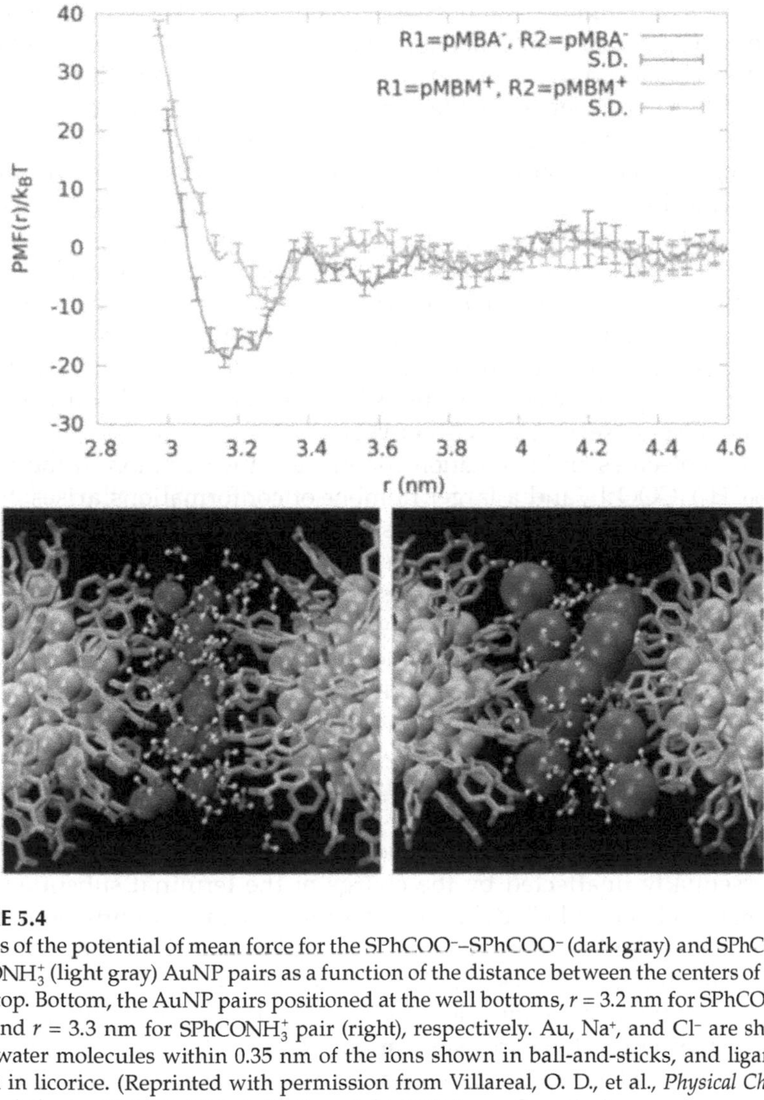

FIGURE 5.4
Profiles of the potential of mean force for the $SPhCOO^-$–$SPhCOO^-$ (dark gray) and $SPhCONH_3^+$–$SPhCONH_3^+$ (light gray) AuNP pairs as a function of the distance between the centers of AuNPs at the top. Bottom, the AuNP pairs positioned at the well bottoms, $r = 3.2$ nm for $SPhCOO^-$ pair (left) and $r = 3.3$ nm for $SPhCONH_3^+$ pair (right), respectively. Au, Na^+, and Cl^- are shown in balls, water molecules within 0.35 nm of the ions shown in ball-and-sticks, and ligands are shown in licorice. (Reprinted with permission from Villareal, O. D., et al., *Physical Chemistry Chemical Physics*, 17: 3680–3688, 2015. Copyright 2015 Owner Societies.)

(trisammonium cation) as potential bridges combined with OH^- or Cl^- anions [6]. The solutions containing only the amount of anions equimolar to the total charge of AuNPs to neutralize the system have also been included in the study. The main findings of these simulations indicate that the presence of smaller counter ions in sufficiently high concentration results into enhanced aggregation of smaller negatively charged AuNPs (~2 nm). Counter ions of aspherical shape eliminate the aggregation of AuNPs, however, they screen the coulombic repulsive interactions at larger distances between AuNPs, whereas ions of the same sign of charge as the sign of AuNPs essentially do not affect the aggregation of AuNPs. The decrease in pH diminishes the number of dissociated carboxylic groups and consequently lowers the susceptibility of AuNPs to aggregate since the number of cationic bridges is reduced.

The structural and dynamic behavior of water molecules in the close vicinity of charged AuNPs have been investigated at the atomistic level as well [5]. The charged AuNPs with the formulas $Au_{144}[S(CH_2)_{11}COO^-]_{60}$ (Na^+ counter ions) and $Au_{44}[S(CH_2)_{11}NH_3^+]_{60}$ (Cl^- counter ions) interact concurrently with the water molecules through hydrogen bonds and electrostatically with the counter ions. That is why the possible conformations of the charged AuNPs simultaneously interacting with the water molecules and counter ions exist. The situation is simpler in the case of $Au_{44}[S(CH_2)_{11}NH_3^+]_{60}$ where only two possible conformations have been reported (Figure 5.5). The feasibility of six to seven contacts between the terminal carboxyl group and the water molecules and Na^+ cation complicates the situation in the case of $Au_{144}[S(CH_2)_{11}COO^-]_{60}$ and a larger number of conformations arises; two of them are illustrated in Figure 5.6. It is not surprising that the carboxyl terminated $AuNP^-$ forms more hydrogen bonds (~404) than the ammonium terminated $AuNP^+$ (~171). Nevertheless, the total number of contacts with counter ions remains more or less the same ~4.4 vs. ~4.7 for $AuNP^-$ and $AuNP^+$, respectively. The stronger binding of Na^+ to two carboxyl oxygens is reflected on the prolonged contact lifetime (~10.1 ps) with respect to Cl^-, which is coordinated only to one ammonium hydrogen (~5.0 ps). Oppositely, since the contact of water molecules with the terminal function is less disturbed by the counter ion in the case of $AuNP^+$ than in the case of $AuNP^-$, the water contact lifetime is about double (~6.5 ps) of the water contact lifetime near $AuNP^-$ (~3.5 ps). Interestingly, the diffusion of interfacial water seems to be essentially unaffected by the charge of the terminal substituent but the more loosely bound Cl^- diffuses faster than the more tightly bound Na^+. There is also a significant preferential orientation found for the water molecules in the first hydration shell of both types of AuNPs.

The mixed monolayers introduced onto NPs have been extensively studied using molecular dynamics simulations. The multiple species presented in the monolayer tend to separate creating microphase structures and thus the knowledge of the role of nanoscale morphology of NPs becomes important. The AuNPs coated with one monolayer constructed of randomly mixed alkanethiols and their terminally derived alcohols of the same lengths have been examined predominantly from the viewpoint of the monolayer wettability [23]. The wettability of the monolayer turns out to be linearly related to the concentration (fraction) of the terminal hydroxyl groups presented in the monolayer. The number of hydrogen bonds formed between the terminal hydroxyl groups and the interfacial water molecules increases with the concentration of hydroxyl groups. This means a possibility to control the interactions between AuNPs and any target molecule or aqueous medium by varying the mixing ratio of terminal hydroxyl groups present in the monolayer.

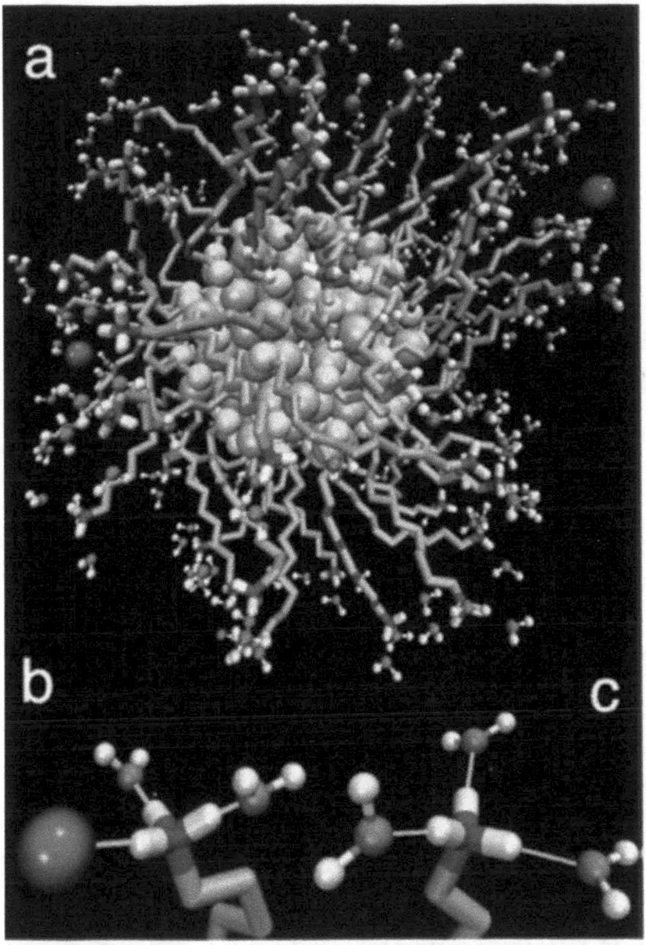

FIGURE 5.5
The solvation shell of AuNP coated with $Au_{44}[S(CH_2)_{11}NH_3^+]_{60}$ and Cl^- counter ions (big spheres) in water within a cutoff distance of 0.34 nm from the center of AuNP (a), NH_3^+ terminal group forming two hydrogen bonds with two water molecules and one contact with Cl^- (b) and three hydrogen bonds with three water molecules (c). (Reprinted with permission from Heikkila, E. et al., *Journal of Physical Chemistry C*, 116: 9805–9815, 2012. Copyright 2012 American Chemical Society.)

The presence of the hydroxyl group in the mixed monolayer of AuNPs is responsible also for the reorientation of the interfacial water molecules. The monolayer formed exclusively by methyl terminated monolayer triggers a broad distribution of the angle between the dipole moment vector of a water molecule and the vector connecting the oxygen atom of a water molecule with the center of mass of the AuNPs, and the maximum of the distribution is located at ~90°. This orientation minimizes the unfavorable interactions between the

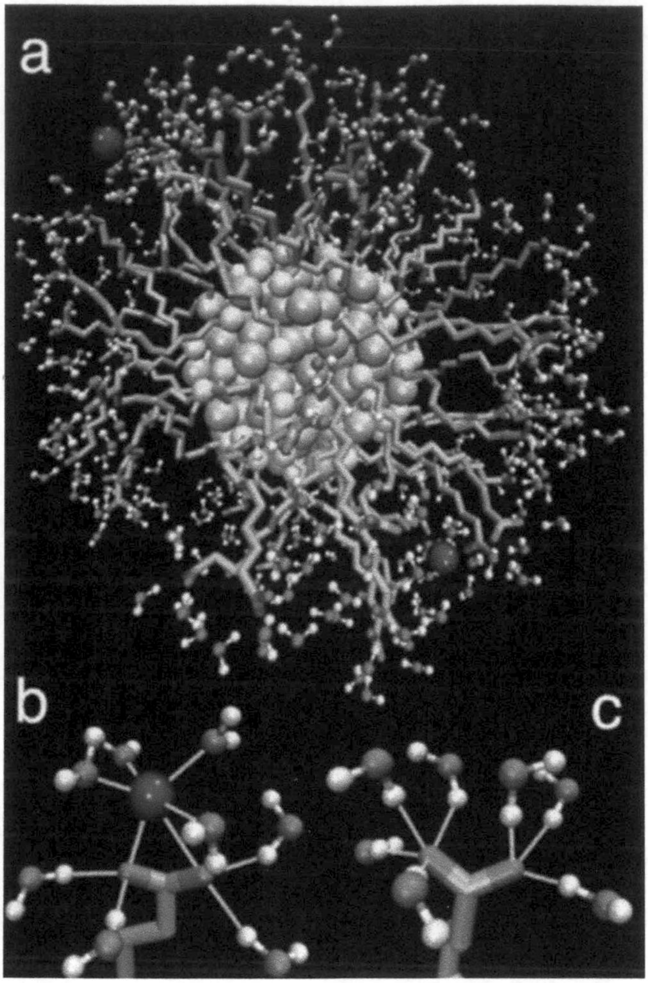

FIGURE 5.6
The solvation shell of AuNP coated with $Au_{144}[S(CH_2)_{11}COO^-]_{60}$ and Na^{-+} counter ion (big spheres) in water within a cutoff distance of 0.36 nm from the center of AuNP (a), Na^+ cation coordinating with four water molecules and COO^- and two carboxylic oxygen atoms forming hydrogen bonds with four water molecules (b), COO^- group forming hydrogen bonds with seven water molecules (c). (Reprinted with permission from Heikkila, E. et al., *Journal of Physical Chemistry C*, 116: 9805–9815, 2012. Copyright 2012 American Chemical Society.)

water molecules and the hydrophobic surface. The introduction of hydroxyl terminal substituents narrows the angle distribution and shifts the maximum to ~40°. With increasing concentration of hydroxyl groups the maximum of the angle distribution becomes higher. This concentration induced trend persistently continues till the homogenous layer of hydroxyl terminated alkanethiols is achieved. The total residence time of the water molecules in the hydration shell of AuNPs increases with the ratio of the terminal hydroxyl groups.

Mixed monolayers constituted of chemically different polymers that also differ in the contour length exhibit interesting characteristics. The AuNPs covered with the 1:1 mixture of hydrophilic 11-sulfanylundecanesulfonate and hydrophobic octane-1-thiol have been studied in water with physiological salt concentration (150 mM) and with additional counter ions to neutralize the system (Figure 5.7) [24]. Striped, regularly mixed, and randomly mixed arrangements of these chains anchored to the AuNPs have been investigated. For comparison, both homogenous monolayers composed only of one of the species have been considered as well. In fact, the structural and electrostatic properties of homogenous monolayer composed of hydrophilic chains are not altered when a certain amount of hydrophilic chains is replaced with shorter hydrophobic chains. The variation of the core diameter affects the tilt angle—for a spherical NP defined as the angle between the vector pointing from the center of mass of the core to the sulfur atom and the vector pointing from the sulfur atom to the end of the

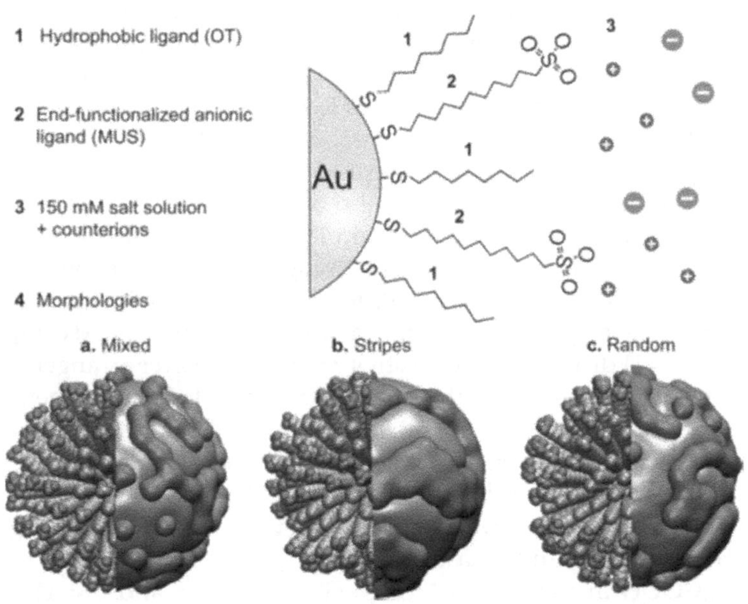

FIGURE 5.7
Schematic illustration of AuNP coated with the 1:1 mixture of hydrophilic 11-sulfanylundecanesulfonate (MUS) and hydrophobic octane-1-thiol (OT) at the top. Mixed (a), striped (b), and random (c) nanoscale morphologies. In the right hemisphere of the AuNPs, the initial, all-trans conformations of the hydrophilic (dark gray) and hydrophobic (light gray) ligands are partially drawn as surfaces to clearly illustrate the difference between the nanoscale domains at the bottom. (Reprinted with permission from Van Lehn, R. C., Alexander-Katz, A., *Journal of Physical Chemistry C*, 117: 20104–20115, 2013. Copyright 2013 American Chemical Society.)

grafted chain, and the fluctuations of tethered chains—and this effect is exceptionally important in the case of NPs with a monolayer comprising species of different contour lengths. In this study, the variation of the core diameter from 2 to 8 nm causes the drop of the tilt angle from ~ 75° to ~ 30°, the latter angle is similar to the tilt angle estimated for homogeneous monolayer on the flat gold surface. The increasing size of the core (decreasing surface curvature) also leads to larger spatial confinement of the chains due to the reduced free space further from the surface of AuNP. Thus, the chains become more rigid and fluctuate less. When the grafted chains constituting the heterogeneous monolayer of the three different arrangements are grafted onto the gold core with the diameter of 4 nm the morphologies and measurable structural characteristics become very similar. This indicates that the extensive chain fluctuations allow the electrostatic repulsion of the charged terminal groups and, thus, their dispersion over the available space in order to maximize their separation. At the same time, the hydrophobic chains are relatively free to adopt conformations avoiding the water exposure. This trend is preserved also when the hydrophobic octane-1-thiols are substituted for longer alkanethiols.

The mixed self-assembled monolayer on AuNPs can be arranged also in a symmetric way to yield the Janus NPs. Such an arrangement, along with the random and striped arrangements, has been analyzed at the atomistic level for 4 combinations of alkanethiols, $S(CH_2)_xCH_3$, with sulfanylalkane-1-carboxylic acids, $S(CH_2)_yCOOH$, of $[x : y] = 11 : 11$, $5 : 5$, $5 : 11$, and $11 : 5$ in water (Figure 5.8) [25]. Based on the tilt angle and radius of gyration of the AuNPs, this study has shown that the structural morphology of mixed monolayers is governed by the chain arrangement, chain length, and the chemical composition of the terminal groups. The different arrangements of the monolayer do not have any impact on the average tilt angle for the monolayer with ligands of $x = 5$ and $y = 5$ but the average tilt angle of the monolayer with the ligand composition of $x = 11$ and $y = 11$ displays small sensitivity to the monolayer arrangement with the larger tilt angle arising from the larger space available for monomers further from the gold surface. In the case of mixing of unequal ligand lengths, the longer chains bend more over the free space above the shorter ligands. Thus, the average tilt angle of the longer chains is again larger and more sensitive to the monolayer morphology than the average tilt angle of the shorter chains regardless of the terminal groups. In the case of Janus type of NPs, the average tilt angle of $S(CH_2)_{11}COOH$ in the mixture composition of $x = 5$ and $y = 11$ remains unchanged when compared with the corresponding value for the Janus type mixture composition of $x = 11$ and $y = 11$ but shortening of the alkanethiols ($x = 5$) brings about small increase in this average tilt angle for the random and striped arrangements. The more dispersed random and striped arrangements of the two kinds of ligands (less clusters of the same species formed) means an enhanced probability of finding shorter chains around a longer one and more free space for a longer chain to bend more. For the random and striped arrangements, the average tilt angles are larger when the longer chains are terminated with carboxyl group than when the terminal group is methyl.

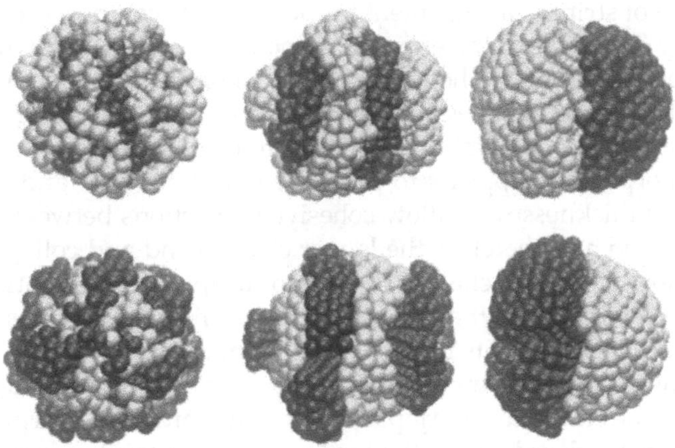

FIGURE 5.8
Equilibrated structures of the AuNPs coated with the mixture of $x = 5, y = 11$ (top row) and $x = 11, y = 5$ (bottom row) at random, striped, and Janus arrangements (from left to right). Color code: methyl terminated alkanethiols (dark gray), carboxyl terminated alkanethiols, (light gray). (Reprinted with permission from Velachi, V., et al., *Journal of Physical Chemistry C*, 119: 3199–3209, 2015. Copyright 2015 American Chemical Society.)

In these morphologies of ligands, the longer chains containing carboxyl terminal group are efficiently involved into the formation of hydrogen bonds with the neighboring water molecules and hence can bend more than when the longer chains are terminated with methyl group. Again, the larger dispersion of the two species of ligands (random > striped > Janus type) increases the isolation probability of longer chains and the carboxyl groups becomes more accessible to water molecules to form hydrogen bonds. The generation of different patterns following from different monolayer arrangements consisting of unequally long ligands is associated also with different hydration of AuNPs. The surface of the AuNP with $x = 5$ and $y = 5$ is practically depleted of water. Anyway, this is not true for AuNP with $x = 11$ and $y = 11$, where the water molecules penetrate into the space between chains, which is bigger at larger distances from the AuNP. The hydrogen bond formation between water molecules, N_{ww}, in the close proximity of AuNP depends on the mixture composition as well as on the monolayer morphology. In the composition mixtures of $x = 5, y = 5$ and $x = 11, y = 11$ the monolayer morphology does not affect N_{ww}, however the monolayer morphology of unequal chain lengths substantially governs N_{ww}. This phenomenon originates from the competing hydrogen bonds formed between water molecules and carboxyl groups. This hydrogen bonding is most striking when the longer chains are functionalized with carboxyl group and this is reflected on the smallest N_{ww}. Findings of this study suggest that the mixing of unequal ligand lengths to generate the striped or Janus type monolayer morphology provides better affinity for any foreign molecule than the random monolayer morphology.

In the case of striped arrangement in the mixtures composed of two types of unequally long chains, the effect of stripe thickness on the structural and hydration properties of the heterogeneous monolayer has been also investigated [26]. The stripe thickness has been varied from one to four chain rows (1-sam, 2-sam, 3-sam, 4-sam). In the mixture of unequal lengths of chains, the stripe morphology appears to be stable only for 3-sam and 4-sam. The smaller stripe thicknesses disallow cohesive interactions between the chains of the same kind and therefore the longer chains bend and collapse over the adjacent stripes of smaller chains. These two collapsed conformations (1-sam, 2-sam) are compared with the preserved stripe conformations (3-sam, 4-sam) in Figure 5.9, which presents the density distributions of 5 and 11 methylene carbon atoms for the mixture of $x = 5$ and $y = 11$ taken from the upper hemisphere and projected into the xy plane. As in the previous morphologies, in the mixture of $x = 5$ and $y = 11$, the longer chains containing carboxyl group bend more than the longer chains containing methyl group in the mixture of

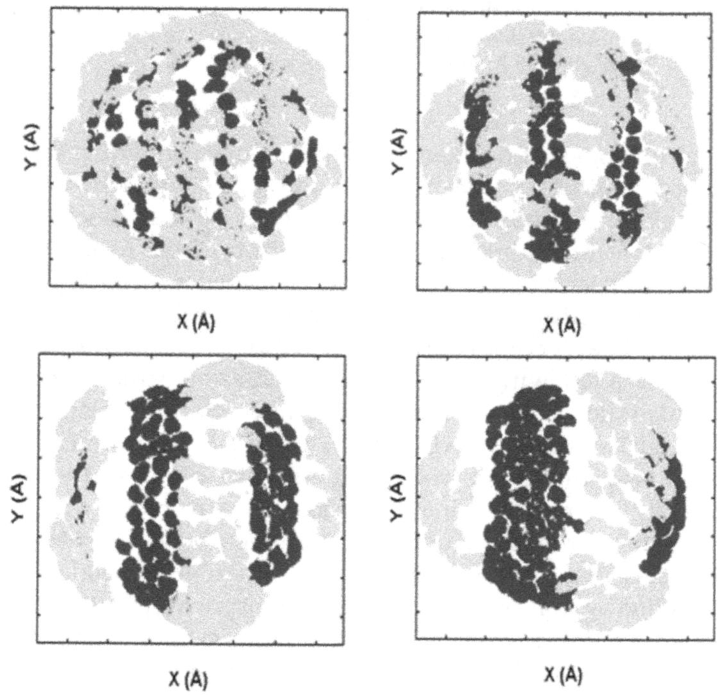

FIGURE 5.9
Density distributions of C5 and C11 carbon atoms from upper half of the AuNP coated with the mixture of $x = 5$, $y = 11$ projected in the xy plane at different stripe thicknesses: 1-sam (upper left), 2-sam (upper right), 3-sam (lower left), and 4-sam (lower right). Color code: methyl terminated carbons (dark gray) and carboxyl terminated (light gray). (Reprinted with permission from Velachi, V. et al., *Journal of Chemical Physics*, 144, 244710, 2016. Copyright 2016 American Institute of Physics.)

$x = 11$ and $y = 5$ for 1-sam and 2-sam arrangements. This again follows from the hydrogen bonding between the carboxyl groups and water molecules in the mixture of $x = 5$ and $y = 11$ and filled space by water molecules over shorter carboxyl terminated ligands, which disfavor the bending of longer methyl substituted chains in the mixture of $x = 11$ and $y = 5$. The information about the binding affinity of the monolayer is obtained from the amount of hydrogen bonds formed between carboxyl groups and water molecules, $N_{w\text{-}CO}$. The mixtures containing longer chains ($y = 11$) terminated with carboxyl groups combined with $x = 5$ or 11 display decreasing trend in $N_{w\text{-}CO}$ with increasing stripe thickness due to the concurrent hydrogen bonds between two carboxyl groups, the number of which increases with the increasing stripe thickness. Opposite tendency is found for striped arrangements of longer methyl terminated ligands ($x = 11$) and shorter carboxyl terminated ligands ($y = 5$). As the thickness of the stripes increases, the shielding of carboxyl groups by longer bent methyl terminated chains is attenuated and the carboxyl groups become more available for water, which increases the value of $N_{w\text{-}CO}$. These simulations have been extended also to other combinations of unequal chain lengths where the difference in the chain length amounts to 2–5 CH_2 units. These results demonstrate that the preservation of the stripe morphology is governed not only by the difference in the chain length but also by the absolute contour length of the chains.

The application of NPs in medical diagnostics, catalysis, energy conversion, or plasmonics often requires the possibility to control the crystalline patterns of NPs. It has been pointed out that binary mixtures of hybrid NPs composed of oligonucleotides and gold core (NA-AuNPs) may form one-, two-, and three-dimensional superlattices whose structure depends on the hydrodynamic size of NA-AuNPs, DNA length, number of DNA connections, and sequence complementarity. The molecular dynamics studies concerning such systems are tractable using a coarse-grained approach and have been carried out for binary mixture of NA-AuNPs of different stoichiometry [27]. Indeed, when taking into account the different size of the components in the binary mixtures, three types of body centered cell crystalline structures were formed, namely: CsCl, AlB_2, and Cs_3Si. For the formation of an equilibrium crystal, however, there is a rather narrow distribution of the number of hybridization pairs per NA-AuNP. Outside this region, the system is either kinetically trapped in an amorphous state or melts.

The atomistic molecular dynamics study of small AuNPs of diameters 1.4 and 2.5 nm coated with a monolayer of antioxidant glutathione in a cell culture have shown that the electrostatic forces help to stabilize the AuNPs pairs of both sizes but they alone are not sufficient to promote association of AuNPs [28]. On the other side, the van der Waals interactions ensuing from cations, anions, and net-neutral polar species promote the dimerization. Anyway, under the same conditions, the ultrasmall AuNPs remain colloidally stable while the larger ones aggregate.

5.4 Silica Hybrid Nanoparticles

In most of the applications, stable suspensions of NPs are required and this can be achieved by solvating a monolayer of grafted chains by the surrounding solvent. For instance, the hydration of a monolayer of chains grafted on NPs in water prevents the NPs from aggregation. The hydration of the grafted monolayer is dictated by the chemical structure, length, and coverage density of grafted chains. The hydration of the silica NPs coated with hydrophobic $Si(OH_3)(CH_2)_3SO_3H$ and hydrophilic $Si(OH_3)(CH_2CH_2O)_nH$ (n = 2, 3, 5) chains has been simulated and estimated from the hydration energies and radial distribution functions [29]. The continuous withdrawal of water molecules from the interstitial space between both types of grafted chains occurs with the increasing coverage density. Interestingly, although the drop in the number of water molecules is steeper, the number of inserted water molecules per functional group turns out to be larger for hydrophobic $Si(OH_3)(CH_2)_3SO_3H$ than for hydrophilic $Si(OH_3)(CH_2CH_2O)_nH$ chains.

Atomistic molecular dynamics simulations have been also adopted to study the velocity-dependent lubrication forces acting on a pair of PEO coated silica NPs (Figure 5.10) [30]. At large separations of NPs, the force exerted on NPs is dominated by a constant resistance to motion due to the solvent and the system is in the viscous drag regime, i.e., the force is linearly dependent on the velocity of NPs. At moderate separation distance, separation-dependent interactions of the coated monolayers dominate the force but the hydrodynamic forces operate as well. As the interparticle separation further decreases and the water molecules are expelled from the space between NPs, the hydrodynamic forces start to prevail.

The investigation of complex behavior of poly(ethylene glycol) (PEG) coated AuNPs in water is too time extensive and thus, a coarse-grained approach is necessary. A systematic study using coarse-grained molecular dynamics simulations has been conducted, and it addressed the influence of PEG coverage density, chain length and surface curvature (size of AuNPs) on the aggregation of PEGylated AuNPs in water [12]. The PEG chains built up of 0–10 monomers have been anchored to the AuNP trough a linker of an alkane chain (8 monomers units) at a coverage density of 50% (see Figure 5.11). The PEG coverage density has covered the range from 0%–100% for PEG of 5 monomers. The size of AuNPs has been varied and taken the values 5 nm, 10 nm, 50 nm, 100 nm, and 500 nm (coverage density 50% and PEG length of 5 monomers). The radius of gyration grows linearly with both the coverage density and the chain length of PEG chains; the latter factor has larger impact on the radius of gyration. The characteristic "mushroom"-"brush" conformational transition of chains grafted on a flat substrate observed when the crowding of chains is enhanced due to the increased chain length or coverage density is not evident in the case of the PEG chains attached to rather small AuNPs (5 nm) upon an increase in the coverage density. On the other hand, this transition is

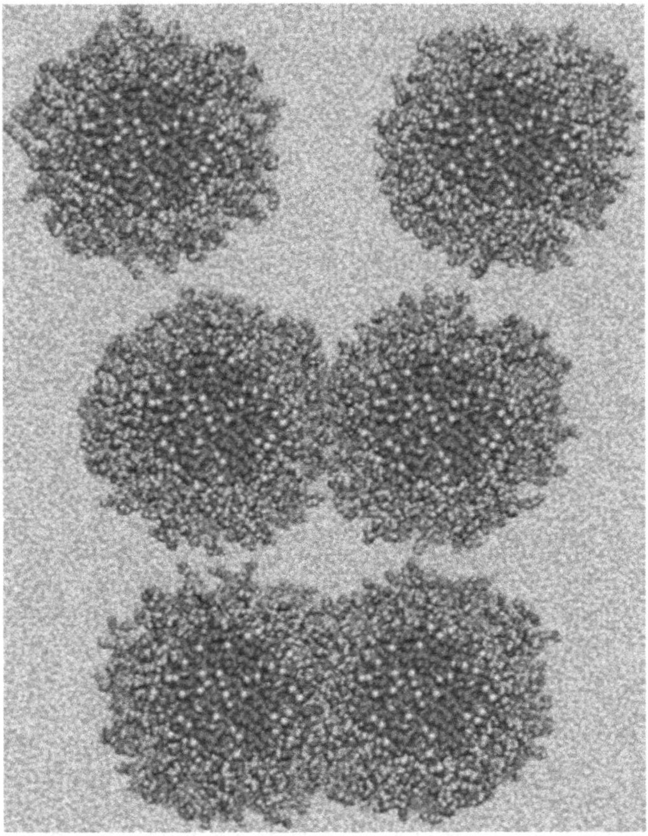

FIGURE 5.10
Silica NPs coated with PEO oligomers in water. The distance between the centers of the NPs is 11.5 nm (top), 8.0 nm (middle), and 6.0 nm (bottom). (Reprinted with permission from Lane, J. M. D. et al., *Physical Review E* 79, 050501, 2009. Copyright 2009 American Physical Society.)

achieved upon an increase of the chain length or flattening of the supporting substrate (increasing core diameter). The dispersion of PEGylated AuNPs is expected when the hydrophobic area on the core surface becomes covered by PEG chains. Relatively short PEG chains and low coverage densities (5 monomers, 40% coverage) prevent the PEGylated AuNPs from aggregation in aqueous medium. The dispersion of the PEGylated AuNPs occurs when the PEG length exceeds 4 monomers at the coverage 50%. Since the PEGylated AuNPs remain still dissolved even at large coverage densities, one would expect that the hydration of PEG layer pertains at these coverages. The manipulation of all three variables, i.e., the coverage density and the length of PEG chains along with the surface curvature reveals that the accessibility of the hydrophobic content of AuNP to a spherical probe of the size of short α-helix is more effectively reduced by the reduction of the surface curvature or increase of the coverage density. For a larger spherical probe of the size of albumin, the desirable

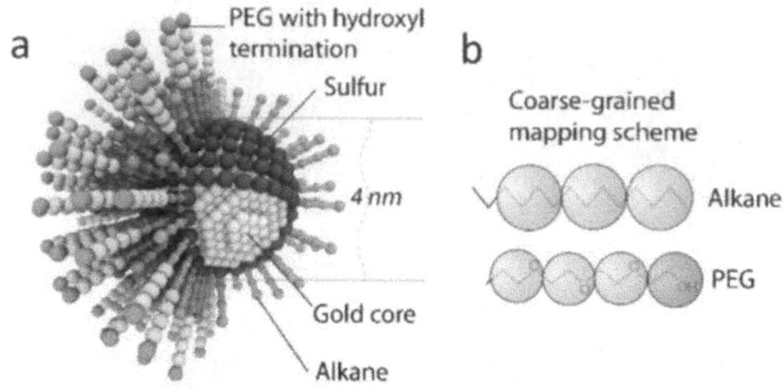

FIGURE 5.11
Schematic structure of the PEGylated AuNPs. PEG chains are attached to the AuNP through alkane linker (a). Coarse-grained to atomistic mapping of PEG and alkane chains (MARTINI force field) (b). (Reprinted from *Journal of Colloid and Interface Science*, 504, Lin, J. et al., PEGylation on Mixed Monolayer Gold Nanoparticles: Effect of Grafting Density, Chain Length, and Surface Curvature, 325–333, Copyright 2017, with permission from Elsevier.)

reduction in accessibility of the surface might be achieved also by the increase of the chain length.

The behavior of thermoresponsive copolymers (PAMAA) composed of acryl amide (AM) and acrylic acid (AA) grafted onto silica NPs has been simulated using molecular dynamics simulations at different temperatures and in the presence of different surfactant additives such as sodium dodecylsulfate, $C_{12}H_{25}SO_3^-Na^+$, dodecyltrimethylammonium bromide, $C_{12}H_{25}(CH_3)_3N^+Br^-$, and alcohol ethoxylate, $C_{12}H_{25}(OC_2H_2)_3OH$ in aqueous medium [31]. For comparison, the pure water as a solvating medium has been also considered. All three surfactant species support stretching of the PAMAA chains compared with the pure water solvent at all investigated temperatures (325 K, 350 K, and 375 K). In the absence of surfactants, the increasing temperature is capable of gradual deswelling of the PAMAA chains as a consequence of disruptions of the hydrogen bonds between water and PAMAA chains at elevated temperature. Also, this study highlights that the small AuNPs do not support the "mushroom"-"brush" conformational transition of PEG chains upon increase of the coverage density, however, a gradual transition is detected with increasing coverage density.

5.5 Conclusions

Coating of nanoparticles with a monolayer of synthetic polymers or biopolymers substantially modifies their structural and dynamic properties and enables fine tuning of the behavior of nanoparticles. In aqueous environment, the specific interactions between water molecules and functional groups of grafted chains determine the colloidal stability as well as the interactions with other chemical species present in the systems. Knowledge of these delicate interactions at the atomistic level is important for understanding both the biological function of coated nanoparticles and estimating their toxicity, biodegradability, and biocompatibility. This is a multidisciplinary task where molecular computer simulations provide insight into the processes and events at the atomistic level and suggest the proper manufacturing of nanoparticles to satisfy demands of desired application. *In silico* experiments are particularly suitable as nondestructive methods for the investigation of biochemical processes triggered by the presence of nanoparticles *in vivo*. Nowadays, the debates arise about possible cytotoxic effects of newly designed, synthesized, and manufactured nanoparticles on DNA, proteins, and membrane structures in solutions and this is the realm where computer simulations remarkably contribute to the elucidation of these questions. With the continuous development of computational techniques, the more detailed and extensive molecular simulations become feasible.

Acknowledgments

The authors acknowledged the financial support from Fundação para a Ciência e a Tecnologia (FCT/MEC) through national funds, and cofinanced by the European Union (FEDER funds) under the Partnership Agreement PT2020, through projects UID/QUI/50006/2013, POCI/01/0145/FEDER/007265, and NORTE-01-0145-FEDER-000011 (LAQV@REQUIMTE). Z. B. also thanks the financial support by grants SRDA-15-0323, VEGA 2/0055/16, VEGA 2/0098/16.

References

1. Terrill, R. H., Postlethwaite, T. A., Chen, C. H., Poon, C. D., Terzis, A., Chen, A. D., Hutchison, J. E., Clark, M. R., Wignall, G., Londono, J. D., Superfine, R., Falvo, M., Johnson, C. S., Samulski, E. T., Murray, R. W. (1995) Monolayers in Three Dimensions: NMR, SAXS, Thermal, and Electron Hopping Studies of Alkanethiol Stabilized Gold Clusters. *Journal of the American Chemical Society* 117: 12537–12548.
2. Zanchet, D., Hall, B. D., Ugarte, D. (2000) Structure Population in Thiol-Passivated Gold Nanoparticles. *Journal of Physical Chemistry B* 104: 11013–11018.
3. Jackson, A. M., Hu, Y., Silva, P. J., Stellacci, F. (2006) From Homoligand- to Mixed-Ligand-Monolayer-Protected Metal anoparticles: A scanning Tunneling Microscopy Investigation. *Journal of the American Chemical Society* 128: 11135–11149.

4. Lopez-Acevedo, O., Akola, J., Whetten, R. L., Gronbeck, H., Hakkinen, H. (2009) Structure and Bonding in the Ubiquitous Icosahedral Metallic Gold Cluster Au-144(SR)(60). *Journal of Physical Chemistry C* 113: 5035–5038.
5. Heikkila, E., Gurtovenko, A. A., Martinez-Seara, H., Hakkinen, H., Vattulainen, I., Akola, J. (2012) Atomistic Simulations of Functional Au-144(SR)(60) Gold Nanoparticles in Aqueous Environment. *Journal of Physical Chemistry C* 116: 9805–9815.
6. Villarreal, O. D., Chen, L. Y., Whetten, R. L., Demeler, B. (2015) Aspheric Solute Ions Modulate Gold Nanoparticle Interactions in an Aqueous Solution: An Optimal Way to Reversibly Concentrate Functionalized Nanoparticles. *Journal of Physical Chemistry B* 119: 15502–15508.
7. Villarreal, O. D., Chen, L. Y., Whetten, R. L., Yacaman, M. J. (2015) Ligand-Modulated Interactions between Charged Monolayer-Protected Au-144(SR)(60) Gold Nanoparticles in Physiological Saline. *Physical Chemistry Chemical Physics* 17: 3680–3688.
8. Hong, B. B., Panagiotopoulos, A. Z. (2012) Molecular Dynamics Simulations of Silica Nanoparticles Grafted with Poly(ethylene oxide) Oligomer Chains. *Journal of Physical Chemistry B* 116: 2385–2395.
9. Ndoro, T. V. M., Voyiatzis, E., Ghanbari, A., Theodorou, D. N., Böhm, M. C., Müller-Plathe, F. (2011) Interface of Grafted and Ungrafted Silica Nanoparticles with a Polystyrene Matrix: Atomistic Molecular Dynamics Simulations. *Macromolecules* 44: 2316–2327.
10. Benková, Z., Szefczyk, B., Cordeiro, M. N. D. S. (2011) Molecular Dynamics Study of Hydrated Poly(ethylene Oxide) Chains Grafted on Siloxane Surface. *Macromolecules* 44: 3639–3648.
11. Cai, L., Lv, W., Zhu, H., Xu, Q. (2016) Molecular Dynamics Simulation on Adsorption of Pyrene-Polyethylene onto Ultrathin Single-Walled Carbon Nanotube. *Physica E* 81: 226–234.
12. Lin, J., Morovati, V., Dargazany, R. (2017) PEGylation on Mixed Monolayer Gold Nanoparticles: Effect of Grafting Density, Chain Length, and Surface Curvature. *Journal Colloid Interface Science* 504: 325–333.
13. Heinz, H., Vaia, R. A., Farmer, B. L., Naik, R. R. (2008) Accurate Simulation of Surfaces and Interfaces of Face-Centered Cubic Metals Using 12–6 and 9–6 Lennard-Jones Potentials. *Journal of Physical Chemistry B* 112: 17281–17290.
14. Pearlman, D. A., Case, D. A., Caldwell, J. W., Ross, W. S., Cheatham, T. E., DeBolt, S., Ferguson, D., Seibel, G., Kollman, P. (1995) AMBER, a Package of Computer Programs for Applying Molecular Mechanics, Normal Mode Analysis, Molecular Dynamics and Free Energy Calculations to Simulate the Structural and Energetic Properties of Molecules. *Computer Physics Communications* 91: 1–41.
15. Jorgensen, W. L., Maxwell, D. S., Tirado-Rives, J. (1996) Development and Testing of the OPLS All-Atom Force Field on Conformational Energetics and Properties of Organic Liquids. *Journal of the American Chemical Society* 118: 11225–11236.
16. MacKerell, A. D., Jr., Bashford, D., Bellott, R. L., Dunbrack, R. L., Jr., Evanseck, J. D., Field, M. J., Fischer, S., Gao, J., Guo, H., Ha, S., Joseph-McCarthy, D., Kuchnir, L., Kuczera, K., Lau, F. T. K., Mattos, C., Michnick, S., Ngo, T., Nguyen, D. T., Prodhom, B., Reiher, W. E., III, Roux, B., Schlenkrich, M., Smith, J. C., Stote, R., Straub, J., Watanabe, M., Wiorkiewicz-Kuczera, J., Yin, D., Karplus, M.

(1998) All-Atom Empirical Potential for Molecular Modeling and Dynamics Studies of Proteins. *Journal of Physical Chemistry B* 102: 3586–2616.

17. Dauber-Osguthorpe, P., Roberts, V. A., Osguthorpe, D. J., Wolff, J., Genest, M., Hagler, A. T. (1988) Structure and Energetics of Ligand Binding to Proteins: Escherichia Coli Dihydrofolate Reductase-Trimethoprim, a Drug-Receptor System. *Proteins: Structure, Function and Genetics* 4: 31–47.
18. Sun, H. (1998) COMPASS: An ab Initio Force-Field Optimized for Condensed-Phase Applications Overview with Details on Alkane and Benzene Compounds. *Journal of Physical Chemistry B* 102: 7338–7364.
19. Sun, H. (1995) Ab Initio Calculations and Force Field Development for Computer Simulation of Polysilanes. *Macromolecules* 28: 701–712.
20. Jorgensen, W. L., Chandrasekhar, J., Madura, J. D. (1983) Comparison of Simple Potential Functions for Simulating Liquid Water. *Journal of Chemical Physics 79*: 926–935.
21. Meena Devi, J. (2014) Aggregation of Thiol Coated Gold Nanoparticles: A Simulation Study on the Effect of Polymer Coverage Density and Solvent. *Computational Materials Science* 86: 174–179.
22. Yang, A.-C., Weng, C.-I. (2010) Structural and Dynamic Properties of Water near Monolayer-Protected Gold Clusters with Various Alkanethiol Tail Groups. *Journal of Physical Chemistry C* 114: 8697–8709.
23. Meena Devi, J. (2015) Molecular Simulations of Mixed Self-Assembled Monolayer Coated Gold Nanoparticles in Water. *Journal of Molecular Modeling* 21: 149–158.
24. Van Lehn, R. C., Alexander-Katz, A. (2013) Structure of Mixed-Monolayer-Protected Nanoparticles in Aqueous Salt Solution from Atomistic Molecular Dynamics Simulations. *Journal of Physical Chemistry C* 117: 20104–20115.
25. Velachi, V., Bhandary, D., Singh, J. K., Cordeiro, M. N. D. S. (2015) Structure of Mixed Self-Assembled Monolayers on Gold Nanoparticles at Three Different Arrangements. *Journal of Physical Chemistry C* 119: 3199–3209.
26. Velachi, V., Bhandary, D., Singh, J. K., Cordeiro, M. N. D. S. (2016) Striped Gold Nanoparticles: New Insights from Molecular Dynamics Simulations. *Journal of Physical Chemistry* 144: 244710.
27. Li, T., Sknepnek, R., Macfarlane, R. J., Mirkin, C. A., de la Cruz, M. O. (2012) Modeling the Crystallization of Spherical Nucleic Acid Nanoparticle Conjugates with Molecular Dynamics Simulations. *Nano Letters* 12: 2509–2514.
28. Hassan, S. A. (2017) Computational Study of the Forces Driving Aggregation of Ultrasmall Nanoparticles in Biological Fluids. *Acs Nano* 11: 4145–4154.
29. Rigo, V. A., de Lara, L. S., Miranda, C. R. (2014) Energetics of Formation and Hydration of Functionalized Silica Nanoparticles: An Atomistic Computational Study. *Applied Surface Science* 292: 742–749.
30. Lane, J. M. D., Ismail, A. E., Chandross, M., Lorenz, C. D., Grest, G. S. (2009) Forces between Functionalized Silica Nanoparticles in Solution. *Physical Review E* 79: 050501.
31. Dai, S. S., Zhang, J. H., Zhang, T. L., Huang, Z. Y., Quan, H. P., Lu, H. S., Zhao, X. W. (2016) Molecular Dynamic Simulations of the Core-Shell Microsphere of Nanosilica Grafted by Acrylamide Acrylic Acid Copolymer PAMAA: Study of Its Microstructure and Interaction between Microsphere and Additives. *New Journal of Chemistry* 40: 5143–5151.

6

Lattice-Boltzmann Modeling of Adsorption Breakthrough in Packed Beds

Nishith Verma

CONTENTS

6.1 Introduction

This chapter introduces the basic procedure of implementing LBM-based 3D adsorption models. The chapter begins with the description of the present LBM-based 3D models for simulating adsorption in porous media, mathematical complexities and challenges in developing such models, and a rigorous mathematical approach for determining adsorption breakthrough in tubular column packed with porous adsorbent particles.

Readers should be aware that apart from packed beds, adsorption and reactive flow have been successfully simulated for a variety of geological processes, using LBM-based models. Common examples are transport of contaminants in soil, flow of crude oil in earthen reservoirs, weathering and dissolution of rocks and concrete degradation [1–6]. Therefore, they should also study such examples in conjunction with those for packed columns. In any case, one of the complexities involved in simulating hydrodynamics or concentration distributions in liquid flow past a solid is the irregular geometrical surface of the solid material in contact with liquid particles. Although most of adsorbers use approximately regular spherical adsorbent

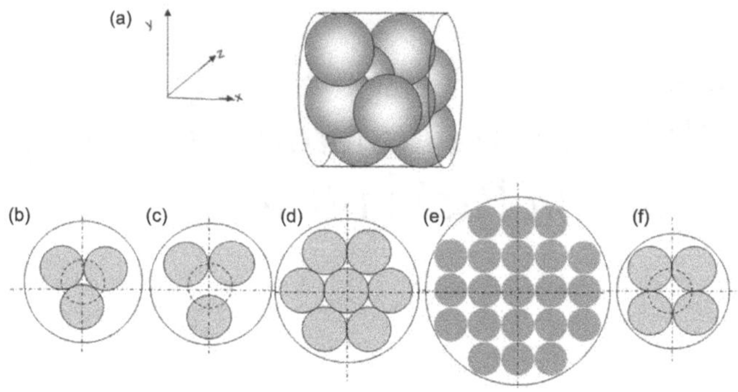

FIGURE 6.1
(a) tubular packed adsorber, (b–f) different packing arrangements (dotted circle represents single sphere placed alternately in axial direction). (From Manjhi, N. et al., *Chem. Eng. Sci.*, 61: 7754–7765, 2006.)

particles, interfacial surface between solute in fluid and solid is curvilinear. Considering that lattice particles move along the straight horizontal and vertical links, LBM simulation of momentum-, mass-, or energy-exchange between fluid and solid particles at curvilinear surface requires special mathematical treatment and is described next in this chapter. It may be mentioned that traditional mathematical models for packed beds consider adsorbent particles to be a point source of mass and thermal flux in the respective conservation equation [7]. Such assumptions are valid for tube-to-particle diameter ratio > 10. However, we re-emphasize that, in a narrow tubular adsorbers, finite size of adsorbent particles should be considered for simulating concentration and temperature profiles, especially in voids between the particles. Figure 6.1 schematically shows different types of the model regular packing arrangements in a tubular bed packed with spherical adsorbent materials of relatively larger size with tube-to-particle diameter ratio < 10.

6.2 Boundary Fitting Approach for Curvilinear Surface

The boundary-fitting concept was originally proposed by Filippova and Hanel [9] for determining particle density distribution functions at curvilinear surfaces. Figure 6.2 schematically shows a curved boundary lying between the computational nodes in the lattice of uniform size ε. As per the proposed method, two computational nodes are defined: "rigid nodes" within the particle and "fluid nodes" in the flow field. For the fluid nodes located adjacent to the solid boundary lying between the nodes of the uniform rectangular lattice, a modified distribution function is used. The distribution function coming to the fluid node r_f from the "rigid" node r_b is computed as follows:

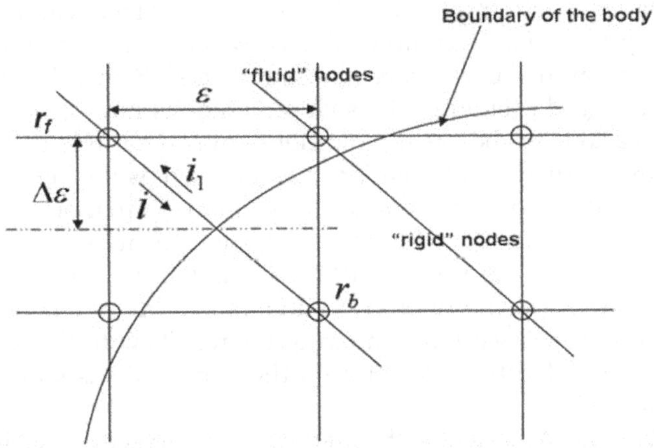

FIGURE 6.2
Computational mesh and geometrical relations for fitting of solid boundaries. (From Verma, N., Mewes, D., *Comp. Math. Appl.*, 58: 1003–1014, 2009.)

$$f_i(t+\delta_t, r_f) = \left[(1-w) f_i(t, r_f) + w\, f_i^{eq}(t, r_f)\right](1-w_1) + w_1\left[a_1 f_i^{eq}(t, r_b) + a_2 f_i^{eq}(t, r_f)\right] \tag{6.1}$$

where r_f and r_b are locations (coordinates) of the adjacent fluid and rigid nodes, respectively; w is the same as $1/\tau$, or the inverse of relaxation factor. The constants, w_1, a_1, and a_2 are calculated as follows:

$$\begin{aligned} w_1 &= w\,(2\Delta - 1),\ a_1 = 1,\ a_2 = 0 \text{ if } \Delta \geq 0.5 \\ &= w\frac{(2\Delta-1)}{1-w},\quad a_1 = 0, a_2 = 1, \text{ if } \Delta < 0.5 \end{aligned} \tag{6.2}$$

where Δ is the Cartesian component of the fraction of distance between fluid node at r_f and rigid node at r_b. The equilibrium distribution function $f_i^{eq}(t, r_b)$ at the solid rigid node is evaluated in the similar way as $f_i^{eq}(t, r_f)$ at fluid node, except the velocity $u_\alpha(t, r_b)$ along α direction at the solid node is determined as:

$$u_\alpha(t, r_b) = \frac{(\Delta - 1)}{\Delta} u_\alpha(t, r_f) \tag{6.3}$$

For the square geometry, which is naturally aligned with the horizontal and vertical grids, $w_1 = 0$. In such case, it can be easily shown that equation (6.1) is reduced to the equation we earlier discussed in Chapter 1 for calculating distribution function f^{eq} for a square geometry.

The above-approach proposed by Filippova and Hanel can be validated by simulating flow of an incompressible Newtonian fluid past a single circular cylinder confined in a square channel, at small Reynolds numbers and blockage ratios (d_t/d_p) (Figure 6.3). Such simulations have been extensively reported in various studies using traditional numerical Finite difference or Finite Volume methods, and therefore, can be used as the benchmark tests [11]. Figure 6.3 also shows the representative LBM simulation results for the velocity texture. Clearly, curvilinear velocity profiles are successfully simulated on rectangular grids used in LBM. Calculations of drag coefficients for circular cylinders also quantitatively validate the proposed model approach for the curvilinear surface. On comparing the results with the reported values in literature [13], difference between the two results is found to be less than 5% (Table 6.1).

Readers must have noted that the mathematical procedure for implementing the LBM boundary condition for a circular geometry is barely different from that for a square geometry. In fact, one common programming code

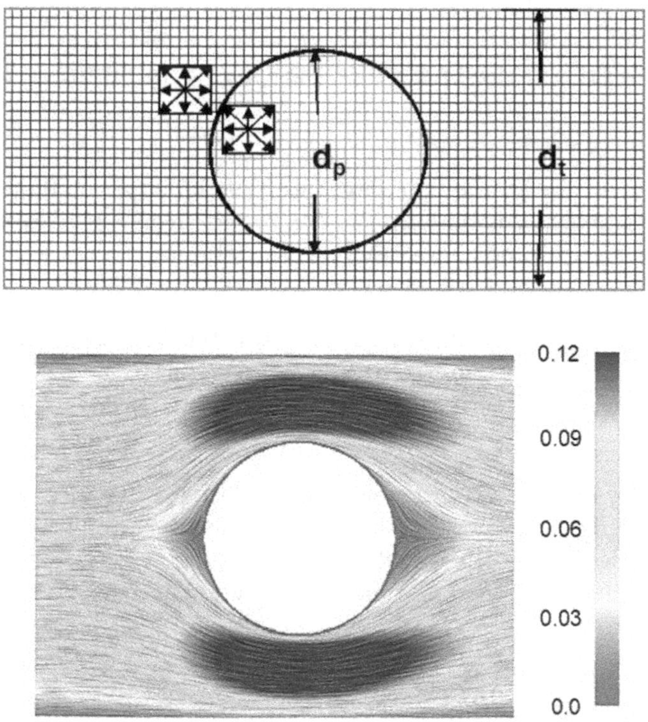

FIGURE 6.3
(Top) LBM rectangular grids mapped on both inside and outside the cylindrical particle. Streaming and collision were allowed along the horizontal and vertical links inside and outside particles, with macroscopic velocity within the particle set at zero; (Bottom) the LBM-simulated velocity texture in the flow at *Re, p* = 5 past a circular long cylinder confined in a channel. (From Verma, N. et al., *Chem. Eng. Sci.*, 62: 3685–3698, 2007.)

TABLE 6.1

Computed Drag Coefficient vs Reynolds Number for Circular Cylinder in a Steady Flow. Velocity Profiles at the Inlet to the Channel Were Assumed to be Flat

	Circular Cylinder						
	This Work[a]			Chakraborty et al.[b] [13]			Filippova and Hanel [9]
Re	$\lambda = 1.54$	2	4	$\lambda = 1.54$	2	4	$\lambda = 4$
0.1	5551.8	2011.6	531.5	5603	2112	551	–
1	611.6	241.2	65.1	631	251	62	–
10	84.3	28.1	10.5	87	27	10	–
20	56.2	17.8	5.8	52	17	5.7	5.52

Source: Manjhi, N. et al., *Chem. Eng. Sci.*, 61: 7754–7765, 2006.

[a] LBM calculations.

[b] Numerical simulation using finite difference/volume method.

may be tweaked to simulate two scenarios, i.e., flow past the rectangular and the circular objects, without requiring major modifications in the code. This is one of the salient features of the lattice Boltzmann modelling. Traditional numerical methods may require extensive numerical modifications in that context.

We have described in details the computational steps for developing higher dimensional models later in this chapter. At this stage, it is noteworthy to mention that LBM can efficiently simulate 3D velocity profiles in a column of circular cross-section using computational grids on a Cartesian coordinates [8]. Figure 6.4 shows the texture plot of the steady-state velocity profiles calculated from the 3D simulation of velocity and concentrations in a tubular column (L = 233, d = 59 nodes). The column is assumed to be packed with equal size spheres (d_p = 19 nodes) on a triangular packing arrangement

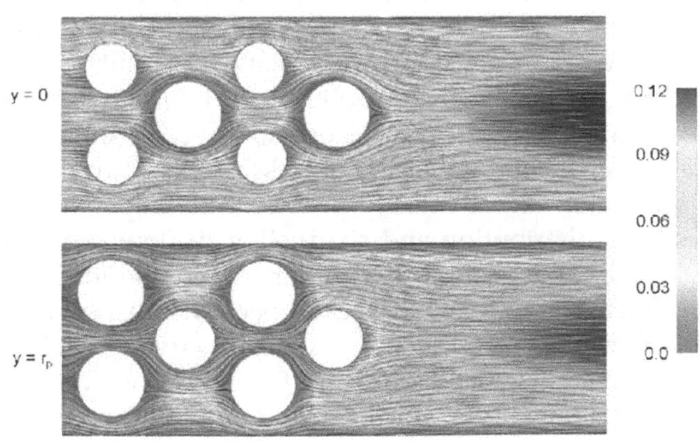

FIGURE 6.4

Velocity texture for x-z planes located at y = 0 and y = rp in 3 spheres traingular packing with 1 sphere places alternately along x-axis. (From Manjhi, N. et al., *Chem. Eng. Sci.*, 61: 7754–7765, 2006.)

schematically shown in Figure 6.1 (top). In such an arrangement, one sphere is placed concentrically between two adjacent planes along the x-axis (flow-direction), whereas the other sphere is kept below the horizontal axis as shown in the schematics for artificially creating non-uniformity in the voids along circumferential or θ-direction. The column was assumed to be packed till middle of the length from the inlet side of the packed column, leaving the remaining half of the column empty. This way velocity profiles in the packed and empty sections may be compared.

The simulated velocity profiles (shown for two different horizontal (x-z) planes, $y = 0$ (center) and $y = r_p$ (half the diameter of the sphere towards the tube's walls in the vertical y direction) clearly show the distinctly different radial and axial flow fields at two planes, with relatively larger magnitudes of velocity calculated at $y = 0$. The simulation profiles also show significant regions of interstitial (macro) voids between the particles where there is expectedly nearly stagnated flow field or relatively smaller fluid velocity (shown by red color in the plots). Another salient observation is a distinct variation in the fluid's velocity along the columns' length (x direction). The radially non-uniform flow profile in the packed section changes to approximately radially uniform in the middle section, to parabolic near the exit of the column. We thus re-emphasize the ability of LBM-based models in successfully simulating different velocity patterns prevalent in packed beds. We will discuss the impact of the non-uniform velocity profiles on concentration distributions during adsorption later in this chapter. It is obvious that 1D models cannot be used to simulate such scenarios.

Freund et al. [14] simulated 3D flow profiles in a tubular adsorber randomly packed with spherical particles with tube-to-particle diameter ratio = 3 – 7. Although local shape was not considered in the model, each sphere was discretized with 30 cells ("voxels") per diameter. Most importantly, the LBM simulation allowed radial variation in bed porosity. The porosity profile was assumed to be strongly oscillating, starting with a porosity of unity near the wall and reaching a minimum at the wall distance of about 0.5 particle diameter, which is in agreement with the experimental observations. The salient feature of the simulation results was the quantitative agreement between the predicted and experimentally measured pressure drops in the packed bed. Further, the simulated radial velocity profiles and snapshot of 3D profiles clearly showed maldistribution and channeling in flow, which are also in agreement with the experimental observations. Such non-ideal scenarios are commonplace during flow through packed beds. It must be reiterated that simple 1D models can not show such non-ideal scenarios prevalent in packed beds. Schure et al. [15] also used LBM simulations to corroborate the effects of variation in radial porosity on flow distribution in packed beds. The study used different types of structured packing arrangements (see Figure 6.1 top) for the simulation.

Staircase approach can also be used to map a circular cross-section onto a rectangular grids [16]. Although such approach is able to simulate flow around curvilinear objects including small rocky or granular particles in porous soil media, required number of computational grids is relatively larger. Manjhi et al. [17] have successfully used the staircase approach to determine particle distribution function in a tube of circular cross-section by mapping the cross-section with the grids generated along the vertical and horizontal directions on a rectangular cross-section (Figure 6.5). The higher the number of grids were, the more accurately the flow was resolved. Grids closest from outside (either horizontally or vertically) to the circumference of the circular cross-section were assigned the boundary points as shown in the figure. These boundary points were matched with different types (vertical plane, horinzontal plane, edge, corner, etc.) of the boundary points for the square or rectangular cross-sectional duct. The authors showed a close agreement between the LBM simulation results for a Poiseuille flow in a tube of circular cross-section with the theoretically calculated 3D velocity profiles.

Modelling at pore scales is critical to accurately predicting adsorption processes including adsorption rate and concentration distributions in the flow fields. In the geological processes such as dissolution of salts from solid particles, where solutes are advected (convected and diffused) in the liquid flow through narrow voids between the particles, resolution of flow as well as concentration distributions of the solute is required at mm or micron-levels. In packed bed columns, solutes are advected in the flow through macropores (voids between adsorption particles). Also, solutes simultaneously diffuse in the pores within the solid particles. Clearly, modelling adsorption in packed bed is even more challenging, for transport occurs at different time scales which may be apart by several orders of magnitude. In other words, there are mathematical challenges in simulating transport over a wide range of Pelect numbers (1–5000).

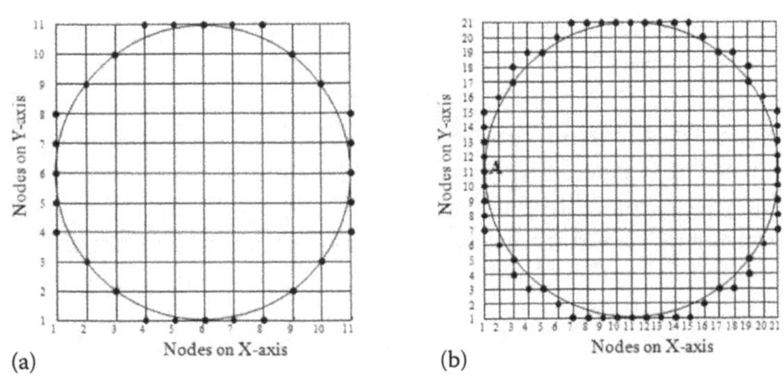

FIGURE 6.5
Mapping from rectangular to circular cross-section by (a) 11 × 11 (b) 21 × 21 grids. (From Manjhi, N. et al., *Chem. Eng. Sci.*, 61: 2510–21, 2006.)

6.3 Macroscopic Conservation Equations

There are two conservations equations: (1) solute balance in the bulk aqueous phase in the column and (2) that within the porous particles [12]. The former considers convective transport, axial and radial dispersions, and diffusion flux at the fluid-solid interface because of adsorption and desorption within the porous particles:

$$\partial_t C_l + \partial_\alpha C_l u_\alpha = D\nabla^2 C_l + \frac{(1-\varepsilon)}{\varepsilon} D_{pore} \nabla C_p \,|_{r=R_p} \times a_p \tag{6.4}$$

where C_l and C_p are solute concentrations in the liquid phase and within pore of the solid, respectively; u_l is the liquid phase velocity in α direction; D is the dispersion coefficient of the solute in the liquid phase; ε is the bed porosity; D_{pore} is the effective pore diffusion coefficient; R_p is the adsorbent particle size and a_p is the geometrical surface area per unit volume of the particle. The last term in equation (6.4) represents the source or sink of solute in the bulk phase because of adsorption or desorption depending upon the "wetting" or "drying" (regeneration) of the adsorbent particles. Therefore, there is a finite contribution from the flux term to the total species concentrations at the computational nodes which cross the particle surface in the packed bed. However, the term becomes redundant at the nodes where there are no particles.

Solute balance within the adsorbent particles considers pore diffusion and surface adsorption/desorption rates at the pore walls of the adsorbents:

$$\partial_t c_p = D_{pore} \nabla^2 C_p - a\partial_t C_s \tag{6.5}$$

where C_s is the solute concentrations at the solid surface. The second term on the right hand side of equation (6.2) represents the rate of change in surface concentration of the adsorbents during adsorption. The coefficient "a" in the equation is the internal surface area to pore-volume ratio of the adsorbents. It may be recalled that quasi steady-state condition was assumed for the adsorption and desorption kinetic rates in Chapter 1, leading to the assumption of an instant surface equilibrium following change in solute concentration within the pores of the adsorbents. The equilibrium is described by a suitable isotherm. The mathematical representation of the above statements is as follows: $\partial_t C_S = \partial_t C_p (dC_S/dC_p)$, where the term in the parenthesis can be recognized as the slope m of the adsorption isotherm. As a consequence of the aforementioned assumptions, equation (5) is re-arranged as:

$$\partial_t C_p = \frac{D_{pore} \nabla^2 C_p}{1 + am} \tag{6.6}$$

Equation 6.6 allows linear as well as nonlinear adsorption systems. The latter situation may arise if concentration changes over a wide range. In such case, slope of the isotherm depends on adsorbent concentrations in the aqueous phase. It would be appropriate to compare the above sets of conservation equations to those for adsorption in non-porous particles. In the latter case, equation (6.5) becomes redundant and the adsorption rate is directly included in the liquid phase balance (equation 6.1) as the source (or sink) term. For the case of "point" particles or small particles ($d_t/dp >> 10$), concentration gradients within the particles are assumed to be insignificant, obviating the requirement of equation (6.2) or (6.3), and the adsorption/desorption rate is included in equation (6.4) as a homogenous source (or sink) term. To sum up, macroscopic conservations equations (6.1–6.6) with appropriate boundary conditions are the required governing equations for describing concentration profiles in the bulk liquid phase of the packed bed and within the porous adsorbent particles. The following liquid-solid interfacial boundary conditions (continuity in concentration and diffusion flux) are applied:

$$C_l|_{Rp+} = C_p|_{Rp-} \tag{6.7}$$

$$-D\nabla C_l|_{R_p+} = -D_{pore}\nabla C_p|_{R_p-} \tag{6.8}$$

At the inlet of adsorber, liquid concentration may be prescribed, whereas at the outlet long tube approximation may be made, i.e., a zero-concentration gradient is assumed to exist. The convective velocity, u_α is *a priori* set at the values calculated from the solution to the macroscopic momentum conservation equation (Navier-Stokes equation) for the flow through the packed bed. Pore diffusion coefficient may be assumed between one to three orders of magnitude smaller than molecular diffusion coefficient of the solute in the liquid phase.

6.4 Mesoscopic Lattice Boltzmann Equations

Similar to the approach described in Chapter 1, a common approach to the development of 3D LBM-based methods corresponding to the macroscopic advective (convection-diffusion) and momentum conservation equations is based on determining the sum (ρ) of and difference ($\Delta\rho$) between densities of two components (liquid and solute) in the flowing mixture, from the popular lattice Bhatnagar-Gross-Krook (BGK) collision rules applied on both, total density distribution function, f and density-difference distribution function, g:

$$f_i(r+e_i,t+1)-f_i(r,t)=\frac{-1}{\tau_m}\left(f_i-f_i^0\right) \tag{6.9}$$

$$g_i(r+e_i,t+1)-g_i(r,t)=\frac{-1}{\tau_d}\left(g_i-g_i^0\right) \tag{6.10}$$

where τ_m and τ_d are the relaxation times to reach local equilibrium $f_i^0\, f_i^0$ and g_i^0, respectively. A *d3q*19 cubic lattice (Figure 6.6) is commonly used for 3D simulations. A suitable choice of equilibrium distribution functions, f_i^0 and g_i^0 corresponding to ρ and $\Delta\rho$ for a binary miscible fluid was defined as per the free energy-based diffusion model proposed by Swift et al. [18]

$$\begin{aligned} f_i^0(u) &= A_0 + C_0\mathbf{u}_{i.}\mathbf{u}, i = 0 \\ &= f_{1i}^0(0)\left[A + B\mathbf{e}_{i.}\mathbf{u} + C(e_{i.}\mathbf{u})^2 + D(u_{i.}\mathbf{u})\right], i = 1\ldots6 \\ &= f_{2i}^0(0)\left[A + B\mathbf{e}_{i.}\mathbf{u} + C(e_{i.}\mathbf{u})^2 + D(u_{i.}\mathbf{u})\right], i = 7\ldots18 \end{aligned} \tag{6.11}$$

Similar expressions are written for g_i^0. The values of Lagrangian coefficients in equation (6.5) for f_i^0 and those for g_i^0 are as follows:

$A_0=\rho(1\text{–}2T), C_0=1/2\,\rho, f_{1i}^0(0)=\rho/18, f_{2i}^0(0)=\rho/36, A=3T, B=3, C=9/12, D=3/2.$

The corresponding coefficients of the distribution function, $g_i^0(u)$ were obtained as follows:

$A_0=\Delta\rho(1-2D'), C_0=-1/2\Delta\rho, f_{1i}^0(0)=\Delta\rho/18, f_{2i}^0(0)=\Delta\rho/36, A=3D', B=3,$
$C=9/2, D=-3/2$

Particle levels constraints on f^0 and g^0 are the same as defined earlier in Chapter 1.

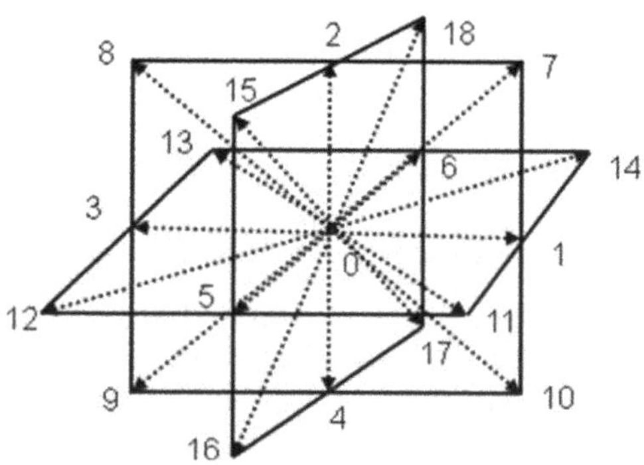

FIGURE 6.6
Schematics of 19-speed lattice molecules used to simulate the concentration profiles within the macro and micro pores of the adsorption tube. (From Verma, N. et al., *Chem. Eng. Sci.*, 62: 3685–3698, 2007.)

We now turn our attention to the numerical scheme used to simulate hydrodynamics and concentration profiles in the macro (inter-particle) voids and concentration profiles within the micro (intra-particle) voids.

6.5 LBM Numerical Scheme

Choice of D (or Peclet number) sets the relaxation time for diffusion (τ_d) independent from kinematics diffusivity or Reynolds number. Advantage of multi-scaling in lattice modelling may be realized in solving concentration fields in the packed bed by segregating computational domains within the particles from those outside the particles. Thus, streaming and collision of lattice molecules are permitted both inside and outside of the particles. However, macroscopic velocity within the particle is set at zero. Considering that lattice mass diffusivity is related to the relaxation coefficient, two different τ_ds (one for micro and the other for macro) may be used to allow for difference between the molecular and pore diffusivity or the corresponding Peclet numbers. The dimensionless lattice mass diffusivity is related to actual diffusivity via grid and time step size as $D_{actual} = D_{lattice}$ $\Delta x^2/\Delta t$. Therefore, inter- and intra particle diffusivities may be related to computational grids.

$$\frac{D_{pore}}{D} = \frac{\tau_{pore} - 1/2}{\tau_m - 1/2} \frac{\Delta x_{pore}^2}{\Delta x^2} \frac{\Delta t}{\Delta t_{pore}} \tag{6.12}$$

Considering that pore diffusion coefficient may be orders of magnitude smaller than molecular diffusion coefficient, lattice modelling allows adjustment of time-step of calculation for concentrations inside the particles independently from the outside region by keeping τ_{pore} different from τ_m. In other words, considering that physical time to reach steady state inside the particles is intrinsically much larger than that outside the pore, several numerical iterations may be carried out for calculating the intra-particle concentration profiles over each single iteration required for calculating concentration profiles in the bulk phase, thus obviating simultaneous computations over each time-step. Two regions of computations, however, exchange the information at the liquid-solid interfaces via macroscopic diffusion flux calculated from equation (6.5), which is included as the source term in the respective microscopic LBGK equations. The mathematical LB procedure of including source term in the species conservation equation is similar to that for including gravity as the momentum generation term in the LBGK momentum equation [19]:

$$g_i(\mathbf{r}+\mathbf{e}_i,t+1)-g_i(\mathbf{r},t)=\frac{-1}{\tau_d}\left(g_i-g_i^0\right)+\Delta t\,h_i(\mathbf{r})a_p \tag{6.13}$$

where $h_i(r)$ is the diffusion flux term along the links at the interface of solid and liquid phases. Knutson et al. [3] have also simulated solute transport with hydrodynamics Δt different from mass transfer Δt, though diffusivity was restricted by $\tau = \tau_d$. Similarly, Sullivan et al. [20] in their LBM calculations of flow in porous media assumed two different time-steps for flow and reaction, enabling low value of diffusivity to be realized. Further, the regions of low intra-particle diffusivity were included by assigning τ_d different from that for large diffusivity so that D_{pore} was made different from D by one-order of magnitude. Their implementation code allowed for the chemical species to be not only diffused but advected based on the local velocity field.

6.6 Refining Mesh Sizes in LBM

Because of potential numerical stiffness, regions around the spherical particles require relatively larger numbers of meshes or fine meshes for large computational accuracy, which could be generated using the method proposed by Succi et al. [21]. The authors applied the method for hydrodynamic simulation, but it can be easily extended for the simulation of concentration profiles. The method uses coarse grids around the spheres divided into n fine grids. Considering that the particle speed is the same on both coarse and fine grids, the time-steps of calculation for the two regions are different: $\Delta x_{\text{fine}} = \Delta x_{coarse}/n$ and $\Delta t_{\text{fine}} = \Delta t_{coarse}/n$. This implies that calculation proceeds n small time-steps on the fine grids during one large time-step on the coarse grids. To achieve the same mass diffusivity (D) on both coarse and fine grids outside the particles, two τ_d are used, one for coarse grids and the other for fine grids:

$$\tau_d(\text{fine}) = [(2\tau_d(\text{coarse})-1)n+1]/2. \tag{6.14}$$

Exchange of density-difference distribution function, g, between coarse and fine grids at the coarse-fine interface requires smoothness of macroscopic variables (concentration) and conservation of the similarity parameters (Peclect numbers), which is implemented using the expressions developed by Succi et al. [21]. Here, we do not reproduce the numerical details for brevity. Considering limitation of CPU time, a refinement factor of two, i.e. $n = 2$ was found to be sufficient for the computation in the

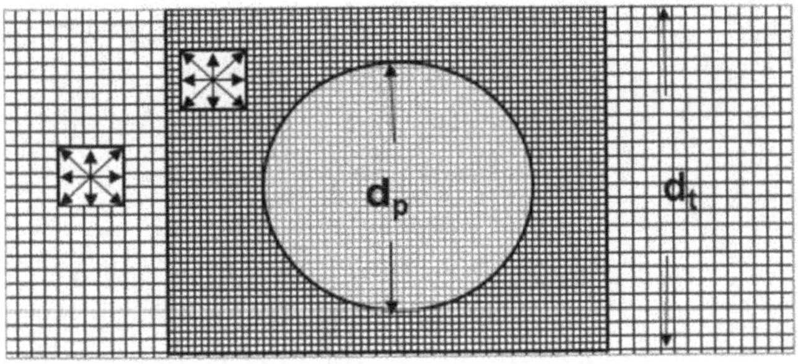

FIGURE 6.7
Scheme using uniform fine grids inside and around the particles, and coarse grids away from the surfaces where there are no particles. (From Verma, N. et al., *Chem. Eng. Sci.*, 62: 3685–3698, 2007.)

study of Manjhi et al. [12]. Figure 6.7 describes the scheme using uniform fine grids inside and around the particles, and coarse grids away from the surfaces where there are no particles to determine concentration profiles. The coarse grids were kept the same as those used for the hydrodynamic calculations.

6.7 Model Validation

One easy way of validating the model for porous adsorbent particles is the simulation using different effective pore diffusivities, and comparing the predicted breakthrough profiles to those for the adsorption in non-porous particles. In other words, under the limiting case of pore-diffusion controlled adsorption (small pore diffusion coefficients), inter-particle concentration distributions are shown to be the same as those corresponding to adsorption in the column packed with non-porous adsorbents, the other operating conditions including hydrodynamics remaining identical.

Figure 6.8 describes breakthrough concentration profiles for different pore-diffusivities ($D_{pore} = 2.5 \times 10^{-7}, 25 \times 10^{-7}, 60 \times 10^{-7}$ m^2/s) for two locations in the adsorber, one near the wall and the other at the center. Simulation was carried out for the identical conditions as used in the previous LBM study [17] on the adsorption in non-porous solids (L × d_t = 200 mm × 50 mm, $d_t/dp = 5$, $Re,p = 22$). For each case of simulation, molecular diffusivity, D, was set constant at 2.5×10^{-5} m^2/s. The simulation results show that both the center and wall breakthrough curves calculated for decreasing values of intra-particle diffusion coefficient asymptotically converged to the model-predicted curve for non-porous adsorbents.

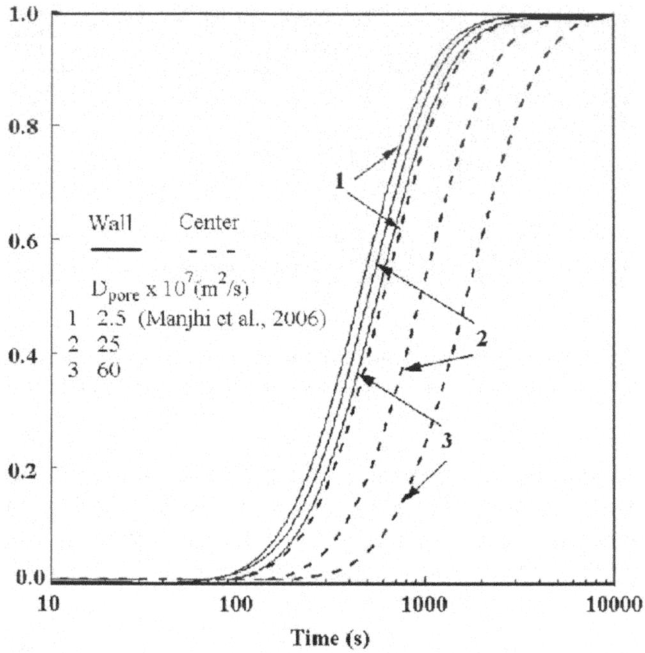

FIGURE 6.8
Model validation for adsorption breakthrough curves for decreasing values of intra-particle diffusivity ($L \times d_t$ = 200 mm × 50 mm, d_t/d_p = 5, Re,p = 22). (From Verma, N. et al., *Chem. Eng. Sci.*, 62: 3685–3698, 2007.)

6.8 Intra-Particle Concentration Distribution

As pointed out earlier in the text, flow field in the inter-particle voids was solved by the LB method and the solutions were superimposed on the concentration fields. The model simulation in the study [12] was carried out assuming viscous flow with Re,p chosen between 1 and 40. D was set constant at 2.5×10^{-5} m^2/s as in the previous case (model validation), whereas D_{pore} was chosen between 0.01 D and D for the particle sizes between 1 and 10 mm. Corresponding to these conditions, macroscopic Peclet number, Pe (vd_p/D) was calculated to be between 1 and 70. Kinematic diffusivity, ν was chosen in the same order of as D.

Figure 6.9 describes the concentration contours across Y-Z plane in a tubular reactor (L = 0.16 m, d_t = 0.04 m) packed with 4-particles (d_t/dp = 3). The particles were placed on the horizontal cross-section of the tube. One particle was placed alternately between two such planes. The 3D simulation was carried out on 164 × 41 × 41 grids, with the number of mesh inside the particles chosen to be 13. The concentration profiles shown in the figure correspond to the section $X = L/2$ containing 4-spheres, at the instance when the concentration in the bulk phase had reached approximately 50% of the

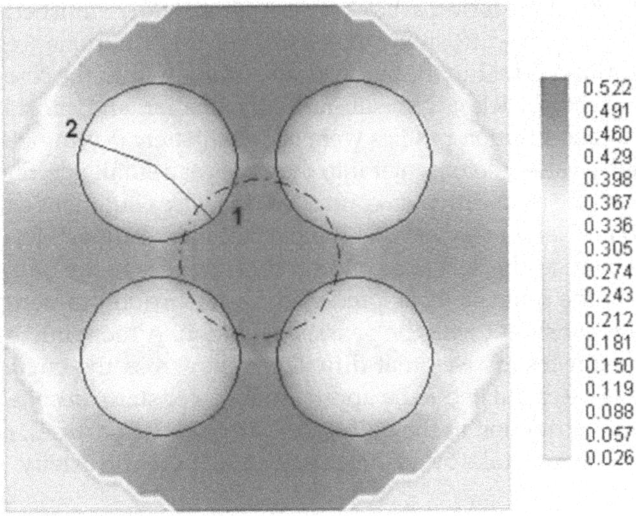

FIGURE 6.9
Concentration contours across *Y-Z* plane in the tubular reactor packed with porous adsorbents ($D/D_{pore} = 10$). (From Verma, N. et al., *Chem. Eng. Sci.*, 62: 3685–3698, 2007.)

inlet concentration. It was assumed that the adsorber initially (i.e. $t = 0$) saturated with very low concentration of the solute (~0.001%) was challenged at $t = 0^+$ by the solute at relatively higher concentration (1%) in the inlet to the tube, whereas the outlet concentration was held constant at 0.001. The results are shown for $D/D_{pore} = 10$. At the outset, three distinct observations may be made, which are consistent with the physics of the simulation. (1) Most of the bulk phase concentration across the tube cross-section, including voids between the particles, are filled with the solute up to 50% of the inlet concentration level. (2) Small intra-particle regions around the circumference of the particles are at relatively higher concentrations (~40%), while the regions inside the particles are deplete with the solute. Concentrations at and near the center of the particles are observed to be in the vicinity of approximately 0.1 only. (3) Concentration profiles inside the particle are non-uniform and θ-unsymmetric. The regions toward the center of the tube have larger concentrations than those in proximity with the periphery of the tube. Concentrations calculated within the spherical particle varies from approximately 0.55 at location *1* to 0.3 at the center, whereas the variation is from approximately 0.4 at location *2* to 0.3 at the center. A close look of the plot reveals that unsymmetricity in θ-direction also exists in the (macro) voids between the spheres and the tube walls, especially around the outer surfaces of the spheres. Readers should appreciate that 3D LBM simulations reveal an interesting aspect of concentration distributions prevalent in the packed column, which 1D models cannot predict.

The LBM simulation was performed for $D/D_{pore} = 100$, keeping the molecular diffusivity and other variables unchanged. The results expectedly showed that un-symmetricity in the intra-particle concentration nearly vanished at relatively smaller value for pore-diffusivity; inter-particle concentrations were relatively lower; concentrations within the particles (including within the core and the circumference of the particles) were substantially smaller; and, most importantly, intra-particle concentration profiles were approximately θ-symmetric. At small pore-diffusivity values, solute enter into the pores at a small rate. Such case may be considered to be pore-diffusion controlled. In fact, calculations carried out for $D/D_{pore} = 100$ and larger showed that the particles were almost devoid of solute even after (quasi) steady-state and concentration profiles in the bulk phase were approximately the same as those predicted for adsorption in non-porous particles, which is consistent with the physical situation. A more interesting aspect of the simulation results was that diffusion flux across the circumference of the particles were calculated to be approximately constant, despite the uneven concentration distribution in the bulk phase around the particles, attributed to small rates of solute uptake by adsorbents at small pore diffusivity.

6.9 Thermal LBM

Adsorption may be accompanied with release or absorption of thermal energy. In such case, significant temperature gradients may exist in the packed beds and the process may be considered to be non-isothermal. The LB simulations of temperature profiles is not difficult. Various lattice models reported in literature for simulating temperature profiles in a fluid flow may be broadly classified into two categories: (1) multi speed approach [22–25], and (2) passive-scalar approach [26–31]. The former approach is apparently an extension of the lattice models for isothermal conditions in which only the density distribution function is used, however, with additional speeds to determine the temperature evolution. Also, equilibrium distribution function must include the higher-order velocity term. In the latter approach, an auxiliary lattice Boltzmann equation is used assuming that temperature field is passively advected. This scheme is relatively simpler to implement, numerically stable, and also permits use of varying Prandtle numbers over a wider range [32]. In this chapter, we restrict ourselves to the implementation aspect of the thermal LBM. The representative simulation results for temperature profiles in a packed bed are qualitatively analogous to those for concentration profiles and can be referred in the study [32].

Drawing analogy with the advection of a species, the lattice Boltzmann equation may be written as:

$$h_i(r+e_i,t+1)-h_i(r,t)=\frac{-1}{\tau_h}\left(h_i-h_i^0\right)+\psi(T) \qquad (6.15)$$

Equation (6.15) contains distribution functions and the local equilibrium distribution function, with the thermal relaxation constant τ_h related to thermal diffusivity in the similar fashion as mass diffusivity and kinamatic viscosity are related to the corresponding relaxation constants of the LBGK transport equations for the mass and momentum transport of a solute in the bulk phase:

$$\alpha = e^2\left(\frac{2\tau_h - 1}{6}\right)\Delta t \tag{6.16}$$

The exothermic heat generation rate is included via $\psi(T)$ in equation (6.15) using the similar approach as used for including gravity as an external body force in the momentum conservation equation. Finally, the development of local equilibrium distribution function h_i^0 is also analogous to that of the equilibrium distribution functions, f_i^0 and g_i^0 corresponding to ρ (sum-density) and $\Delta\rho$ (difference-density) of a binary miscible fluid as per the diffusion model proposed by Swift et al. [18]:

$$\begin{aligned} h_i^0(u) &= A_0 + C_0 u.u, i = 0 \\ &= h_{1i}^0(0)\left[A + Be_i.u + C(e_i.u)^2 + D(u.u)\right], i = 1\ldots9 \\ &= h_{2i}^0(0)\left[A + Be_i.u + C(e_i.u)^2 + D(u.u)\right],\ i = 10\ldots18 \end{aligned} \tag{6.17}$$

In the case of porous adsorbent particles, the intra-particle thermal diffusivity is set at values different from those in the bulk region (voids between the particles) by segregating two domains of computations, one for the inside and the other for the outside, and using two different τ_h related to α_{macro} and α_{micro} as per equation (6.16). The numerical scheme for solving temperature profiles is identically the same as that used for solving concentration profiles in the adsorber (both macro and micro) and described in the previous study [12].

6.10 Lattice Thermal Boundary Conditions

There are two types of commonly used boundary conditions: one is the constant wall temperature and the other is the constant heat flux. The flux may be set at zero for the thermally insulated walls, i.e. adiabatic condition may be assumed to prevail. In the case of fixed wall temperature, the scheme originally proposed by Yean and Schaefer [29] for 2D geometry may be adopted and extended for 3D case.

First, distribution functions after streaming for the known links are determined. For example, referring to the 3D lattice molecule of Figure 6.1, the distribution functions which are known at the bottom plane of a tube are: *h0, h1, h3-6,* h9-14, *h16,* and *h17.* The unknown functions are *h2, h7, h8, h15,* and *h18.* Then, an intermediate quantity, say T′ is calculated from the known distribution functions and the prescribed wall temperature T_{wall}:

$$T' = T_{wall} - (h0 + h1 + h3 + h4 + h5 + h6 + h9 + h10 + h11 + h12 + h13 + h14 + h16 + h17) \tag{6.18}$$

The unknown distribution functions are temporarily set at their equilibrium values given by equation (6.17) with temperature set at T′. Thus, realizing that $\sum hi(i = 1..18) = T_{wall}$, the intermediate temperature is accordingly corrected as:

$$T_{correct} = T' \times T_{wall} / \left(h_2^0 + h_7^0 + h_8^0 + h_{15}^0 + h_{18}^0\right) \tag{6.19}$$

Finally, the unknown distribution functions are calculated as:

$$h2 = h_2^0 / T_{wall} \times T_{correct}; h7 = h_7^0 / T_{wall} \times T_{correct}, \text{etc.} \tag{6.20}$$

In the case of wall flux boundary condition, the second-order finite difference scheme based on the discretization of three points may be carried out:

$$\left.\frac{\partial T}{\partial Y}\right|_{i,1} = (4T_{i,2} - T_{i,3} - 3T_{i,1})/(2\Delta y) \tag{6.21}$$

Thus, by prescribing the wall flux, the wall temperature $T_{i,1}$ is calculated. The after-procedure is the same as that used for the constant wall temperature case. The implementation of the boundary conditions (continuity in temperature and heat flux across the surface of the sphere) is performed exactly in the same way as in the case of (mass) diffusion in the porous spherical particles:

$$T|_{R_p+} = T|_{R_p-}, \text{ and}$$

$$-k\nabla T|_{R_p+} = -k_{pore}\nabla T|_{R_p-} \tag{6.22}$$

Before we close this chapter, we want to recommend readers, especially those who wish to learn and practice the LBM-based modelling of adsorption, that new methods continue to evolve enormously on all fronts. There are new theoretical approaches continuously developed with a view to increasing accuracy of the model predictions. Then there are new studies from the perspective of speeding up the computer programming code so that CPU time of calculations is reduced. There are also emerging applications of LBM, for example, flow in microchannels, flow of Non-Newtonian fluids, and also, biomedical transport. In fact, at the time of writing this chapter, several novel LB models for adsorption have already been reported in literature which are worthy of studying and applying to practical applications [33–35]. Therefore, it is in the interest of readers to keep themselves abreast with new developments and improvise the limitations and constraints of the present LBM-based models.

References

1. O'Brien, G. S., Bean, C. J., McDermott, F. (2002) A Comparison of Published Experimental Data with a Coupled Lattice Boltzmann-Analytic Advection-Diffusion Method for Reactive Transport in Porous Media. *J. Hydrol.* 268: 143–157.
2. Zhang, X., Ren, L. (2003) Lattice Boltzmann Model for Agrochemical Transport in Soils. *J. Contaminant Hydrol.* 67: 37–42.
3. Knutson, C. E., Wert, C. J., Valocchi, A. J. (2001) Pore-Scale Modelling of Dissolution from Variably Distributed Nonaqueous Phase Liquid Blobs. *Water Resource Res.* 37(12): 2951–2963.
4. Manz, B., Gladden, L. F., Warren, P. B. (1999) Flow and Dispersion in Porous Media: Lattice-Boltzmann and NMR Studies. *AIChE J.* 45(9): 1845–1854.
5. Tan, M.-L., Qian, Y. H., Goldhirsch, I., Orszag, S. (1995) A Lattice-BGK Approach to Simulating Granular Flows. *J Stat Phys* 91(172): 87–92.
6. Kang, Q., Lichtner, P. C., Zhang, D. (2007) An Improved Lattice Boltzmann Model for Multicomponent Reactive Transport in Porous Media at the Pore Scale. *Water Resource Res.* 43: 1–12.
7. Yang R. T., *Gas Separation by Adsorption Processes* (1997) Imperial College, UK.
8. Manjhi, N., Verma, N., Salem, K., Mewes, D. (2006) Simulation of 3D Velocity and Concentration Profiles in a Packed Bed Adsorber by Lattice Boltzmann Methods. *Chem. Eng. Sci.* 61: 7754–7765.
9. Filippova, O., Hanel, D. (1998) Boundary-Fitting and Local Grid Refinement for Lattice-BGK Models. *Int. J. Mod. Phys.* 9(8): 1271–1279.
10. Verma, N., Mewes, D. (2009) Lattice Boltzmann Methods for Simulation of Micro and Macro Transport in a Packed Bed of Porous Adsorbents under Non-Isothermal Condition. *Comp. Math. Appl.* 58: 1003–1014.

11. Breuer, M., Bernsdorf, J., Zeiser, T., Durst, F. (2000) Accurate Computation of the Laminar Flow Past a Square Cylinder Based on Two Different Methods: Lattice-Boltzmann and Finite-Volume. *Int. J. Heat Fluid Flow* 21: 186–196.
12. Verma, N., Salem, K., Mewes, D. (2007) Simulation of Micro and Macro Transport in a Packed Bed of Porous Adsorbents by Lattice Boltzmann Methods. *Chem. Eng. Sci.* 62: 3685–3698.
13. Chakraborty, J., Verma, N., Chhabra, R. P. (2004) Wall Effects in Flow Past a Circular Cylinder in a Plane Channel: A Numerical Study. *Chem. Eng. Processing* 43(12): 1529–1537.
14. Freund, H., Zeiser, T., Huber, F., Klemm, E., Brenner, G., Durst, F., Eming, G., (2003) Numerical Simulations of Single Phase Reacting Flows in Randomly Packed Fixed-Bed Reactors and Experimental Validation. *Chem. Eng. Sci.* 58: 903–910.
15. Schure, M. R., Maier, R. S., Kroll, D. M., Davis, H. T. (2004) Simulation of Ordered Packed Beds in Chromatography. *J. Chromat. A* 1031: 79–86.
16. Succi, S. (2001), *The Lattice Boltzmann Equation for Fluid Dynamics and Beyond*. Oxford Press, Oxford, U.K.
17. Manjhi, N., Verma, N., Salem, K., Mewes, D. (2006) Lattice Boltzmann Modelling of Unsteady-State 2D Concentration Profiles in Adsorption Bed. *Chem. Eng. Sci.* 61: 2510–21.
18. Swift, M. R., Orlandini, E., Osborn, W. R., Yeomans, J. M. (1996) Lattice Boltzmann Simulation of Liquid-Gas and Binary Fluid Systems. *Phys. Rev. E* 54(5): 5041–5046.
19. Zou, Q., Hou, S., Chen, S., Doolen, G. D. (1995) Improved Incompressible Lattice Boltzmann Model for Time-Independent Flows. *J Stat Phys* 81: 35–38.
20. Sullivan, S. P., Sani, F. M., Jones, M. L., Gladden, L. F. (2005) Simulation of Packed Bed Reactors Using Lattice Boltzmann Methods. *Chem. Eng. Sci.* 60: 3405–3418.
21. Succi, S., Filippova, O., Smith, G., Kaxiras, E. (2001) Applying the Lattice Boltzmann Equations to Multiscale Fluid Problems. *Computing Sci. Eng.* November/December, 26–37.
22. Alexander, F. J., Chen, S., Sterling, J. D. (1993) Lattice Boltzmann Thermo-hydrodynamics. *Physical Review E* 47: 2249–2252.
23. Shan, X., Doolen, G. (1996) Diffusion in a Multicomponent Lattice Boltzmann Equation Model. *Physical Review* 54: 3614–3619.
24. McNamara, G., Garcia, A. L., Adler, B. J. (1995) Stablization of Thermal Lattice Boltzmann Models. *J. Statistical Physics* 81: 395–408.
25. Chen, Y., Ohashi, H., Akiyama, M. (1997) Two-Parameter Thermal Lattice BGK Model with a Controllable Prandtl Number. *J Scientific Computation* 12: 169–277.
26. Eggels, J. G. M., Somers, J. A. (1995) Numerical Simulation of Free Convection Flow Using the Lattice Boltzmann Scheme. *J. Heat and Fluid Flow* 16: 357–361.
27. Shan, X., (1997) Simulation of Rayleigh-Benard Convection Using a Lattice Boltzmann Method. *Physical Review E* 55: 2780–2788.
28. Inamuro, T., Yoshino, M., Inoue, H., Mizuno, R., Ogino, F. (2002) A Lattice Boltzmann Method for a Binary Miscible Fluid Mixture and Its Application to a Heat-Transfer Problem. *J. Computational Phys.* 179: 201–215.
29. Yuan, P., Schaefer, L. (2006) A Thermal Lattice Boltzmann Two-Phase Flow Model and Its Application to Heat Transfer Problems—Part 1. Theoretical Foundation. *J. Fluid Eng.* 128: 142–150.

30. Yuan, P., Schaefer, L. (2006) A Thermal Lattice Boltzmann Two-Phase Flow Model and Its Application to Heat Transfer Problems—Part 2. Integration and Validation. *J. Fluid Eng.* 128: 151–156.
31. Kao, P.–H., Yang, R.–J. (2007) Simulating Oscillatory Flows in Rayleigh-Benard Convection Using the Lattice Boltzmann Method. *Int. J. Heat and Mass Transf.* 50: 3315–328.
32. Verma, N., Mewes, D. (2008) Simulation of Temperature Fields in a Narrow Tubular Adsorber by Thermal Lattice Boltzmann Methods. *Chem. Eng. Sci.* 63: 4269–4279.
33. Vanson, J.-M., Boutin, A., Klotz, M., Coudert, F.-X. (2017) Transport and Adsorption under Liquid Flow: The Role of Pore Geometry. *Soft Matter* 13: 875–885.
34. Ma, Q., Chen, Z., Liu, H. (2017) Multiple-Relaxation-Time Lattice Boltzmann Simulation for Flow, Mass Transfer, and Adsorption in Porous Media. *Physical Review E* 96: 013313.
35. Vanson, J.-M., Coudert, F.-X., Klotz, M., Boutin, A. (2017) Kinetic Accessibility of Porous Material Adsorption Sites Studied through the Lattice Boltzmann Method. *Langmuir* 33: 1405–1411.

7

Improved Removal of Toxic Contaminants in Water by Green Adsorbents: Nanozeolite and Metal-Nanozeolite for the Removal of Heavy Metals and Phenolic Compounds

Thi-Huong Pham and Byeong-Kyu Lee

CONTENTS

7.1 Introduction

7.1.1 Concepts of Green Adsorbents

This chapter describes the concepts and conditions that green adsorbents should satisfy in order to act as an ideal adsorbent.

The following summarizes some conditions proposed to become ideal green adsorbents:

1. Precursors and manufacturing processes
   - Precursors (or sources) of green adsorbents should be abundant and easy accessible from nature and/or markets with lower collection and transportation costs and without making adverse environmental impacts. In many cases, if the precursors are biomass, materials with higher carbon density per unit mass are better than those with lower carbon density.
   - Synthesis or manufacturing processes should be easy or simple with high product yields, lower consumption of materials (precursors, chemicals) and energy (or heat or electricity), and fewer requirements for complicated or expensive equipment or instruments.
   - During and/or after being synthesized or prepared, they should not produce significant amounts of toxic chemicals or hazardous materials that require difficult or expensive treatments.
2. Physicochemical Properties
   - The green adsorbents should have a more porous structure or large pore volume, high surface area, large amount of active sites for adsorption, and easy or proper desorption and regeneration characteristics, as these are considered among the most important factors of adsorbents. In addition, the properties of the synthesized or prepared green adsorbents should be almost inert or neutral, chemically and thermally stable, and structurally durable even after repeated use.

3. Separation
    - After being used, they should be easily separated from water and/or air pollutant treatment or purification systems and their disposal methods after final use should be simple or environmentally sustainable with reduced generation of toxic or hazardous substances.
4. Costs
    - Their total preparation or manufacturing costs, including environmental impacts, should be low enough to compete with other currently available adsorbents.

7.1.2 Surface Characteristics Modification for New Zeolite Development

Zeolites have been widely used for various industries, particularly in gas separation, ion exchange, and catalysis areas [1,2]. These applications demand the development of highly porous solids, such as new zeolites with smaller size, increased surface area, metal-modified surface chemistry, and improved adsorption selectivity [3,4]. The decrease in zeolite crystal size from the micrometer to the nanometer scale greatly enhances zeolite properties, such as increased surface area and decreased diffusion path lengths. As the zeolite crystal size is decreased, the external surface area of the prepared zeolite is greatly enhanced. The large improvement of the external surface area, which has not been expected for micron-sized zeolites, can be a distinct feature as compared to the internal pore surface improvement. This is because active sites are incorporated onto the external surface characteristics of zeolites, and thus particularly affect the surface reactivity, resulting in improved catalysis [5–7]. The crystal size reduction can also lead to a significant reduction in diffusion path lengths of adsorbates toward adsorbents. Many studies have been devoted to evaluating the adsorption and catalytic functions of nanozeolites for the removal of inorganic and organic contaminants such as heavy metals and phenols from aqueous solutions [8–11]. The following scheme summarizes the four main application areas of zeolites, which are mostly based on their surface characteristics and reaction activities, as shown in Figure 7.1.

The adsorption and separation characteristics of chemicals to be treated or disposed of are both determined by the different migration speed of various compounds along the adsorbent surface. The chemical migration features depend on their interaction features with the adsorbent surface and also the steric effects, including the size effects of zeolite pores and the molecular size effects of chemicals to be treated. The shape-selective properties of zeolites greatly affect their molecular level adsorption features (preferentiality or excluding) of adsorbates. Thus one of the research directions for new and efficient zeolite development is how to find or modify the surface characteristics of zeolites, including their pore size change and surface functionality change.

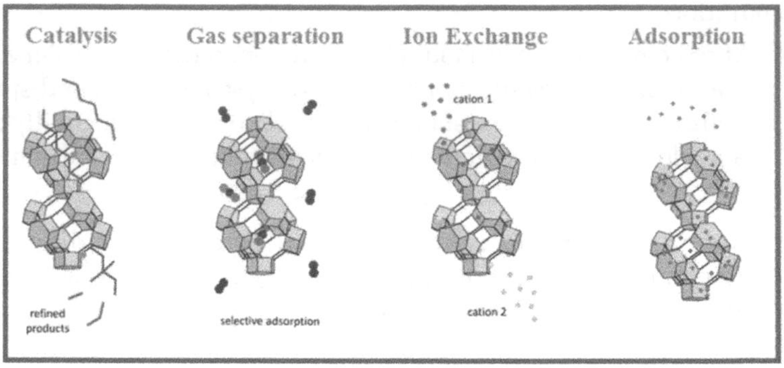

FIGURE 7.1
The main applications of zeolite.

7.1.3 Metals-Modified Zeolites

The metal-modified zeolites benefit from the introduction of different metals from the metals located in the zeolite pores, which extends the application areas of conventional zeolites [12–14]. The two most common methods for metal introduction into the pores are ion-exchange and wet impregnation techniques. In the ion-exchange process, a zeolite is repeatedly immersed in a solution containing the target metal ions to obtain the maximum degree of substitution by the target metal cations [15,16], followed by extensive washing to remove the excess of the metal salt existing inside the zeolite pores. In many cases, however, ion-exchange methods do not provide successful results in terms of sufficient metal loadings, resulting in low exchange capacity, which is typical for high silica zeolites, zeolites without ion exchange sites and the use of neutral or negatively charged metal complexes [17,18]. Thus, the wet impregnation technique is commonly applied to obtain the desired degree of metal loading. In the wet impregnation techniques, zeolite is added into small volumes of solutions containing the target amount of metal, resulting in a zeolite-metal slurry. This slurry needs to equilibrate with a uniform distribution of the solution across the zeolite surface. The modified zeolite is obtained by evaporating the solvent under reduced pressure [19,20]. Partial replacement of Si^{4+} by Al^{3+} in the zeolite structure can produce an excessive negative charge which disturbs the effective adsorption of organic pollutants. The metals exchanged in the zeolite can decrease the excess negative charge of zeolite and thereby improve their organic pollutants adsorption capacity [21].

7.1.4 Heavy Metals Contaminants

Many industries are suffering from the release of heavy metal ions into the environment and the resulting serious water pollution. The most common toxic metal ions in water environments are Cr(VI), Pb(II), Cd(II), Cu(II), Hg(II), Ni(II), and As(III), which are persistent environmental contaminants. These heavy metals cannot be degraded or destroyed naturally. Such heavy metal ions improperly discharged from industries have become a great threat to human health and have been implicated in serious diseases such as cancer [22,23].

7.1.5 Phenolic Contaminants

Phenolic compounds are almost ubiquitous contaminants because they have been widely discharged into wastewater from oil refineries, the paint, drug and pesticide industries, and as various process intermediates [24,25]. The potential toxicity of phenol derivatives has received increasing attention, with some having been proven to be toxic. Chlorophenols may cause inflammation in the digestive system and increased blood pressure. Nitrophenols can reduce the ability of the blood to carry oxygen to tissues and organs. Because of their toxicity, they must be removed from industrial effluents before being discharged into the water stream.

7.2 Preparation of Nanozeolite and Metal-Nanozeolites as Green Adsorbents

Many researchers have investigated the use of various types of zeolites to remove heavy metals and organic pollutants released into the environment. Many nano-sized zeolites have been prepared by collaborating scientists from interdisciplinary fields including chemistry and nano technology. In nanozeolites, the substantial change in their size and surface properties can greatly expand their adsorption capacity as compared to conventional micro-sized zeolites. The great increase in the outer surface area of nano-crystalline zeolite can improve the surface properties, such as specific surface area, effective functional groups, hydrophobic/hydrophilic property, and surface charge. For example, copper, iron, silver, or zinc are used to modify the surface properties of metal-loaded nanozeolites. The addition of iron ions into the nanozeolite structure can form hydroxyl groups on the surface of the metal particles during the preparation process of iron-modified nanozeolite [26–28], which assists in the removal of heavy metals and organic contaminants in aqueous solutions.

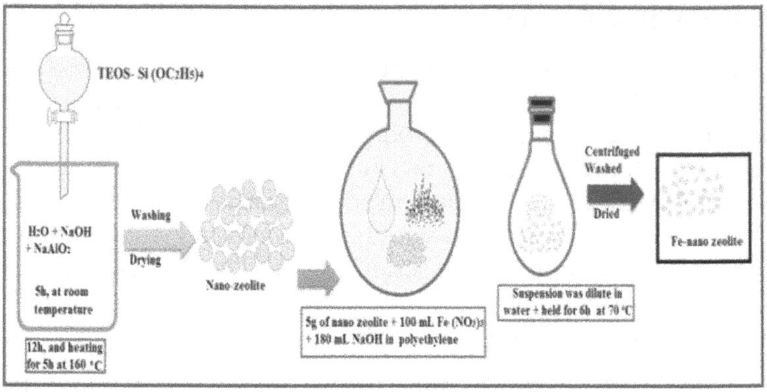

FIGURE 7.2
Process for synthesis of nanozeolite and iron-modified nanozeolite.

The synthesis processes of nanozeolite and metal-modified nanozeolite are shown in Figure 7.2.

Sodium aluminate salt (>63% Al_2O_3) and silica sol can be used as aluminum and silicon sources for zeolite preparation, respectively. For the nanozeolite preparation, 0.35 g NaOH and 0.147 g sodium aluminate salt in H_2O are mixed together and then aged for 5 h at 20°C with magnetic stirring. After that, 6.6 g of silica sol is added dropwise and then the resulting mixture is stirred at room temperature for 12 h to give a homogenous mixture followed by heating for 24 h at 180°C under autogenous pressure. The solid product is centrifuged and washed with deionized water until its pH reaches 5.0. After drying the synthesized crystalline nano, it is immersed into $Fe(NO_3)_3$ solutions with a concentration of 1M to obtain Fe ions loaded onto nanozeolite. The additionally obtained or modified functional groups from the modification processes can be beneficial for the removal of heavy metals and organic contaminants. In particular, the hydroxyl groups formed on the surface of the iron particles added during the iron-nanozeolite preparation can greatly improve the removal capacity of heavy metals and organics in water [29].

7.3 Physicochemical Characterization of the Green Adsorbents

7.3.1 Morphological Analysis

The SEM images of the nanozeolites and iron-modified nanozeolite are shown in Figure 7.3(a) & 7.3(b). The surface morphology of the nanozeolites exhibited very small nanocrystals with a size range of 90–100 nm and a tetrahedral structure, which was different from that of the Fe-nanozeolite. The surface of the iron nanozeolite, modified by Fe $(NO_3)_3$, was covered with

FIGURE 7.3
SEM analysis of (a) nanozeolite and (b) Fe-nanozeolite.

some small particulates, indicating that some iron ions were doped on the nanozeolite surface.

7.3.2 TEM Analysis

Transmission electron microscopy (TEM) images of nanozeolite and Fe-nanozeolite were collected in order to determine the size and morphology of the nanocrystals (Figure 7.4(a) & 7.4(b)). The nanoparticles of nanozeolite had a spherical shape with an average particle size of 100 nm. However, not all the particles were spherical in the Fe-nanozeolite, which was attributed to the Fe ions doped on the nanozeolite [30].

7.3.3 XRD Analysis

X-ray diffraction (XRD) was used to analyze the crystallinity of nanozeolite and Fe-nanozeolite (see in Figure 7.5(a) & 7.5(b)). The peaks at 2θ values of 23.5, 30.1, and 54. 3° showed the aluminate of the nanozeolite (NZ) and

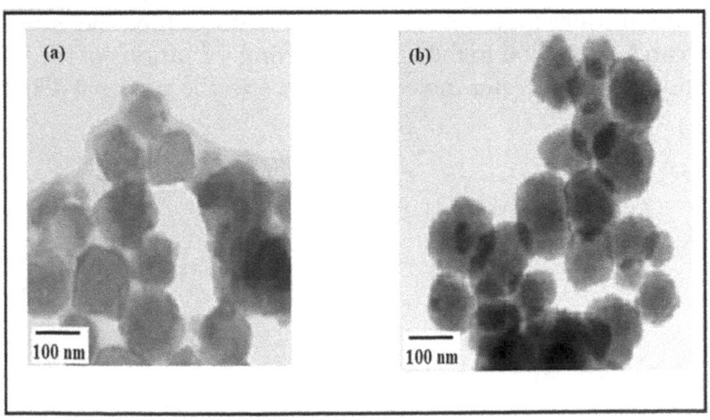

FIGURE 7.4
TEM analysis (a) nanozeolite and (b) Fe-nanozeolite.

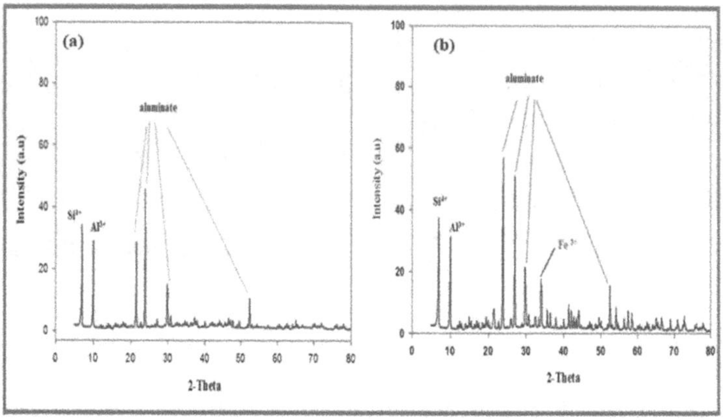

FIGURE 7.5
XRD analysis (a) nanozeolite and (b) Fe-nanozeolite.

Fe-nanozeolite. The peaks at $2\theta = 7–10°$ correspond to the specific peaks of Al^{3+} and Si^{4+} of the adsorbents. The high intensity of the peak at $2\theta = 33.8°$ represented the Fe^{3+}-loaded nanozeolite (Fe-NZ). The characteristic peaks shown in the iron-nanozeolite structure were almost similar to those of nanozeolite ZMS-5 [32,33].

7.3.4 Infrared Spectra Analysis

To identify the functional groups of nanozeolite and Fe-nanozeolite, the Fourier transform infrared (FTIR) spectra were analyzed (Figure 7.6(a) & 7.6(b)). The characteristic band at 1080–1090 cm^{-1}, observed in both nanozeolite and Fe-nanozeolite, was attributed to the Si-O stretching of the Si-O-Si structure. The absorption band at 1600–1645 cm^{-1} is assigned to H-O-H bending vibration of water in nanozeolite and Fe-nanozeolite. The broad band at 3432 cm^{-1} can be assigned to the O-H stretching vibrations of the hydrogen-bonded and adsorbed water molecules. The specific peaks at 480–560 cm^{-1}

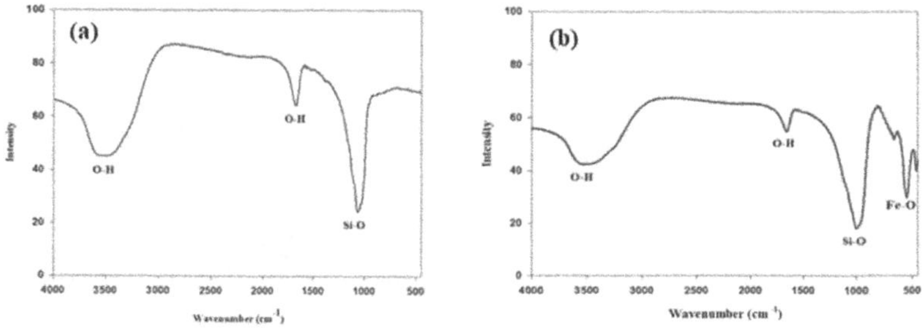

FIGURE 7.6
FTIR analysis (a) nanozeolite and (b) Fe-nanozeolite.

TABLE 7.1

BET Analysis of Nanozeolite and Fe-Nanozeolite

Adsorbents	Surface Area (m^2/g)	Pores Volume (cm^3/g)
Nanozeolite	692	0.36
Fe-nanozeolite	886	0.47

well represent the characteristics of Fe-NZ (Figure 7.6(b)), which consists of Fe-O group [31].

7.3.5 BET Analysis

The surface area and pore volume of the nanozeolite and Fe-nanozeolite were identified by Brunauer–Emmett–Teller (BET) analysis, as shown in Table 7.1. The specific surface areas of nanozeolite and Fe-nanozeolite were 692 and 886 m^2/g, respectively. Generally, the increased surface area and pore size of iron-modified nanozeolite (FE-NZ) also facilitates the adsorption process of heavy metals and organic pollutants, having even large molecular size. Thus Fe-NZ can bind to the adsorption sites more easily, which improves the adsorption capacity of the adsorbates [34].

The characterization of the nanozeolites using physicochemical parameters, such as SEM, TEM, XRD, EDX, FT-IR, and BET data, offers insights into the properties of surface or crystal morphology, size, structure, elementals, functional groups, and surface area and pore information, respectively, before and after modification of the nanozeolites. The surface area and pore volumes of modified nanozeolite were significantly increased by the iron addition. These physicochemical analyses help to explain the adsorption performance of pollutants onto adsorbents such as nanozeolites and iron-modified nanozeolites.

7.4 Removal of Heavy Metals and Organic Compounds

Based on the physicochemical properties of prepared nanozeolite and Fe-nanozeolite, this study investigated their feasibility as green adsorbents for removing organic compounds such as phenol (Ph), 2-chlorophenol (2-CP), and 2-nitrophenol (2-NP) and heavy metals such as cadmium ions (Cd (II)) and lead ions (Pb (II)) in waste water. Various adsorption parameters or characteristics of the prepared adsorbents for the removal of the target pollutants, including solution pH, reaction (contact) time, adsorbent dose, initial concentration, adsorption isotherm and kinetics, and regeneration and removal mechanisms, are analyzed and compared to the previously reported values. Furthermore, this study also examines the economic factors for the removal of heavy metals and phenolic pollutants by using synthesized nanozeolite and iron-modified nanozeolite. The economic consideration is based on simple estimates of manufacturing and regeneration costs, electricity and utility costs, and environmental impacts/costs for the practical use of these adsorbents.

7.4.1 Removal of Heavy Metals

7.4.1.1 Effect of Solution pH

The pH of the solution containing the adsorbate greatly affects the adsorption process between the adsorbent surfaces and the heavy metals ions [35]. The pH at the point of zero charge (pH_{pzc}) also plays an important role in the adsorption process. At pH > pH_{pzc}, the surface of NZ and Fe-NZ is negatively charged, giving a strong electrostatic attraction between the surface groups and metal ion species, which can contribute to increasing the metal adsorption. The decrease in the adsorption amount observed at pH less than pH_{pzc} (when the surface of the adsorbent is positively charged) was attributed to increased competition between protons and metal ion species for the adsorption sites. The effect of solution pH on the removal of Cd (II) and Pb (II) ions using NZ and Fe-NZ was investigated within the pH range of 2.0 to 6.0. An adsorption of Cd (II) and Pb (II) at pH greater than 6.0 was avoided due to the presence of insoluble cadmium and lead hydroxide precipitates in the solution. Figure 7.7 shows the effect of solution pH on the removal efficiency of Cd (II) and Pb (II) ions of 200 mg/L concentration and 200 mL volume. The removal efficiencies of Cd (II) and Pb (II) ions were relatively lower in strong acid solution (pH = 2.0–4.0) than at higher pH (5.0–6.0). In the more acidic

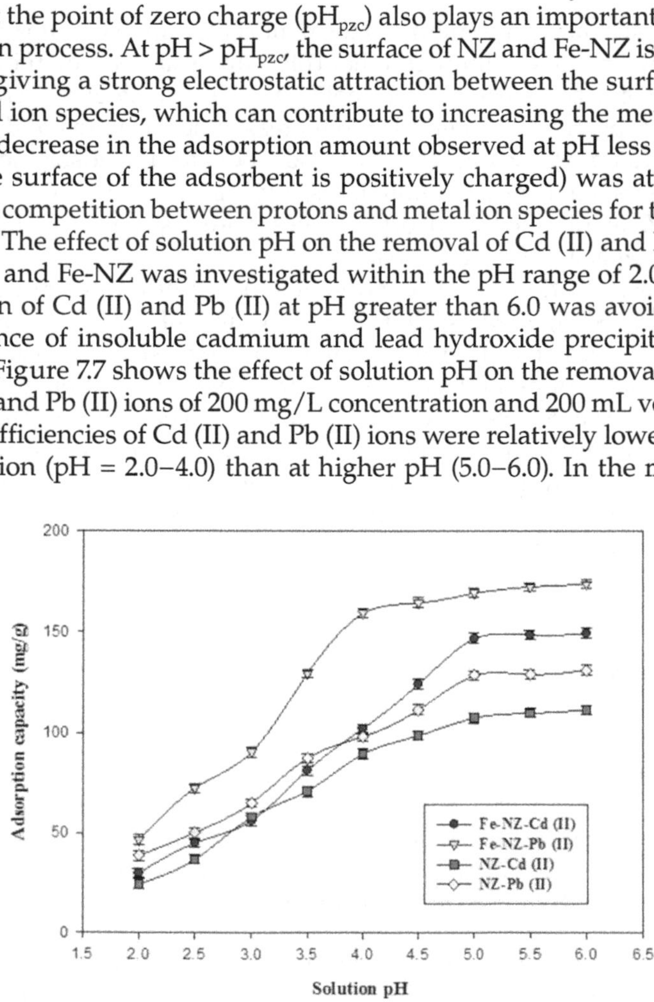

FIGURE 7.7
Effect of solution pH on the removal of Cd (II) and Pb (II) ions using NZ and Fe-NZ.

conditions (pH = 2.0–4.0), the increased competition between H^+ and Cd (II) or Pb (II) ions for the adsorption sites on the adsorbent surface reduced the removal efficiencies of Cd (II) and Pb (II) by Fe-NZ. In the weakly acidic conditions (pH = 5.0–6.0), the amount of H^+ ions in the solution is reduced due to the relative increase in OH^- anions, which weakens the competition effect between the metals ions and H^+ ions and leads to the development of more negative charges on the adsorbent surface as compared with the more acidic pH range of 2.0–4.0 [26,27]. However, the values of the pH_{pzc} were 5.0 and 4.78 for Fe-NZ and NZ, respectively. Thus the adsorbent surfaces start to build a partial negative charge at pH higher than pH_{pzc}. The electrostatic attraction between the positive charge of the metal ions and the negative charge on the absorbent surface increased, which increased the removal efficiencies of Cd (II) and Pb (II) ions at weakly acidic pH values (pH = 5.0–6.0). The adsorption capacities of Cd (II) and Pb (II) ions using NZ and Fe-NZ at pH 6.0 were 111.9 and 151.6 mg/g, and 130.8 and 174.5 mg/g, respectively. Therefore, other experiments to perform Cd (II) and Pb (II) adsorption by NZ and Fe-NZ were conducted at pH 6.0.

7.4.1.2 Effect of Contact Time

Adsorption experiments were performed in 200 mL of single-cation solution containing Cd (II) or Pb (II) ions and mixed-cation {Cd (II)/[Pb (II) + Cd (II)] and Pb (II)/[Pb (II) + Cd (II)]} solution of 200 mg/L with contact times ranging from 10 to 300 min. Figure 7.8(a) & 7.8(b) show the effect of different contact times on the adsorption of the Cd (II) and Pb (II) solution having single- and mixed-cations systems. As time elapsed, the adsorption capacities of Cd (II) and Pb (II) ions slightly increased when the contact time was increased to 300 min. The adsorption rates of Cd (II) or Pb (II) ions onto the adsorbent surface were high in the beginning of the adsorption process because there are more adsorption sites initially available for the adsorption removal of Cd (II) and Pb (II) ions. After 120 min, only a small fraction of active or vacant sites are available for further metal ions removal because the increased repulsion, against the new or free metal ions approaching toward the adsorbent, could be developed by the Cd (II) or Pb (II) ions already adsorbed on the surface of Fe-NZ. Competitive adsorption occurred to a certain extend between the Cd (II) and Pb (II) ions. The adsorption capacities of Cd (II) and Pb (II) at the equilibrium time from mixed-cation solution were decreased (9.8 and 11.7% using NZ, and 6.2 and 8.4% using Fe-NZ, respectively) as compared to the single-cation solution. However, the removal efficiency of Pb (II) ions was higher than that of Cd (II) ions in the single- and mixed-cation systems. This was attributed to the effect of electronegativity, the hydrated ionic radius, and the first hydrolysis constants of heavy metals. In general, the lower the first hydrolysis constant, the greater the proportion of MOH^+ species that can exist. Among the various metal species in solution, therefore, MOH^+ species can get stronger

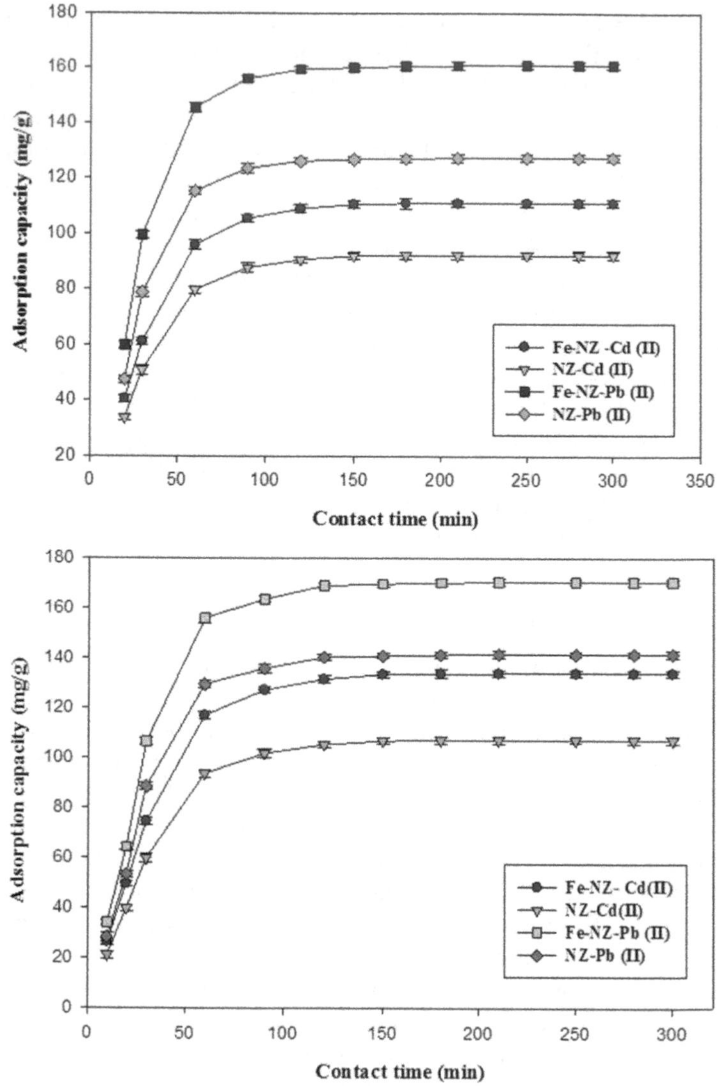

FIGURE 7.8
Effect of reaction time for the removal of Cd (II) and Pb (II) ions in (a) single-cation and (b) mixture-cation system using nanozeolite and Fe-nanozeolite.

adsorption than M^{2+} species. The hydrated ionic radius and electronegativity also affect the interaction between metals ion and adsorbent. The metal ion with high electronegativity and smaller hydrated ionic radius can easily be adsorbed on the adsorbent surface. Lead has a higher electronegativity (2.33), lower first hydrolysis constant (7.71) and smaller hydrated ionic radius (0.261 nm) than cadmium, and thus has a higher adsorption capacity and removal efficiency [28].

7.4.1.3 Adsorption Kinetics

In order to predict the adsorption kinetics of an adsorbate, the pseudo-first-order and pseudo-second-order models are commonly applied to the experimental data.

The pseudo-first-order reaction is expressed by the following equation (7.1):

$$\mathrm{Log}(q_e - q) = \log\ q_e - k_1 t/2.303 \tag{7.1}$$

where q_e and q_t are the amount of Cd (II) or Pb (II) sorbed per unit weight of sorbent (mg/g) at equilibrium and at any time t, respectively, and k_1 is the rate constant of pseudo-first-order sorption (min^{-1}).

The pseudo second-order kinetic model is described by the following reaction (7.2):

$$t/q = 1/k_2 q_e^2 + t/q_e \tag{7.2}$$

where k_2 is the rate constant of the pseudo-second-order sorption (g/mg min^{-1}).

The pseudo-first order model did not provide good fitting (R^2 = 0.67-0.75 and 0.72-0.58 for Cd (II) and Pb (II) ions removal in both single- and mixed-cation systems using NZ and Fe-NZ, respectively) for the experimental data obtained (Table 7.2(a) & 7.2(b)).

In the first-order kinetic model, the adsorption equilibrium capacity (q_e) and the experimental data differed widely. These results indicated that the adsorption data of the heavy metals did not well follow the first-order kinetic model for either single- or mixed-cation system. The calculation adsorption capacities of Cd (II) and Pb (II) in the single- and mixed-cation systems based on the second-order kinetic were 94.04 and 89.42 mg/g, and 98.12 and 93.57 mg/g using NZ, and 93.52 and 82.64 mg/g for Cd (II) and 97.73 and 91.17 mg/g for Pb (II), using Fe-NZ, respectively, and their values were very similar with the experimentally obtained adsorption capacity of

TABLE 7.2(a)

Comparison of the First- and Second-Order Adsorption Rate Constants and Calculated and Experimental q_e Values of Cd (II) and Pb (II) Ions Using NZ

	First-Order Kinetic Model			Second-Order Kinetic Model	
Heavy Metal	**q_e, exp (mg/g)**	**q_e, cal (mg/g)**	**R^2**	**q_e, cal (mg/g)**	**R^2**
Pb (II)	95.10	38.88	0.75	98.12	0.98
Cd (II)	94.27	40.96	0.81	94.04	0.95
Pb (II)/[Pb (II) + Cd (II)]	91.46	36.38	0.69	93.57	0.91
Cd (II)/[Pb (II) + Cd (II)]	87.05	30.20	0.67	89.42	0.93

Note: exp and cal stand for experimental and calculated, respectively.

TABLE 7.2(b)

Comparison of the First- and Second-Order Adsorption Rate Constants and Calculated and Experimental q_e Values of Cd (II) and Pb (II) Ions Using Fe-NZ

	First-Order Kinetic Model			Second-Order Kinetic Model	
Heavy Metal	**q_e, exp (mg/g)**	**q_e, cal (mg/g)**	**R^2**	**q_e, cal (mg/g)**	**R^2**
Pb (II)	99.30	48.88	0.72	97.73	0.97
Cd (II)	96.02	43.96	0.71	93.52	0.94
Pb (II)/[Pb (II) + Cd (II)]	94.21	40.38	0.63	91.17	0.89
Cd (II)/[Pb (II) + Cd (II)]	89.32	36.20	0.58	82.64	0.83

Note: exp and cal stand for experimental and calculated, respectively.

Cd (II) and Pb (II) ions. In addition, the experimental data were well fitted by the pseudo-second order kinetic model, with high determination coefficient (R^2) values ranging from 0.98 to 0.83 in both the single- and mixed-cation systems. These results suggest that the adsorption of Cd (II) and Pb (II) ions by NZ and the Fe-NZ followed the second-order kinetic model rather than the first-order model [29,30].

However, the adsorption efficiencies of cadmium and lead cations were significantly decreased in the mixed-cation systems compared to those in the single-cation systems because of the adsorption competition between Cd (II) and Pb (II) ions for the limited active sites on the adsorbent surface in the mixed-cation system.

7.4.1.4 Effect of Initial Concentration

Figure 7.9(a) & 7.9(b) shows the adsorption capacity of Cd (II) and Pb (II) ions in the single- and mixed-cation systems onto NZ and Fe-NZ as a function of the initial metal concentrations.

At the low and medium initial concentrations (10–500 mg/L), a large surface area and many active sites were available for the adsorption of heavy metals. Thus the adsorption capacities of Cd (II) and Pb (II) ions in both single- and mixed-cation systems were increased with increasing initial metals concentrations.

As the initial concentrations increased up to 800 mg/L, the adsorption capacity of Cd (II) and Pb (II) ions in the single-cation systems slightly improved to approach almost constant values. This was attributed to the occupation by Cd (II) or Pb (II) ions of all the active sites that are necessary for further adsorption. Further increases in the concentration of Cd (II) or Pb (II) ions to a given amount of NZ or Fe-NZ did not further increase Cd (II) or Pb (II) ions adsorption. The adsorption capacities of Pb (II) ions in the single- and mixed-cation systems at an initial concentration of 1000 mg/L were 224.26 and 209.44 mg/g using NZ, and 278.86 and 236.48 mg/g using Fe-NZ, respectively. However, the adsorption capacities of Cd (II) ions in the single- and mixed-cation systems were 166.29 and

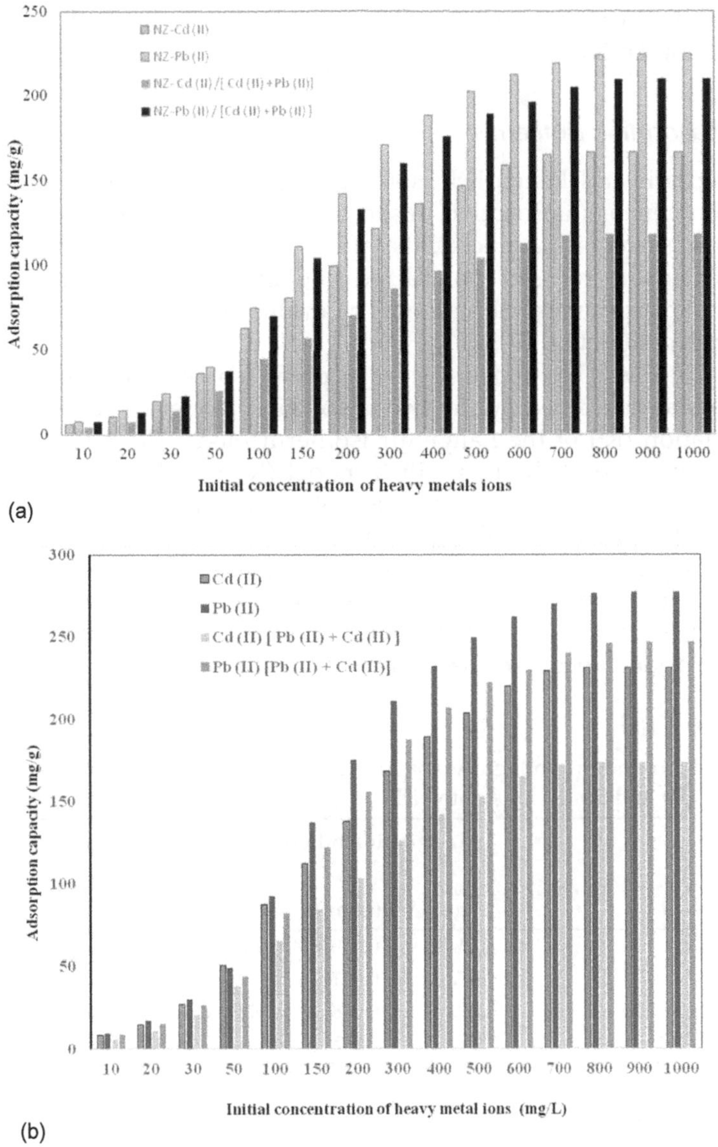

FIGURE 7.9
Effect of initial heavy metals concentration using (a) NZ and (b) Fe-NZ.

117.84 mg/g using NZ, and 230.96 and 173.29 mg/g using Fe-NZ, respectively. The greater adsorption capacity of Pb (II) than that of Cd (II) was due to the greater competitive power of Pb (II) ions for the adsorption sites. On the other hand, Fe-NZ showed much higher adsorption capacity for Cd (II) and Pb (II) in single and mixture-cation systems than those of NZ. The result indicated that the addition of iron ions to nanozeolite was helpful for the adsorption of heavy metals, which may be due to the hydroxyl groups generated on Fe-NZ.

7.4.1.5 Adsorption Isotherms

The Langmuir and Freundlich isotherms corresponding to homogeneous and heterogeneous adsorption, respectively, were used to investigate the adsorption systems. In general, the affinity and surface properties of the particular adsorbent used are utilized to distinguish the main adsorption regime or adsorption isotherm model by evaluating values of the respective adsorption parameters [31,32].

Table 7.3(a) & 7.3(b) shows the adsorption isotherm results with their experimental data fitted by the two adsorption models. The coefficients of determination (R^2) of the Langmuir isotherm model (R^2 = 0.99-0.92) were higher than those of the Freundlich model (R^2 = 0.81-0.64). Based on the Langmuir equation, the maximum adsorption capacities of Cd (II) and Pb (II) ions were 167.68 and 224.26 mg/g using NZ, and 231.82 and 279.96 mg/g using Fe-NZ in the single-cation systems, respectively. In the mixed-cation or coexisting system, they were 117.82 and 173.68 mg/g for Cd (II) ions, and 209.45 and 236.94 mg/g for Pb (II) ions, using NZ and Fe-NZ, respectively. The adsorption capacity of heavy metals in the mixed-cation system was

TABLE 7.3(a)

Langmuir and Freundlich Adsorption Isotherms for the Adsorption of Cd (II) and Pb (II) Ions onto NZ

	Langmuir Isotherm			Freundlich Isotherm		
Adsorbed	**q_{max} (mg/g)**	**b(L/mg)**	**R^2**	**1/n**	**K_f ((mg/g) $(L/mg)^{1/n}$)**	**R^2**
Pb (II)	224.26	0.013	0.96	0.22	4.32	0.73
Cd (II)	167.68	0.021	0.97	0.41	6.1 8	0.68
Pb (II)/[Pb (II) + Cd (II)]	209.45	0.048	0.92	0.34	5.04	0.72
Cd (II)/[Pb (II) + Cd (II)]	117.82	0.057	0.94	0.25	6.52	0.76

TABLE 7.3(b)

Langmuir and Freundlich Adsorption Isotherms for the Adsorption of Cd (II) and Pb (II) Ions onto Fe-NZ

	Langmuir Isotherm			Freundlich Isotherm		
Adsorbed	**q_{max} (mg/g)**	**b(L/mg)**	**R^2**	**1/n**	**K_f ((mg/g) $(L/mg)^{1/n}$)**	**R^2**
Pb (II)	279.96	0.026	0.99	0.28	8.02	0.81
Cd (II)	231.82	0.018	0.98	0.15	5.39	0.72
Pb (II)/[Pb (II) + Cd (II)]	236.94	0.023	0.95	0.22	6.24	0.76
Cd (II)/[Pb (II) + Cd (II)]	173.68	0.016	0.92	0.14	4.97	0.64

TABLE 7.3(c)

Comparison of the Adsorption Capacity of Heavy Metals Among Various Adsorbents

Adsorbents Fe-Nano	Adsorption Capacity of Pb (II) (mg/g)	Adsorption Capacity of Cd (II) (mg/g)
Zeolite (Australia)	9.9	6.7
Iron- zeolite	11.1	7.2
ZSM-5 zeolite	20.1	12.3
Coconut fiber	–	31.1
Sulfur-functionalized silica	–	30.7
Polyving alcohol/polyacrylic acid	62.2	115.8
Polyving alcohol/polyacrylic acid	194.9	6.6
Mesoporous carbon stabilized alumina (MC/Al_2O_3)	9.1	249.8
AC- Fe_3O_4 –NH_2.	335.5	–
Fe_3O_4 nanomaterials	104.2	–
Magnetic biochar composite	36	–
Ethylenediamine functionalized AC	25.3	–
GO	360	–
CNT	217.6	
AC	302	121.6
Fe nano particle	138.3	158.4
α-Fe_2O_3 nanoparticles coated volcanic rocks	–	127.2
AC	–	158
Fe-nano zeolite	–	231.8

lower than that in the single-cation system. The reduced adsorption capacity in the mixed-cation or coexisting ion system was due to the increased competition by the coexisting counterpart metal ions. However, the smaller hydrated ionic radius and stronger electronegativity of Pb (II), compared to those of Cd (II), were relatively less influential in the adsorption reduction due to the coexisting counter metal ions. On the other hand, we compared the adsorption capacity of Pb (II) and Cd (II) ions among various adsorbents (see in Table 7.3(c)). The result revealed that Fe-nanozeolite had a higher adsorption capacity of heavy metals as compared with the other adsorbents.

These findings support the promise of Fe-nanozeolite as a superior adsorbent for heavy metals removal.

7.4.1.6 Regeneration of Heavy Metal-Loaded Adsorbents

Adsorbent reusability is an important characteristic for proper operation of adsorbents for industrial applications. In addition, after an adsorbate is loaded onto a selected sorbent, the sorbent regeneration characteristic based on sorption-desorption performance can assist in determining the basic field

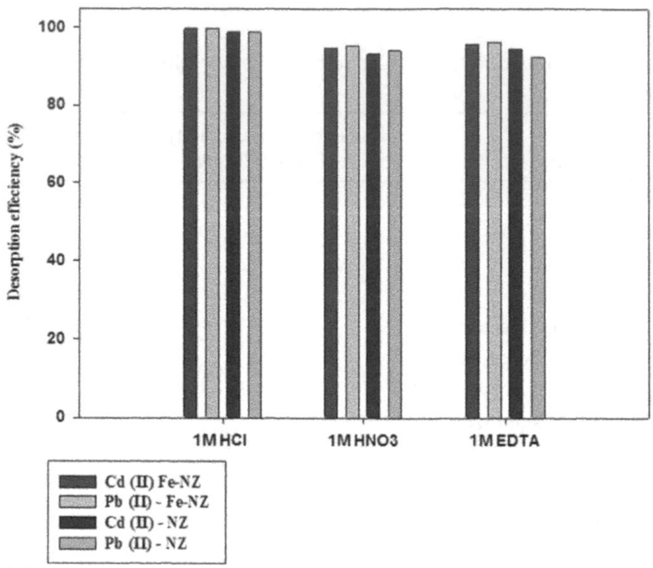

(a)

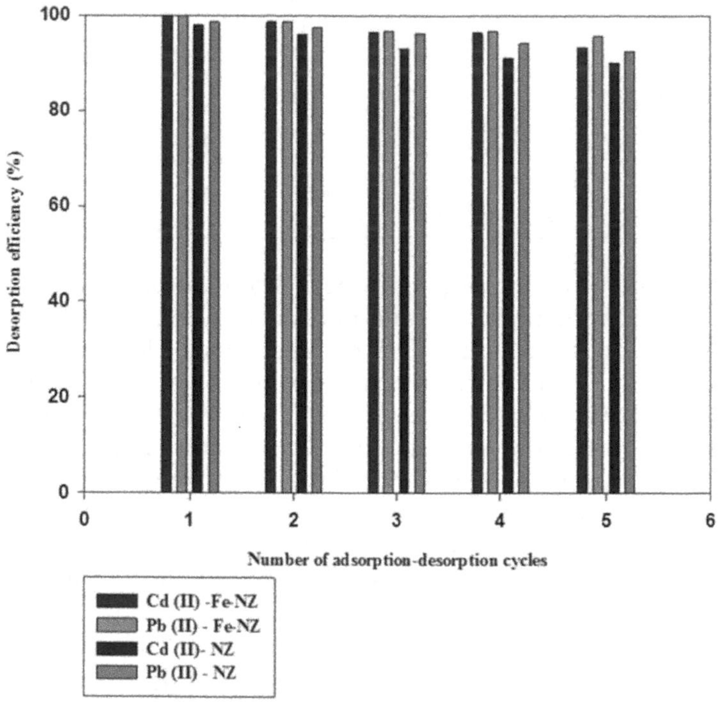

(b)

FIGURE 7.10
(a) Effect of eluents in the desorption of Cd (II) and Pb (II) ions and (b) desorption efficiency of heavy metals-loaded NZ and Fe-NZ by 1M HCl.

application feasibility and can provide useful data for selecting an appropriate sorbent for the target metal. Before the reusability experiment, several desorption reagents, including 1M HCl, 1M HNO_3, and 1M EDTA, were investigated after the heavy metals were loaded onto NZ and Fe-NZ. Among the three tested desorption reagents, 1M HCl was the best eluent with the highest desorption efficiency of around 100% towards Cd (II) and Pb(II) ions (Figure 7.10(a) & 7.10(b)). Even after the adsorption-desorption cycles were repeated five times, the desorption efficiencies for Cd (II) and Pb (II) ions were not significantly decreased and the adsorption capacities of Cd (II) and Pb (II) were decreased by only 8.12 and 6.38% from Cd (II) and Pb (II)-loaded Fe-NZ, and 8.57 and 10.26% from Cd (II) and Pb (II)-loaded NZ, respectively, as compared to the original adsorption capacity. This indicates that 1M HCl was the best desorption reagent for Pb (II)- or Cd (II)-loaded Fe-NZ and NZ. This result also suggests that NZ and Fe-NZ are stable adsorbents even after 5 adsorption-desorption cycles without significant loss of adsorption capacity of Cd (II) or/and Pb (II).

7.4.1.7 Proposed Removal Mechanism

The pH on the metal solution can affect not only the activity of the functional groups contained in the sorbent but also the metal species in the solution. Therefore, it is very important to assess the adsorption behavior of Cd (II) and Pb(II) ions by Fe-NZ under different pH conditions [33].

In this study we propose a mechanism for the removal of Cd (II) and Pb (II) ions at optimum pH 6.0 for the adsorption process.

At pH 6.0, the adsorbent surface is negatively charged and heavy metals ions exist in the form of Cd (II), Cd $(OH)^+$, Pb (II), and $Pb(OH)^+$ cations. The adsorption of Cd (II) and Pb (II) ions by Fe-NZ is followed by both physical and chemical sorption. Physisorption proceeds via adsorption by surface area, active sites, van der Waals forces, and electrostatic outer-sphere complexes (ion-exchange), whereas chemisorption proceeds via short-range interactions that include inner-sphere complexation (ligand ex-change and covalent bonding). The surface functional group plays a role in the adsorption process [34,35]. The major surface functional groups in Fe-NZ are OH groups coordinated to Si^{4+} and Al^{3+} and also FeOH with the O coordinated with Fe^{3+} ions. The Cd (II) and Pb (II) ions can interact with the surface functional groups of Fe-NZ by complexation. This complexation can be followed by outer- and inner-sphere complexes (see in Figure 7.11). Outer-sphere complexes involve electrostatic interactions, which are weak compared to inner-sphere complexes in which the binding is covalent. Adsorption by outer-sphere complexation occurs strongly at pH 6.0 because the surface of Fe-NZ is negatively charged and the adsorbed Cd (II) and Pb (II) ions are positively charged. However, the adsorption kinetic and adsorption isotherm results imply that Cd (II) and Pb (II) ions can also be removed by inner-sphere complexation with bonding of oxygen-Cd (II)/or Pb (II) ions.

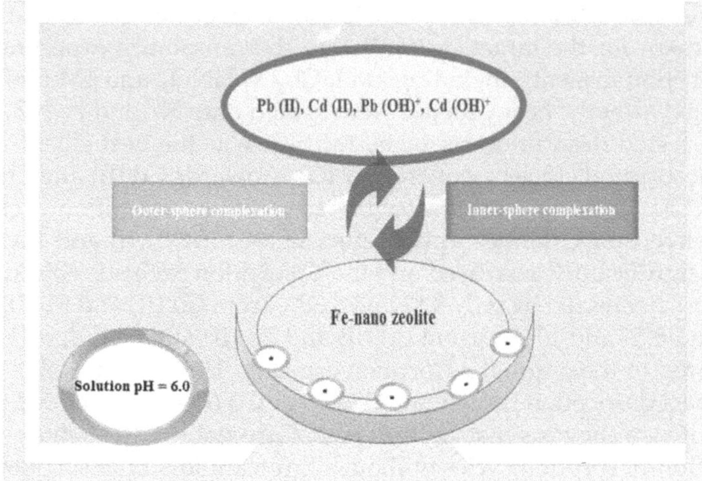

FIGURE 7.11
Proposed sorption mechanism for the removal of Cd (II) and Pb (II) ions by Fe-NZ.

7.4.1.8 Adsorption Cost Estimation

The adsorption cost for Cd (II) and Pb (II) removal using Fe-NZ was calculated based on the heavy metal adsorption capacity and adsorbent preparation cost, as shown in Equation 7.3:

$$\text{Adsorption cost}_{(\text{US\$/g})} = \frac{\left[\text{Chemical purchase cost}\left(\frac{\text{US\$}}{\text{g}}\right) + \text{Energy cost}\left((\text{US\$})\frac{\text{Kw}}{\text{g}}\right)\right]}{\text{Adsorption capacity}\left(\frac{\text{mg}}{\text{g}}\right)\times 10-3\left(\frac{\text{g}}{\text{mg}}\right)} \tag{7.3}$$

The preparation costs of NZ and Fe-NZ were simply estimated based on the purchase costs of their precursor chemicals in Korea and the energy cost required for synthesizing the adsorbent. The chemicals costs were based on the purchase prices of $Fe(NO_3)_3$, NaOH, H_2O, $NaAlO_2$, Al_2O_3, and $Si(OC_2H_5)_4$ in Korea. The energy cost required for synthesizing Fe-NZ was based on the electricity cost used for heating and drying during the synthesis processes in Korea [4, 23, and 28]. These simple adsorption cost estimates for the removal of a unit of mass of Cd (II) and Pb (II) ions in water using NZ and Fe-NZ are shown in Table 7.4(a) & 7.4(b) for single- and mixed-metal systems. Even though the two heavy metals or both [Pb (II) or/and Cd (II)] and an organic pollutant (phenol) existed simultaneously in the system, Fe-NZ could be applied for the removal of contaminants. These results support the promising potential

TABLE 7.4(a)

Adsorption Cost for the removal of Cd (II) and Pb (II) Ions Using NZ

Heavy Metal	Adsorption Capacity (mg/g)	Preparation Cost (US$/g)	Adsorption Cost (US$/g)
Cd (II)	167.29	0.009	0.053
Pb (II)	224.26	0.009	0.040
Cd (II) [Pb (II) + Cd (II)]	117.8	0.009	0.076
Pb (II) [Pb (II) + Cd (II)]	209.45	0.009	0.042

TABLE 7.4(b)

Adsorption Cost for the Removal of Cd (II) and Pb (II) Ions Using Fe-NZ

Heavy Metal	Adsorption Capacity (mg/g)	Preparation Cost (US$/g)	Adsorption Cost (US$/g)
Cd (II)	231.82	0.0118	0.050
Pb (II)	279.96	0.0118	0.042
Cd (II) [Pb (II) + Cd (II)]	173.68	0.0118	0.067
Pb (II) [Pb (II) + Cd (II)]	236.94	0.0118	0.049

of Fe-NZ as a green and high capacity adsorbent for the removal of heavy metals and organic pollutants at low-cost.

7.4.2 Removal of Phenolic Compounds

7.4.2.1 Effect of Solution pH

As already mentioned in the previous section, the solution pH strongly affects the adsorption process between the adsorbent surfaces and the adsorbate (phenolic compounds) [18]. The removal efficiencies were increased from 31 to 86.4% for Ph, 32.7 to 89.8% for 2-CP, and 41.7 to 97.2% for 2-NP as the solution pH was increased from 2.0 to 5.0. The sorption removal peaked at pH 5.0 and then decreased by 58.18, 27.83, and 32.38% at pH 10 for Ph, 2-CP, and 2-NP, respectively (Figure 7.12). Fe-NZ showed pH_{pzc} at pH 5.03 ± 0.1. In acidic solution (pH ≤ 5.0), increasing pH decreased the hydronium ion concentration. Thus more electrostatic attraction can be developed between the positively charged surface of Fe-NZ and the negative charge of the phenolic compounds, thereby increasing the removal efficiencies. The surface of the adsorbent starts to build a partial negative charge as the pH exceeds the pH_{pzc} of pH 5.03. Therefore, increasing the solution pH strengthens the repulsion on the surface of Fe-NZ with the increased number of hydroxyl anions. However, the pKa values of Ph, 2-CP, and 2-NP were 9.3, 8.5, and 7.2, respectively. When the pH increased above their pKa, the phenolic compounds existed in anionic form with greater negative charge, which increased the repulsion with the anionic charge of the adsorbent and thus decreased the removal efficiencies. In particular, as the

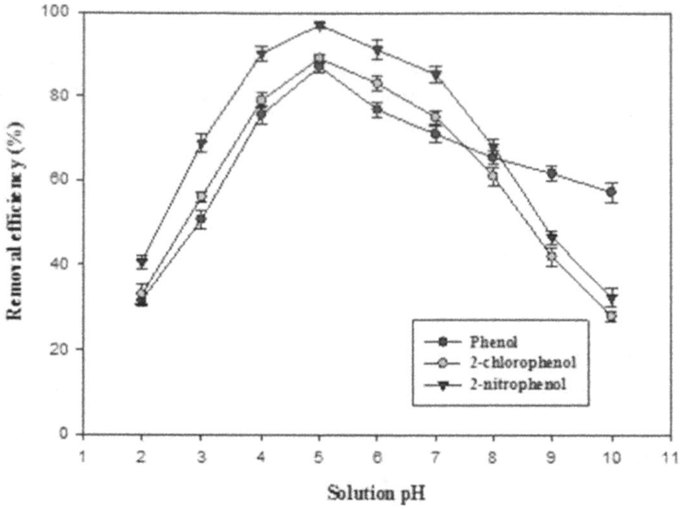

FIGURE 7.12
Effect of solution pH for the removal of phenolic compounds.

pH exceeded pH 8.5, the removal efficiencies of 2-CP and 2-NP, which existed in anionic form, were sharply decreased [32,33]. However, Ph, with relatively more positive ions than 2-CP and 2-NP in the alkaline pH range (pH > 9.0), suffered a smaller decrease in the removal efficiency (Figure 7.12).

7.4.2.2 Effect of Contact Time

Adsorption experiments were performed in 100 mL of Ph, 2-CP, and 2-NP solution of 200 mg/L with contact time varying from 10 to 300 min. Figure 7.13 shows the effect of different reaction times on the adsorption of Ph, 2-CP, and 2-NP onto Fe-NZ. The adsorption initially proceeded rapidly and was almost completed after 230 min, when adsorption equilibrium was almost attained. The removal efficiencies at equilibrium of Ph, 2-CP and 2-NP using Fe-NZ were 82.6, 91.8 and 97.9%, respectively.

The Ph, 2-CP, and 2-NP adsorption rates were initially high because more vacant or active sites are initially available for the phenolic compound adsorption [36]. Subsequently, however, the remaining vacant surface sites of the adsorbent were not easily available for further Ph, 2-CP, and 2-NP removal due to blocking and the repulsion by the molecules already adsorbed on the surface of Fe-NZ. Thus, 230 min was chosen as the optimum contact time for further adsorption experiments.

7.4.2.3 Effect of Initial Concentration

Figure 7.14 shows the adsorption capacities of Ph, 2-CP, and 2-NP using Fe-NZ as a function of the initial concentrations. At low initial concentrations less

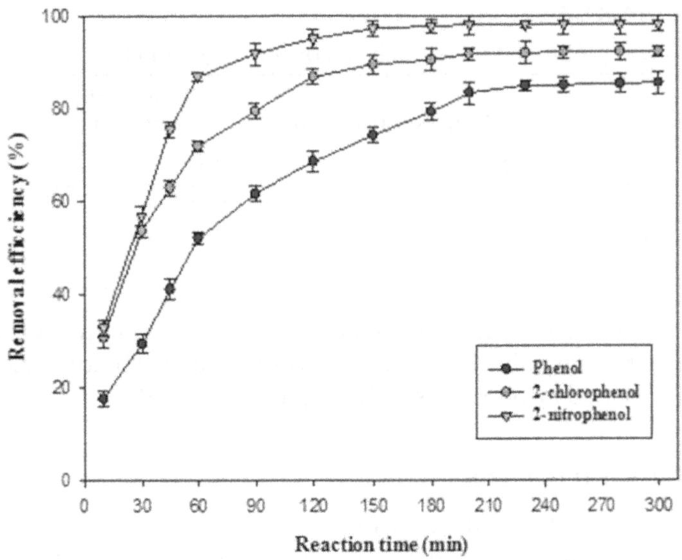

FIGURE 7.13
Effect of reaction time on the removal of phenolic compounds.

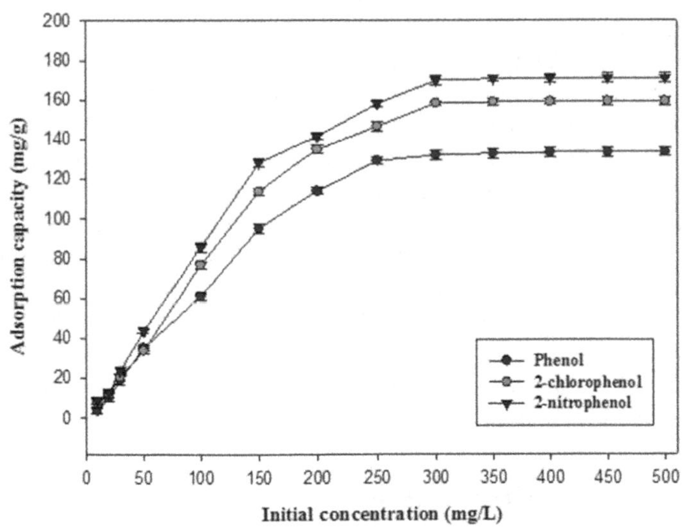

FIGURE 7.14
Effect of initial concentration on the removal of phenolic compounds.

than 250 mg/L, more unoccupied surface areas or active sites were available for the removal of phenolic compounds. The increase in initial concentrations increased the adsorption capacity. At initial concentrations above 300 mg/L, any further increase only slightly increased their adsorption capacities. This can be explained because all active sites for further adsorption were almost

TABLE 7.5
Langmuir and Freundlich Adsorption Isotherms

	Langmuir Isotherm			Freundlich Isotherm		
Adsorbed	Q_{max} **(mg/g)**	**b**	R^2	**1/n**	K_f	R^2
Ph	138.7	4.38	0.98	0.23	64.2	0.86
2-CP	158.9	5.41	0.97	0.21	65.8	0.87
2-NP	171.2	8.32	0.99	0.14	73.1	0.79

occupied or saturated by the adsorbates [37–40]. This greatly weakened the force driving the phenolic compounds toward the active binding sites; therefore, any further adsorption was only slightly increased.

7.4.2.4 Adsorption Isotherm

The Langmuir and Freundlich adsorption isotherms models were used to simulate the adsorption systems. Table 7.5 shows the adsorption isotherm results with their data fitting to the two adsorption models. The coefficients of determination (R^2) of the Langmuir isotherm model were much higher (R^2 = 0.99-0.97) than those of the Freundlich model (R^2 = 0.87-0.79). Based on the Langmuir equation, the maximum adsorption capacities of Ph, 2-CP, and 2-NP were 138.7, 158.9 and 171.2 mg/g, respectively. The adsorption capacities increased in this order due to the different properties of dipole moment and functional groups (OH, Cl, and NO_2) of Ph, 2-CP, and 2-NP. The dipole moment of 2-NP (3.01) is much larger than that of 2-CP (1.13) and Ph (1.22). Thus 2-NP has a stronger electrostatic attraction to Fe-NZ, which gives it the highest adsorption capacity among the three phenolics. The adsorption capacity of 2-CP was also higher than that of Ph because 2-CP has one more functional group (Cl), which has more electronegativity and is thus available for further interaction with the positive charged components in Fe-NZ (hydrogen in OH groups and Fe ions), as compared Ph, which increases the adsorption capacity [40–43]. However, we compared the adsorption capacity of phenolic compounds using Fe-nanozeolite and other adsorbents (Table 7.6(a)–(c)).

7.4.2.5 Regeneration of Loaded Adsorbents

In order to minimize the costs of the adsorption process, the adsorbed Ph, 2-CP, and 2-NP should be desorbed and then the adsorbent (Fe-NZ) should be regenerated for further cycles of adsorption-desorption processes [44,45]. Figure 7.15 shows the desorption efficiencies of Ph, 2-CP, and 2-NP from these phenolic compound-loaded Fe-NZ samples. Even after 10 adsorption-desorption cycles, the desorption efficiencies of Ph, 2-CP, and 2-NP were 46.2, 50.9, and 55.7%, respectively. These high reusability rates and facile

TABLE 7.6(a)

Comparison of Phenol Adsorption Capacity among Various Adsorbents

Adsorbents	**Adsorption Capacity of Phenol (mg/g)**
Porous Clay	15.4
Activated carbon power and granular from Eucalyptus Wood	60.0
Natural Zeolitcs	32.63
Zeolite	74.6
Polymeric Adsorbent	123.20
Bentonite and Caloinite	25.57
Polymeric adsorbents with amide groups	198.87
N-methylacetamide-modified hypercrosslinked resin	160.0
Power activated carbon treatment (PACT)	135.7
Power activated carbon (PAC)	32.2
Dried activated sludge	86.1
Fe-nano zeolite	**138.7**

TABLE 7.6(b)

Comparison of 2-Chlophenol Adsorption Capacity among Various Adsorbents

Adsorbents	**Adsorption Capacity of 2-Chlorophenol (mg/g)**
Paper mill sludge	**0.34**
Red mud	**17.3**
Rice husk	**14.4**
Activated Carbon	**102.4**
Rice husk char	**36.2**
Activated carbon fibers (ACFs)	**141**
Surfactant-modified bentonite	**50**
Graphene-zirconiumoxide nanocomposite	**140**
Surfactant-modified natural zeolite	**12.7**
amino-modified ordered mesoporous silica	**400**
Bagasse Fly Ash (BFA)	**52.52**
Hydrothermal treatment BFA	**78.80**
Fusion treatment BFA	**85.65**
Fe-nano zeolite	**158.9**

TABLE 7.6(c)

Comparison of 2-Nitrophenol Adsorption Capacity among Various Adsorbents

Adsorbents	Adsorption Capacity of 2-Nitrophenol (mg/g)
Fe/Zn -AC	**157.89**
Fe-AC	**122.2**
Zn- AC	**117.8**
Activated jute stick char	**39.38**
Sugar fly ash	**0.76–1.15**
Wood fly ash	**134.9**
Fly ash	**7.80–9.68**
Samla coal	**51.54**
Rice husk	**15.31**
Rice husk char	**39.21**
Petroleum coke	**11.06**
Coke breeze	**4.64**
Bagasse fly ash	**8.3**
Zeolite	**41.02**
Pyrolysed oil shale	**4.895**
$ZnCl_2$ pyrolysed oil shale	**6.026**
KOH pyrolysed oil shale	**0.895**
Fe-nano zeolite	**171.2**

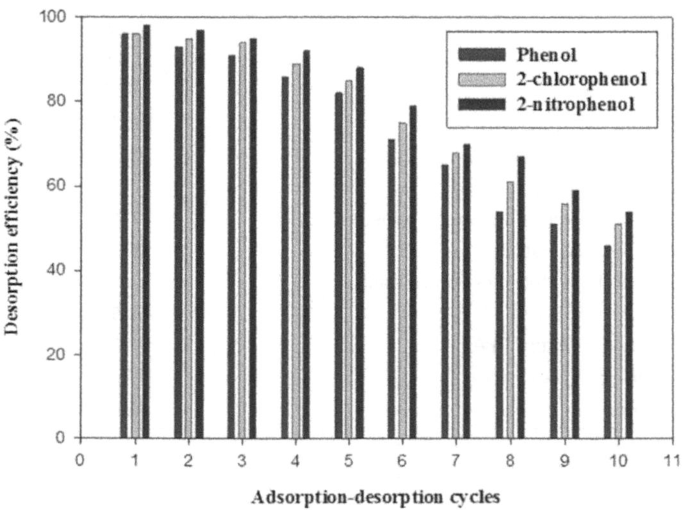

FIGURE 7.15
Regeneration of adsorbent.

regeneration characteristics support the potential use of Fe-NZ as an adsorbent for the removal of phenolic compounds from aqueous solution.

7.4.2.6 Adsorption Mechanisms

The adsorption mechanism for Ph, 2-CP, and 2-NP is dependent on the physical and chemical characteristics of the adsorbent and the adsorbate. In addition, the removal efficiency of organic pollutants was highly dependent on the solution pH. In the weakly acidic condition ($4.0 \leq pH \leq 7.0$), the adsorbates mostly exist in their molecular forms because the pKa values of Ph, 2-CP, and 2-NP were 9.3, 8.5, and 7.2, respectively. Thus their removal mechanism includes both physisorption and chemisorption.

1. Physisorption proceeds via adsorption by surface areas, active sites, and van der Waals forces between the adsorbates and adsorbent. Fe-NZ has a high surface area, numerous pores, and active sites. Thus the adsorbates near the surface and the active sites of Fe-NZ can get encapsulated into the pores of the adsorbent, leading to physisorption.
2. Chemisorption proceeds via short-range interactions through additional complexation between H^+, Si^{4+}, Al^{3+}, and Fe^{3+} ions on Fe-NZ and the functional groups such as (OH, Cl, and NO_2) on the Ph, 2-CP and 2-NP molecules (Figure 7.16).

In the basic condition, both the adsorbate and adsorbent have more negative charges, and this increased electrostatic repulsion decreases the adsorption

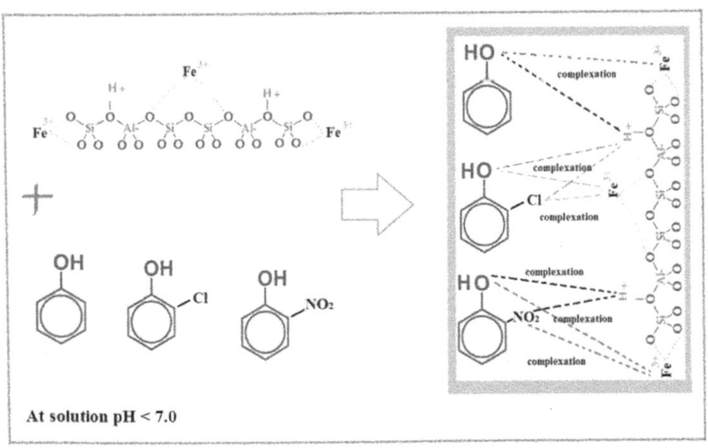

FIGURE. 7.16
Removal mechanism of phenolic compounds.

TABLE 7.7

Comparison of Adsorption Cost Required Using Fe-Nanozeolite and AC to Treat Ph, 2-CP, and 2-NP in Wastewaters

		Adsorption Cost Required (US$/g)		
Adsorbent	**Preparation Cost (US$/g)**	**Ph**	**2-CP**	**2-NP**
Fe-nanozeolite	**0.12**	**0.86**	**0.75**	**0.70**
AC	0.23	1.84	1.75	1.68

capacity. In the strongly acidic condition (pH ≤ 3.0), the greater number of hydronium ions produced can compete with the phenolic compounds for the adsorbent's adsorption sites, which further decreases their removal efficiency (as described above in the Effect of pH section).

7.4.2.7 Adsorption Cost Estimation

The simply estimated adsorption costs, based on adsorbent precursor purchase and preparation costs in Korea and their adsorption capacity, required for the removal of a unit weight of Ph, 2-CP, and 2-NP using Fe-NZ were 0.86, 0.75, and 0.70 US$/g, respectively, which were much lower than those (1.81, 1.75, and 1.68 US$/g) by using activated carbon (AC) (Table 7.7). However, the adsorption costs required to treat Ph, 2-CP, and 2-NP at 425, 327, and 368 mg/L concentrations using Fe-NZ were 246.5, 134.1, and 128.8 US$/1000 kg–wastewaters, respectively, and were much lower than those using AC, which were 782, 549, and 485.7 US$/1000 kg-wastewaters. In addition, Fe-NZ can be reused even after 10 regeneration cycles, which further reduces its total treatment costs for wastewaters containing Ph, 2-CP, and 2-NP, as compared to AC.

7.5 Summary

This chapter has demonstrated the potential for using nanozeolite and Fe-nanozeolite as promising, efficient, economic, and environmentally friendly adsorbents for the removal of aqueous contaminants. Based on the experimental results presented herein, the following points can be concluded:

1. The SEM, TEM, BET, XRD, and FTIR analyses confirmed the successful loading of iron onto the nanozeolite, which is essential for improving the environmental properties of nanozeolite.

2. The removal of heavy metals was optimized at pH = 6.0 and the maximum adsorption capacities were 279.96 and 236.94 mg/g for Pb (II), and 238.82 and 173.68 mg/g for Cd (II), in single- and mixed-cation systems using Fe-NZ, respectively. In the regeneration of Cd (II) and Pb (II) ion-loaded NZ and Fe-NZ, even after 5 adsorption-desorption cycles, the desorption efficiencies remained as high as 90.4–95.68% and 92.8–97.72%, respectively. The major removal mechanisms for both heavy metals included outer- and inner-sphere complexes. The adsorption costs required for removing metal ions using Fe-NZ were 0.050 and 0.042 US$/g for Pb (II), and 0.067 and 0.049 US$/g for Cd (II), in the single- and mixed-cation systems, compared to 0.040 and 0.042 US$/g for Pb (II), and 0.053 and 0.076 US$/g for Cd (II) when using NZ, respectively. When nanozeolite and Fe-nanozeolite were applied to remove phenolic compounds, Fe-NZ showed maximum adsorption capacities of 138.5, 158.9, and 171.2 mg/g for Ph, 2-CP, and 2-NP, respectively. Fe-NZ also showed high reusability with a desorption efficiency remaining above 46% even after 10 adsorption-desorption cycles. Physisorption proceeds via adsorption by surface areas, active sites, and van der Waals forces, whereas chemisorption proceeds via short-range interactions arising from additional complexation between H^+, Si^{4+}, Al^{3+}, and Fe^{3+} ions on Fe-NZ and the functional groups such as OH, Cl, and NO_2 on the Ph, 2-CP, and 2-NP molecules. This combination of physisorption and chemisorption formed the basic mechanism responsible for the removal of phenolic compounds using Fe-nanozeolite.
3. The cost for the adsorption treatment of wastewaters containing the three tested phenolic compounds was much lower using Fe-NZ than that using AC. Along with this low cost, the technical and economic feasibility, environmental benefits, high adsorption capacity, easy recovery, and eco-friendly characteristics of nanozeolite and Fe-nanozeolite support their commercial application as promising and green adsorbents for improved removal of aqueous pollutants.

Acknowledgments

This work (C0453909) was supported by Business for Cooperative R&D between Industry, Academy, and Research Institute funded Korea Small and Medium Business Administration in 2017.

References

1. Abrishamkar, M., and Izadi, A. (2013) Nano-ZSM-5 zeolite: Synthesis and application to electrocatalytic oxidation of ethanol. *Microporous Mesoporous Mat.* 180: 56–60.
2. Pham, T. H., Lee, B. K., Kim, J., and Lee, C. H. (2016) Enhancement of CO_2 capture by using synthesized nano-zeolite. *Journal of the Taiwan Institute of Chemical Engineers* 64: 220–226.
3. Larsen, S. C. (2007) Nanocrystalline zeolites and zeolite structures: Synthesis, characterization, and applications. *J. Phys. Chem. C* 111(50): 18464–18474.
4. Yan, Y., Jiang, S., Zhang, H., and Zhang, X. (2015) Preparation of novel Fe-ZSM-5 zeolite membrane catalysts for catalytic wet peroxide oxidation of phenol in a membrane reactor. *Chem. Eng. J.* 259: 243–251.
5. Moamen, O. A., Ismail, I. M., Abdelmonem, N., and Rahman, R. A. (2015) Factorial design analysis for optimizing the removal of cesium and strontium ions on synthetic nano-sized zeolite. *Journal of the Taiwan Institute of Chemical Engineers* 55: 133–144.
6. Abrishamkar, M., and Izadi, A. (2013) Nano-ZSM-5 zeolite: Synthesis and application to electrocatalytic oxidation of ethanol. *Microporous Mesoporous Mat.* 180: 56–60.
7. Abramova, A. V., Slivinskii, E. V., Goldfarb, Y. Y., Panin, A. A., Kulikova, E. A., and Kliger, G. A. (2005) Development of efficient zeolite-containing catalysts for petroleum refining and petrochemistry. *Kinet. Catal.* 46(5): 758–769.
8. Marcilly, C. R. (2000) Where and how shape selectivity of molecular sieves operates in refining and petrochemistry catalytic processes. *Top. Catal.* 13(4): 357–366.
9. Olsbye, U., Svelle, S., Bjørgen, M., Beato, P., Janssens, T. V., Joensen, F., Bordiga, S., and Lillerud, K. P. (2012) Conversion of methanol to hydrocarbons: How zeolite cavity and pore size controls product selectivity. *Angew. Chem. Int. Ed.* 51(24): 5810–5831.
10. Choi, M., Na, K., Kim, J., Sakamoto, Y., Terasaki, O., and Ryoo, R. (2009) Stable single-unit-cell nanosheets of zeolite MFI as active and long-lived catalysts. *Nature* 461(7261): 246–252.
11. Climent, M. J., Corma, A., Iborra, S., and Primo, J. (1995) Base catalysis for fine chemicals production: Claisen-Schmidt condensation on zeolites and hydrotalcites for the production of chalcones and flavanones of pharmaceutical interest. *J. Catal.* 151(1): 60–66.
12. Akhigbe, L., Ouki, S., and Saroj, D. (2016) Disinfection and removal performance for Escherichia coli and heavy metals by silver-modified zeolite in a fixed bed column. *Chem. Eng. J.* 295: 92–98.
13. Rivera-Garza, M., Olguın, M. T., Garcıa-Sosa, I., Alcántara, D., and Rodrıguez-Fuentes, G. (2000) Silver supported on natural Mexican zeolite as an antibacterial material. *Microporous Mesoporous Mat.* 39(3): 431–444.
14. Valkaj, K. M., Katovic, A., and Zrncevic, S. (2011) Catalytic properties of Cu/13X zeolite based catalyst in catalytic wet peroxide oxidation of phenol. *Ind. Eng. Chem. Res.* 50(8): 4390–4397.

15. Wang, J., Park, J. N., Jeong, H. C., Choi, K. S., Wei, X. Y., Hong, S. I., and Lee, C. W. (2004) Cu^{2+}-exchanged zeolites as catalysts for phenol hydroxylation with hydrogen peroxide. *Energy Fuels* 18(2): 470–476.
16. Kwakye-Awuah, B., Williams, C., Kenward, M. A., and Radecka, I. (2008) Antimicrobial action and efficiency of silver-loaded zeolite X. *J. App. Microbiol.* 104(5): 1516–1524.
17. Song, H., Wan, X., Dai, M., Zhang, J., Li, F., and Song, H. (2013) Deep desulfurization of model gasoline by selective adsorption over Cu–Ce bimetal ion-exchanged Y zeolite. *Fuel Process. Technol.* 116: 52–62.
18. Yang, M., Lin, J., Zhan, Y., Zhu, Z., and Zhang, H. (2015) Immobilization of phosphorus from water and sediment using zirconium-modified zeolites. *Environ. Sci. Pollut. Res.* 22(5): 3606–3619.
19. Mhamdi, M., Khaddar-Zine, S., and Ghorbel, A. (2009) Influence of the cobalt salt precursors on the cobalt speciation and catalytic properties of H-ZSM-5 modified with cobalt by solid-state ion exchange reaction. *Appl. Catal. A: General* 357(1): 42–50.
20. Valle, B., Gayubo, A. G., Aguayo, A. T., Olazar, M., and Bilbao, J. (2010) Selective production of aromatics by crude bio-oil valorization with a nickel-modified HZSM-5 zeolite catalyst. *Energy Fuels* 24(3): 2060–2070.
21. Srinivasan, D., Rao, R., and Zribi, A. (2006) Synthesis of novel micro-and mesoporous zeolite nanostructures using electrospinning techniques. *J. Electronic Mat.* 35(3): 504–509.
22. Abrishamkar, M., and Izadi, A. (2013) Nano-ZSM-5 zeolite: Synthesis and application to electrocatalytic oxidation of ethanol. *Microporous Mesoporous Mat.* 180: 56–60.
23. Pham, T. H., Lee, B. K., Kim, J., and Lee, C. H. (2016) Nitrophenols removal from aqueous medium using Fe-nano mesoporous zeolite. *Mat. Des.* 101: 210–217.
24. Mosayebi, A., and Abedini, R. (2014) Partial oxidation of butane to syngas using nano-structure Ni/zeolite catalysts. *J. Ind. Eng. Chem.* 20(4): 1542–1548.
25. Larsen, S. C. (2007) Nanocrystalline zeolites and zeolite structures: Synthesis, characterization, and applications. *J. Phys. Chem. C* 111(50): 18464–18474.
26. Moamen, O. A., Ismail, I. M., Abdelmonem, N., and Rahman, R. A. (2015) Factorial design analysis for optimizing the removal of cesium and strontium ions on synthetic nano-sized zeolite. *Journal of the Taiwan Institute of Chemical Engineers* 55: 133–144.
27. Abrishamkar, M., and Izadi, A. (2013) Nano-ZSM-5 zeolite: Synthesis and application to electrocatalytic oxidation of ethanol. *Microporous Mesoporous Mat.* 180: 56–60.
28. Pham, T. H., Lee, B. K., Kim, J., Lee, C. H., and Chong, M. N. (2016) Acid activation pine cone waste at differences temperature and selective removal of Pb2+ ions in water. *Process Saf. Environ. Prot.* 100: 80–90.
29. Trgo, M., Perić, J., and Medvidović, N. V. (2006) A comparative study of ion exchange kinetics in zinc/lead—Modified zeolite-clinoptilolite systems. *J. Hazardous Mat.* 136(3): 938–945.
30. Yagub, M. T., Sen, T. K., Afroze, S., and Ang, H. M. (2014). Dye and its removal from aqueous solution by adsorption: A review. *Adv. Colloid and Interface Sci.* 209: 172–184.

31. Machida, M., Mochimaru, T., and Tatsumoto, H. (2006) Lead (II) adsorption onto the graphene layer of carbonaceous materials in aqueous solution. *Carbon* 44(13): 2681–2688.
32. Demirbas, A. (2008) Heavy metal adsorption onto agro-based waste materials: A review. *J. Hazardous Mater.* 157(2-3): 220–229.
33. Huang, Y., and Keller, A. A. (2015) EDTA functionalized magnetic nanoparticle sorbents for cadmium and lead contaminated water treatment. *Water Res.* 80: 159–168.
34. Maher, A., Sadeghi, M., and Moheb, A. (2014) Heavy metal elimination from drinking water using nanofiltration membrane technology and process optimization using response surface methodology. *Desalination* 352: 166–173.
35. Chen, T., Liu, F., Ling, C., Gao, J., Xu, C., Li, L., and Li, A. (2013) Insight into highly efficient coremoval of copper and p-nitrophenol by a newly synthesized polyamine chelating resin from aqueous media: Competition and enhancement effect upon site recognition. *Environ. Sci. Technol.* 47(23): 13652–13660.
36. Ngah, W. W., and Hanafiah, M. A. K. M. (2008) Removal of heavy metal ions from wastewater by chemically modified plant wastes as adsorbents: A review. *Bioresource Technol.* 99(10): 3935–3948.
37. Cheng, T. W., Lee, M. L., Ko, M. S., Ueng, T. H., and Yang, S. F. (2012) The heavy metal adsorption characteristics on metakaolin-based geopolymer. *Appl. Clay Sci.* 56: 90–96.
38. Bernard, E., Jimoh, A., and Odigure, J. O. (2013) Heavy metals removal from industrial wastewater by activated carbon prepared from coconut shell. *Res. J. Chem. Sci.* 3: 3–9.
39. Ma, X., Cui, W., Yang, L., Yang, Y., Chen, H., and Wang, K. (2015) Efficient biosorption of lead (II) and cadmium (II) ions from aqueous solutions by functionalized cell with intracellular $CaCO_3$ mineral scaffolds. *Bioresource Technol.* 185: 70–78.
40. Lim, A. P., and Aris, A. Z. (2014) Continuous fixed-bed column study and adsorption modeling: Removal of cadmium (II) and lead (II) ions in aqueous solution by dead calcareous skeletons. *Biochem. Eng. J.* 87: 50–61.
41. González, M. A., Pavlovic, I., and Barriga, C. (2015) Cu (II), Pb (II), and Cd (II) sorption on different layered double hydroxides. A kinetic and thermodynamic study and competing factors. *Chem. Eng. J.* 269: 221–228.
42. Gupta, S. S., and Bhattacharyya, K. G. (2012) Adsorption of heavy metals on kaolinite and montmorillonite: A review. *Phys. Chem. Chem. Phys.* 14(19): 6698–6723.
43. Nguyen, T. C., Loganathan, P., Nguyen, T. V., Vigneswaran, S., Kandasamy, J., and Naidu, R. (2015) Simultaneous adsorption of Cd, Cr, Cu, Pb, and Zn by an iron-coated Australian zeolite in batch and fixed-bed column studies. *Chem. Eng. J.* 270: 393–404.
44. Poursani, A. S., Nilchi, A., Hassani, A. H., Shariat, M., and Nouri, J. (2015) A novel method for synthesis of nano-γ-Al 2 O 3: Study of adsorption behavior of chromium, nickel, cadmium, and lead ions. *Int. J. Environ. Sci. Technol.* 12(6): 2003–2014.
45. An, B., Lee, H., Lee, S., Lee, S. H., and Choi, J. W. (2015) Determining the selectivity of divalent metal cations for the carboxyl group of alginate hydrogel beads during competitive sorption. *J. Hazardous Mat.* 298: 11–18.

8

Abiotic Removal with Adsorption and Photocatalytic Reaction

Robert Chang-Tang Chang, Bor-Yann Chen, Ke-Fu Zhou, Qiao-Jie Yu, Xiao-Dan Xie, Mridula P. Menon, and Arun Kumar Subramani

CONTENTS

8.1 Introduction

With increasing industrial and agricultural productions, new chemical compounds are continuously emerging. Therefore, types and quantities of pollutants entering surface water through various channels are also increasing day by day. It is difficult to remove these substances from water without improving the existing water treatment methods. At present, the issue of water pollution has caused widespread attention. Processing technique combining chemical oxidation and adsorption is promising to be in practical applications (as shown herein).

8.1.1 Definition of Adsorption—General Perspective

Adsorption is the process in which pollutants migrate and transform in the aqueous environment. For a certain adsorption medium, adsorption of pollutants is mainly related to their hydrophobicity, polarity, polarizability, and their spatial configuration. Due to different structures of various pollutants that contain some special functional groups and constituents, their adsorption behavior is different.

8.1.2 Definition of Photocatalytic Reaction—General Perspective

Advanced oxidation processes (AOPs) are well suited for the degradation of contaminants, especially those that are non-biodegradable or have low biodegradability, durability, and high chemical stability. AOPs for wastewater treatment include photolysis, photocatalysis, ozone oxidation, Fenton, light-Fenton, ultrasonic radiation, sonication, electrochemical oxidation, and humid air oxidation. In general, all AOPs share the characteristics of the reactive oxygen species that can react with contaminants in wastewater. Reactive oxygen or free radicals are strong oxidizing agents that can mineralize contaminants into non-toxic small molecules. These free radicals are derived from atoms or molecules that contain one or more unpaired electrons (e.g. hydroxyl radicals (HO·), superoxide anion radicals (O^{2-}·), peroxy hydroxyl radicals (HO^{2}·), and alkoxide free radical (RO·)). Among them, free radical HO· is the most notable one. The high activity of HO· is attributed to its non-selectivity and high oxidation potential (2.8 eV), which could react with many pollutants. The reaction rate constant is usually ca. 106–109 mol/(L·s). In this chapter, we will take antibiotics as an example to describe the adsorption and photocatalysis techniques.

8.1.3 Definition of Antibiotics

In recent years, pharmaceuticals have received widespread attention as emerging pollutants because of their increasing use and induced environmental problems. An antibiotic is defined as a substance that selectively inhibits the growth of bacteria. In fact, the popularly used antibiotics that are mainly applied to treat microbial infections have become major pollutants for environmental protection. Antibiotics could be divided into many categories according to their molecular structure and pharmacological functions. The main categories are sulfonamides, macrolides, β-lactams, and aminoglycosides, which have saved numerous lives since their discovery and application as medicine. The current global production and consumption of antibiotics are still gradually increasing. In the United States, there were more than 250 million prescriptions each year in particular antibiotics [1]. In Europe, total consumption of human medicinal antibiotics increased gradually from 2004 to 2008. Because of concern for the gradually evolving antibiotic-resistant microbes, total amount of antibiotic consumption in most European countries slightly changed from 2008 to 2013. In addition, the consumption of penicillin was the largest, accounting for 48.4% of the total in European countries. Tetracycline antibiotics consumption accounted for 9.8% [2]. The annual antibiotic production in China is as high as 210,000 tons. The annual average amount of antibiotics used per capita is 138 g, far more than 10 times the average level in the United States. In the meantime, there is a serious abuse of antibiotics because of popular applications to various infections in human population.

Antibiotics also play a crucial role in agriculture and animal husbandry. They can be used as veterinary drugs and sterilants in growing crops and feed additives for livestock [3]. Approximately 46% of the annual antibiotic production was used for livestock husbandry [4]. However, antibiotics used in agriculture and animal husbandry, in particular as growth promotants for food animals, easily lead to the microbial evolution and selection of antibiotic-resistant bacteria from the ecosystem [5]. For example, food animals and their breeding grounds have become reservoirs of antibiotic-resistant strains. Regarding routes of transport, they could be released from animal excrement into the environment, and then gradually spread to the surrounding habitat [6–8]. Therefore, many developed countries have taken top-priority measures to prevent the wide use of antibiotics causing resistance to human pathogens. As a matter of fact, EU banned the use of all antibiotics as growth promoters in animal feeding in 2006 [9]. The U.S. Food and Drug Administration has also enacted regulations to reduce the use of antibiotics as growth promoters (FDA, 2005) [10].

Antibiotics have found extensive use in promoting growth, treatment, and prevention of various diseases both in humans and in animals. The abuse and overuse of antibiotics in animal husbandry as well as aquaculture have led to the presence of antibiotics in food products for human consumption.

According to the reports to date, more than 70 antibiotic traces have been observed in various human foods such as chicken, beef, pork, and fishes. The intake of such contaminated food products causes the exposure of antibiotics to human body. Removing antibiotics from the environment is a necessary step towards rejuvenation of the environment. Highlighting this issue in the society is of great importance because of various reasons. One among them is its adverse effect on human health. The entry of antibiotics into the human body is via consumption of contaminated animal products, contaminated dust inhalation, drinking water contaminated with traces of antibiotics, and drug intake. The adverse effects of antibiotics to human health include allergic reactions, nephrotoxicity, childhood obesity, and tooth discoloration. In addition to human adverse effects, antibiotics also cause alterations in the genes of bacteria, which lead to the formation of antibiotic resistant bacteria. Such antibiotics-resistant bacterial strains are exposed to human beings through the food chain. According to an estimation by WHO in 2014, > 1,000,000 deaths are likely to occur annually by 2050 due to antibiotic resistance.

In the study conducted by Wang et al. [11], it was reported that the presence of antibiotics of various concentrations was detected in the urine of Chinese children. Similarly, 18 antibiotics were detected in the urine of healthy children from China. In the same study, it was reported that the concentration of ampicillin in the urine of children reached > 40,000 ng/mL. Therefore, it is highly significant to address the importance of antibiotic removal from the environment for improving the quality of human lives.

8.1.4 Focus on Tetracycline

Popularly used antibiotics such as tetracycline (tetracycline, TC), oxytetracycline (oxytetracycline, OTC) and chlortetracycline (chlortetracycline, CTC) are three kinds of tetracycline antibiotics (tetracycline antibiotics, TCs) that are widely used as broad-spectrum antibacterial agents and feed additives in human medical treatments, animal husbandry, and fisheries. Approximately 50% to 80% of TCs is released as the non-metabolic prototype via excrements of the human or animal bodies, and drained into the sewage system or directly into the water environment [12]. Their typical molecular structures and chemical properties are shown in Table 8.1.

Thus, TCs are often found in sewage from hospitals, pharmaceutical plants, agriculture, and fisheries. Around 400 ng/L of TC was detected in groundwater in midwestern Illinois, whereas 32 ng/L of OTC and less than 690 ng/L of CTC were detected in surface runoff [13]. In the U.S. survey of groundwater and surface water samples, the average TC, OTC, and CTC concentrations were found to be 0.11, 1.34, and 0.15 μg/L, respectively [14].

Because of the excessive misuse of such antibiotics, contamination of water has become a common problem. Most of the antibiotics after uptake are not fully absorbed by animal or human body and are directly excreted into the water bodies. Because of the serious environmental impact caused by such

TABLE 8.1
Chemical Properties of TCs

TCs	Relative Molecular Mass	Molecular Formula	Solubilitymol/L [Daghrir et al., 2013]	$\log K_{OW}$ [Daghrir et al., 2013]	pK_a [Zhou et al., 2017]	Molecular Structure
TC	444.4	$C_{22}H_{24}N_2O_8$	0.008	−0.62	$pK_{a1} = 3.30$ $pK_{a2} = 7.68$ $pK_{a3} = 9.70$	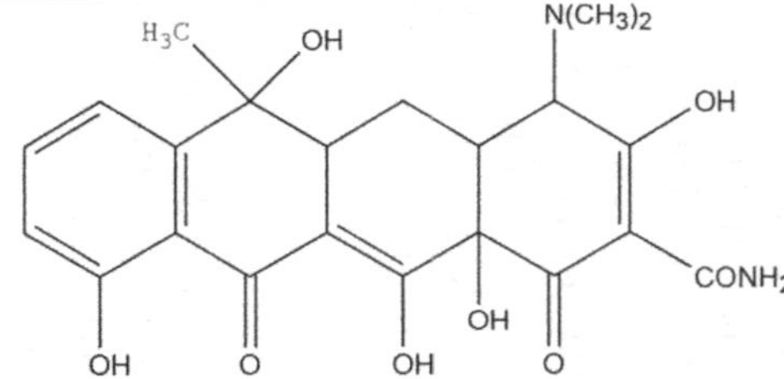
OTC	460.4	$C_{22}H_{24}N_2O_9$	0.041	−1.25	$pK_{a1} = 3.57$ $pK_{a2} = 7.53$ $pK_{a3} = 9.88$	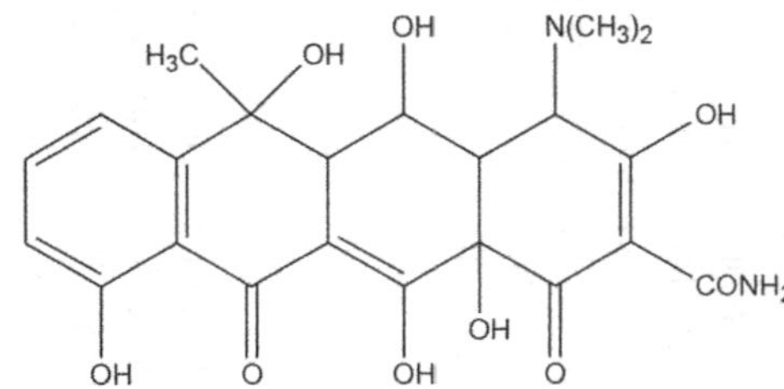
CTC	478.9	$C_{22}H_{23}ClN_2O_8$	0.062	−1.12	$pK_{a1} = 3.30$ $pK_{a2} = 7.55$ $pK_{a3} = 9.30$	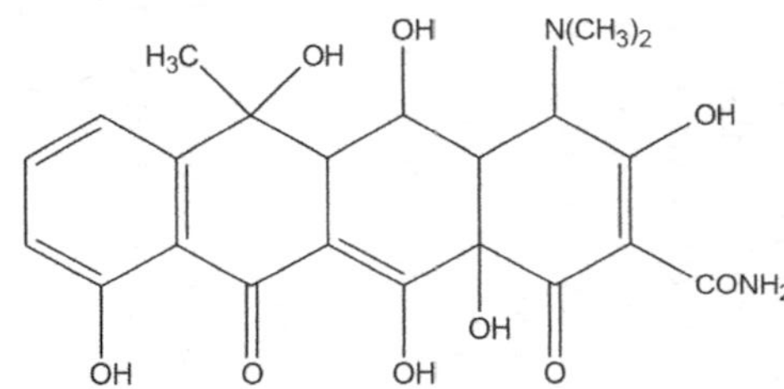

Source: Data from Zhou, K. et al., *Applied Surface Science*, 416: 248–258, 2017.

antibiotics, they can be considered as typical pollutants that are not appropriate to be biotreatment in the environment. With the extensive migration in aqueous phase, these substances can accumulate in human bodies through the food chain without dispute. It results in the spread of drug-resistant genes and causes even serious consequences (e.g. teratogenicity and mutagenesis). Therefore, the removal of antibiotic substances, represented by tetracyclines, has become a pressing scientific research problem [16]. If antibiotics are needed to be treated, abiotic degradation should be the sole selection and upon requirement, biotreatment may be used afterwards for environmental friendliness.

8.1.5 Sources

Antibiotics entering into human bodies and animals cannot be completely absorbed and metabolized. A significant portion of this is excreted in the form of prototypes or metabolites through urine and feces. It can be continuously discharged into the ecosystem. The status of antibiotic contamination in different environments is presented in Tables 8.2–8.5 [17]. These drugs and their metabolites may be directly or indirectly discharged into surface water or groundwater through sewage treatment plants, landfill leachate, drainage ditches, and so forth [18]. In particular, large amounts of antibiotics supplemented to the water body in the aquaculture industry would likely contaminate the surface water. Antibiotics used in human medicine inevitably enter the environment mainly because conventional wastewater treatment processes and sewage treatment plants could not effectively degrade antibiotics [19]. Application of manure is the main measure of soil pollution caused by antibiotics. After entering the soil, antibiotics can enter the surface water through surface runoff, leaching in groundwater and drinking water, and indirectly into the aquatic environment [20,21]. Moreover, contamination can also take place during processing and storing antibiotics. Antibiotics-bearing feed and manure could diffuse into the atmosphere as dust, even as they might have less environmental impact [22].

Antibiotics and their metabolites in the environment could be gradually deposited in soil, surface water, and groundwater by partial conversion and bioaccumulation. The migration and transformation of these compounds in the environment are controlled by a variety of biological, physica,l and chemical processes in the soil-water system (e.g. stability, adsorption, leaching, and degradation). It depends upon the physico-chemical properties of the antibiotics, the nature of the soil, the weather conditions, and so forth [12,23]. Antibiotics and their metabolites in the environment could inevitably be absorbed by animals and humans and thus have persistent effects upon human populations through drinking water and food webs. The source and migration pathways of antibiotics in the environment is shown in Figure 8.1. The widespread use of antibiotics around the globe has led to the release of a large number of antibiotics to the environment. Therefore, the increasingly serious problem of antibiotic contamination needs to be solved urgently as top-concern pollutants.

TABLE 8.2

Typical Characteristics of Some Antibiotics-Bearing Wastewater

Sample Type	Time/ Year	Sampling Location	Main Pollutants	Concentration/(μg·L^{-1})		References
				In	Out	
Sewage treatment plant	2008	Madrid, Spain	Ciprofloxacin	0.16–13.625	<LOQ–5.692	[Rosal et al., 2010]
			Ofloxacin	0.848–5.286	<LOQ–1.651	
Sewage treatment plant	2009	Spain	Amoxicillin	46.0	30.0	[Hijosavalsero et al., 2011]
Sewage treatment plant	2008	Korea	Lincomycin	2.73–25.0	1.06–45.7	[Sim et al., 2011]
Sewage treatment plant	2008	Hong Kong, China	Cephalexin	0.7–5.64	0.17–5.07	[Leung et al., 2012]
			Ofloxacin	0.14–7.9	0.096–7.87	
Sewage treatment plant	2010	Beijing, China	Ofloxacin	0.439–3.1	0.148–1.155	[Gao et al., 2012]
			Sulfadiazine	0.383–2.035	0.118–0.56	
Livestock farms	2008	Korea	Lincomycin	46.2–3005	13.2–2458	[Sim et al., 2011]
			Sulfathiazole	7.44–403	0.028–170	
			Sulfadimethyl pyrimidine	1.76–189	0.011–25.4	
Livestock farms	2009	Jiangsu Province, China	Oxytetracycline	9.14(72.9)^{c}	3.91(11.1)^{c}	[Wei et al., 2011]
			Sulfadimethyl pyrimidine	2.29(211)^{c}	0.75(169)^{c}	

Note: a: limit of quantification; b: not detected; c: median (maximum).

TABLE 8.3

Concentration Profiles of Antibiotics Present in Surface Water

Time/Year	Sample Location	Main Pollutants	Concentration/ ($ng{\cdot}L^{-1}$)	References
2008	Bohai Bay, China	Norfloxacin	460	[Zou et al., 2011]
		Ofloxacin	390	
2009	Huangpu River, China	Sulfamethazine	313.44(5.45)	[Jiang et al., 2011]
2010	Haihe River, China	Sulfamethoxazole	137	[Gao et al., 2012]

Note: a: dry season (flood period) concentration; b: limit of quantification.

TABLE 8.4

Concentration Figures of Antibiotics in Groundwater and Drinking Water

Sample Type	Time/ Year	Sample Location	Main Pollutants	Concentration/ ($ng{\cdot}L^{-1}$)	References
Drinking water	2006	Finland	Ciprofloxacin	20	[Vieno et al., 2007]
Drinking water	2006–2007	United States	Sulfamethoxazole	3.0	[Benotti et al., 2009]
Drinking water	2007	North Carolina, United States	Erythromycin	4.9	[Ye et al., 2007]
			Tylosin	4.2	
			Acrylic acid	4.0	
Ground water	2007–2008	Tianjin, China	Sulfadoxine	78.3	[Hu et al., 2010]
			Ciprofloxacin	31.8-42.5	
			Chloramphenicol	5.8-28.1	

Note: a: not detected.

Sewage of hospital, pharmaceutical industry, farms, and fisheries often contains TCs. In addition, because of the antibacterial property and stability, the traditional biological wastewater treatment could not be effectively used for TC removal [24]. The detectable residual concentration of TCs varied from ng/L to mg/L in the effluent of the sewage treatment plant. It has been found that, in a conventional activated sludge treatment plant where TC was only removed by 44%, CTC was removed by 85%, and OTC was only removed by 24%, the influent concentration was varied from 53.5 to 353 ng/L. The residual concentrations up to 2,683, 105 and 1,079 μg/kg for oxytetracycline, tetracycline, and chlortetracycline, respectively, have been detected in soil (in winter) [25].

Biological treatment of traditional waste water treatment plants (WWTPs) can not effectively and completely remove TCs from wastewater due to their antibacterial properties and stability. Residual TCs would still be detected in effluent from WWTPs at concentrations ranging from ng/L to μg/L [24]. Two activated sludge plants in the south of England monitored TC and OTC effluent concentrations and found that concentrations of effluent antibiotics varied widely, with influent concentrations ranging from 112 to 3028 ng/L

TABLE 8.5

Concentration Data of Antibiotics in Soil, Sludge and Sediment

Sample Type	Time/Year	Sample Location	Main Pollutants	Concentration/(ng·L^{-1}) Dry Weight	References
Sediment	2008	Yellow River, China	Norfloxacin	ND[a]-25 (ND-142)[b]	[Zhou et al., 2011]
			Ofloxacin	ND-8.65 (<LOD[c]-124)[b]	
			Oxytetracycline	ND-2.68 (ND-184)[b]	
Sediment	2008	Liaohe, China	Norfloxacin	ND-120 (ND-177)[b]	[Zhou et al., 2011]
			Oxytetracycline	ND-76.6 (ND-653)[b]	
Sediment	2010	Haihe River, China	Norfloxacin	3.9-141	[Gao et al., 2012]
Sludge	2006–2007	Ohio, United States	Ciprofloxacin	<LOD-46.4	[Spongberg et al., 2008]
			Clarithromycin	<LOD30.2	
Sludge	2007	Japan	Levofloxacin	3110(6310)[e]	[Okuda et al., 2009]
Sludge	2008–2010	Beijing, China	Ofloxacin	550–21000	[Gao et al., 2012]
			Norfloxacin	430–4900	
Soil	2006–2007	Turkey	Oxytetracycline	20–500	[Jacobsen et al., 2009]
Soil	2007	Italy	Oxytetracycline	127–216	[Brambilla et al., 2007]
Soil	2007–2008	Tianjin, China	Oxytetracycline	<LOD (124–2683)[b]	[Hu et al., 2010]
			Chlortetracycline	ND (33.1–1079)[b]	
Soil	2008	Paris, France	Nalidixic acid	<LOD-22.1	[Tamtam et al., 2011]

Note: a: not detected; b: summer (winter) concentration; c: limit of quantification; d: concentration in nitrification sludge (original sludge); e: concentration in primary sludge out of limit; g: soil concentration of sandy soil (sandy soil).

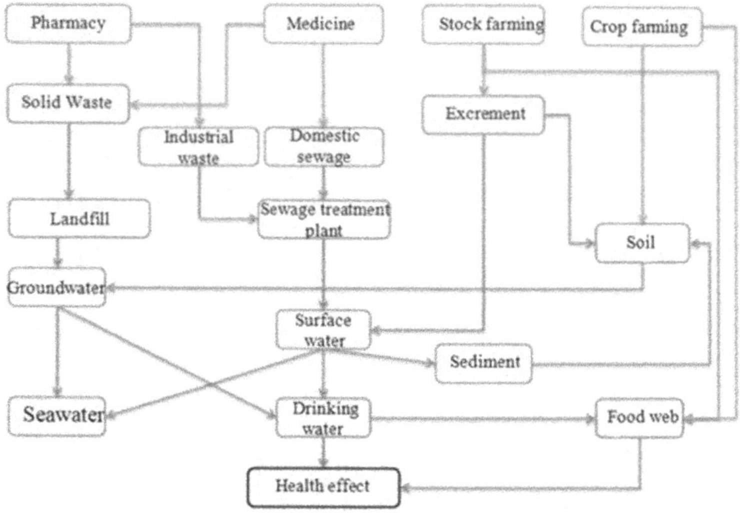

FIGURE 8.1
Sources and pathways of antibiotics in the environment. (Data from Zhou, K. et al., *Applied Surface Science*, 416: 248–258, 2017.)

and effluent concentrations ranging from 17 to 602 ng/L [26]. The TCs inflow and outflow concentrations were studied at an activated sludge treatment plant in Hong Kong. When the influent concentration was between 53.5 and 353 ng/L, the CTC removal rate was 85%, whereas the TC removal was only 44% and the OTC removal was only 24% [27]. The concentration of TC detected in secondary effluent of a sewage treatment plant in Beijing was 195 ng/L, which was apparently higher than the TC concentration in rivers into which effluent inflow were observed [28].

8.2 Removal Technology

8.2.1 Conventional Treatment

Tetracycline substances could not be completely absorbed by the human body and animals. It could be excreted by the body into the city sewer pipes. Therefore, two or three treatment processes of general urban sewage treatment plant would be required to reduce the concentration of tetracycline antibiotics.

Prado et al. [29] studied the treatment of tetracycline-bearing wastewater by Microbiomass Membrane Reactor (MBR). Prado and co-workers showed that the sludge residence time was reduced from 10 to 3 days and the removal rate decreased from 85 to 78%. They also found that the sludge retention time increased to more than 30 days, and the removal rate that could be achieved was more than 89%. Batt et al. [30] also believe that Sludge Retention Time (SRT) is an important factor affecting the removal of tetracycline. When the concentration of tetracycline in sewage is relatively high, coagulation and hydrolysis and acidification could be used for pre-treatment. Moreover, Fan et al. [31] used this technique to degrade oxytetracycline containing wastewater and the removal performance was significantly efficient. The use of this traditional process can effectively remove tetracycline. However, due to the low pertinence of these processes, the further treatment of water bodies could not be avoided. There will still be a certain amount of tetracyclines present because of their migration ability into the environment through the drainage line. Although the aforementioned results could provide some basic data for water treatment plants to remove antibiotic pollution, further study on the removal capabilities of each processing unit is still needed. For example, the activated carbon adsorption method has not been widely adopted in China yet. In China, conventional water treatment process used in the coagulation sedimentation, quartz sand filtration, disinfection process of antibiotic contamination removal, and other issues should be further studied.

8.2.2 Membrane Separation Method

The use of nanofiltration membrane and reverse osmosis membrane to remove micro-pollutants in water is an effective way to purify water. Four nanofiltration and reverse osmosis membranes are used for the treatment of tetracycline at a concentration of 10 mg/L, while the other components of the solution simulated mimic drainage from veterinary drug manufacturers. The findings showed that the four kinds of membrane to treat tetracycline were at retention rate of 99%, most likely due to the tetracycline antibiotic molecular weight. The tetracycline-charged material as well as the hydrophobic nature of the membrane filtration almost had no significant effect on the performance. Although membrane treatment method has a good removal efficiency of tetracyclines, the cost of membrane blocking is still high. Evidently, this is an important factor affecting the popularization and application of membrane treatment.

8.2.3 Chemical Oxidation

Chemical oxidation implied that the antibiotic could be transformed and degraded by the reaction-generating oxidized antibiotics or strong oxidants (e.g. hydroxyl radicals) for remediation. Chemical oxidation could be used for nearly all of the degradation of pollutants. Commonly used oxidants are mainly O_3, $KMnO_4$, ClO_2, and so forth.

As known, capabilities of ozone oxidation are strong. In acidic solution, its redox potential is E_0 = 2171 V and oxidation could be preceded only by fluorine (E_0 = 2187 V). In basic solution E_0 = 1124 V, this potential would be slightly lower than the oxidation of chlorine (E_0 = 1136 V). Meanwhile, ozone could directly react with antibiotics. However, it could also decompose to indirectly produce hydroxyl radical for reactor. The direct reaction strongly depends upon reaction selectivity. The reaction of protonated amino group and double bond is faster, while hydroxyl radical owned non-selective rapid response. The specific reaction pathway is related to the pH value and the water quality of the water body. Usually, it is directly reacted in pure water under acidic conditions (pH < 4) and mainly under indirect conditions under alkaline environment (pH > 10). Both approaches are of great importance for the treatment of both groundwater and surface water (pH ≈ 7). For special wastewater, indirect oxidation is important even at pH = 2, strongly depending on the amount and the characteristics of pollutants. For antibiotics, a small amount of ozone was found to be capable to achieve a promising removal effect under laboratory conditions or waterworks experimental conditions. It was also found that ozonation of sewage was a feasible method to reduce Chemical Oxygen Demand (COD) and Total Organic Carbon (TOC) by 34% and 24%, respectively. It is noteworthy that when ozone was used as a pre-treatment, increasing the amount of ozone can significantly augment the removal rate. Increasing the oxidation time not only decreased the rate of removal, but also reduced the antibiotic removal efficiency. This may be because of an excess of ozone oxidation leading to intermediate accumulation body. Thus, the degradation performance would be significantly attenuated.

Advanced oxidation technology based on ozone (e.g. H_2O_2/O_3, UV/O_3, and Fenton system) can effectively promote the absorption of O_3 in water and generate a large amount of hydroxyl radicals with reduction of the O_3 dose. For substance, recalcitrant to O_3 oxidation, hydroxyl radicals might have relatively promising removal capabilities. O_3 and $O_3/H_2O_2/O_3$ has been significantly used to oxidatively degrade two human antibiotics and one kind of animal antibiotic in wastewater. When 21126 mg/L ozone alone was added, the BOD_5/COD (biodegradability) of the two human antibiotics increased from 0 to 0.1 and 0.127, while the BOD_5/COD of the antibiotic wastewater increased from 0.077 to 0.38. The COD_{Cr} removal rate of human antibiotic wastewater was almost 100% after H_2O_2/O_3 was added at a concentration of 0.013 mol/L. The efficiency of UV/O_3 in this regard has been examined by various researchers globally.In a study, UV/O_3 has been used to treat mice with 5 antibiotics, 5 β-inhibitors, 4 anti-inflammatory agents, 2 lipid metabolites, antiepileptic drugs, ketones, and other pharmaceutical waste water at the dosage of ozone of 15 mg/L. After 18 min exposure, the investigators found that all residual drug concentrations were below LC/MS/MS limits of detection.

Advanced oxidation processes (AOPs) were mainly based on the generation of more hydroxyl radicals to achieve better removal. It is also effective and feasible to augment H_2O_2 and use ultraviolet light to improve the pH of ozone alone. Studies have shown that the basic ozone and advanced oxidation processes have nearly identical removal of COD and residual antibiotics.

The O_3 absorption rate increased by 20% compared to no fight control. Ozone (O_3) could greatly improve the removal rate of aromatics and the biodegradability of wastewater. In particular, advanced oxidation technology own the advantages of relatively shorter reaction time and nearly complete antibiotic degradation. However, where large-scale treatment is concerned, the cost of practical operation would inevitably be relatively high. Thus, it is more appropriate to combine with other technologies. That is to say, synergistic assistance with other technologies to guarantee promising technical feasibility of antibiotic degradation would be necessarily required

8.2.4 Biodegradation

Most antibiotics are considered to be non-biodegradable under aerobic conditions and their degradation should not be satisfactory in lab testing simulations. The biodegradation of 16 kinds of antibiotics when studied under laboratory conditions exhibited that only penicillin G could be completely degraded. In the simulated wastewater treatment experiments with Penicillin G, ceftriaxone, and trimethoprim, it was observed that the biodegradation rate of penicillin G was about 25% within 21 days, while ceftriaxone and trimethoprim seemed to be completely non-biodegradable. In addition, the significant inhibitory effect of lincomycin on nitrification in municipal wastewater was identified in 2006. At present, it found that adsorption should be the main mechanism of tetracycline removal in activated sludge. However, no studies on the biodegradability of tetracycline has been mentioned in literature. According to an environmental survey some of the antibiotics in soils and sediments are non-biodegradable. Therefore, long-term persistence of antibiotics in soils and sediments seemed to have not been prevented. Many studies have reported that some antibiotics have longer half-lives in soils and sediments. However, certain antibiotics can still be partially degraded. For example, ampicillin, doxycycline, oxytetracycline, and thiamphenicol could be significantly degraded in marine sediments, but jasamycin was determined to be non-degradable.

8.2.5 Photolysis

Compared with other AOPs, semiconductor-mediated photocatalytic removal of persistent pollutants in wastewater is relatively promising. The ideal photocatalyst should be chemically inert, photoactive, photostable, economical, non-toxic, and could be excited by visible light and ultraviolet (UV). Although various kinds of semiconductor photocatalysts such as ZnO, ZnS, Fe_2O_3, CdS, CeO_2, WO_3, and SnO_2 exist, they do not satisfy all the characteristics of the ideal photocatalyst. In 1972, the advantages of using photoelectrochemical cells consisting of a rutile TiO_2 photoanode and a Pt electrode were explored in electrolyzing water This discovery has promoted the new application of TiO_2, to water purification, air purification, organic synthesis, sterilization, and anti-cancer drugs. Many literatures have studied the purification of various organic target pollutants in water, such as dyes, pesticides, and drugs. TiO_2-containing catalysts for the photocatalytic degradation are usually chosen because of high photostability, safety, economy, photoreactivity, and chemical and biological inertia of TiO_2 [32].

In fact, the use of Ti-containing catalyst for photocatalytic degradation seemed to be popular [33–37]. Combining Ti to other materials can improve the overall performance of the catalyst. For example, MCM-41 was a kind of porous material with high adsorption capability. Doping Ti into MCM-41 could provide both adsorption property and photocatalytic performance. Ti-containing MCM-41 mesoporous materials with regular 35 Å pore diameters can be synthesised by direct hydrothermal synthesis method. In this method, Ti is in framework positions and no extra framework TiO_2 was detected. Out of several sources of Ti, $(NH_4)_3[Ti(O_2)F_5]$ is used as the titanium source to synthesis Ti-containing MCM-41. Subsequently, Ti-containing MCM-41 became a hot research spot because of excellent absorption capability of larger pore size and catalytic ability of Ti. Application of Ti-containing MCM-41 has been widely mentioned in myriad of professional fields, such as epoxidation [38,39], hydroxylation selective oxidation, oxidative desulfurization, and photocatalysis [40–42]. Hydrothermal method is the most commonly used to synthesis Ti-MCM-41. The duration required for its preparation using this method spans hours to days. However, the traditional hydrothermal method is relatively time-consuming and requires high temperatures, resulting in high cost of operation and maintenance (O/M). Some studies use microwave-assisted methods to shorten the synthesis time [43,44]. A few studies also controlled the synthesis of Ti-MCM-41 at a low temperature [14]. Further technologies to improve such capabilities would be inevitably required.

8.2.6 Adsorption Removal Method

In fact, an adsorption is a method widely used in various fields for practice. The higher the hydrophobicity of tetracycline, the greater the van der Waals force of the adsorbent. The tetracycline molecular plane structure would enhance van der Waals forces. On the other hand, the tetracyclines contain carbonyls and amino groups. It would be easy for the complex to react with the benzene ring and other structures on the adsorption material to facilitate the antibiotic adsorption. The various methods used for the adsorption of antibiotics are explained in detail in the following sections.

8.2.6.1 Ion Adsorption

A typical example of ion adsorption is magnetic Ion Exchange (MIEX). MIEX is Australia's Orica company-developed anion exchange resin. MIEX is well known for its ability to remove 80% of Dissolved Organic Carbon (DOC). MIEX-based adsorption achieves much higher removal efficiencies than coagulation. They have high surface area (3–5 times than conventional resin). They also exhibit rapid exchange kinetics and can be continuously recycled. MIEX resins are suitable for removing hydrophilic and hydrophobic components. Various studies describe the usage of MIEX resins for antibiotic removal. The use of MIEX to remove tetracycline antibiotics in water indicated that the removal rate of MIEX to chlortetracycline was more than 90%. The removal rate of oxytetracycline was about 80% and the removal rate of tetracycline was about 75%. They also used the effluent of sewage treatment plant for comparison, achieving the removal of chlortetracycline from 68% to 98% with MIEX dosage from 0.5 mL/L to 5.0 mL/L. The oxytetracycline was removed from 20% to 60%, and tetracycline removal rate is 5% ~ 50%. The findings indicated that MIEX was favorable for limited adsorption of dissolved organic matter in water compared to tetracyclines.

8.2.6.2 Carbon Material Adsorption

Carbon adsorption materials included granular activated carbon, powdered activated carbon, and carbon nanomaterials. They revealed a good tetracycline removal effect. Ji et al. [45,46], selected multi-walled carbon nanotubes (MWNTs), single-walled carbon nanotubes (SWNTs), and common activated carbon as adsorbents for comparative assessment of tetracycline adsorption. They found that water-soluble substances (such as humic acid, Cu, Ca, and Na) had a certain impact on the adsorption. Humic acid inhibited tetracycline adsorption by MWNT; however, it had less effect on tetracycline adsorption by SWNT. This was due to the fact that pore size of the carbon material was influenced by the adsorption of humic acid. The original size of the tetracycline molecular aperture was apparently not suitable for adsorption. However, the aperture ratio of MWNT, which is more appropriate for adsorbing tetracyclic rings, is larger than that of SWNT. Therefore, the adsorption effect is more effective. As Ji et al. [47] mentioned, the pore size of the adsorbent has a greater effect on adsorption, whereas the sorbent with a higher microporosity ratio seemed to be less effective. Meanwhile, the presence of Cu in water could synergistically enhance the adsorption of carbon nanomaterials. MWNT with macropores could complex with Cu and tetracycline to enhance the adsorption effect.

The carbon adsorbent materials find promising application in the treatment of trace amounts of tetracycline substances in water. However, there are still some problems in the popularization and application of this treatment method. Due to the adsorption characteristics of tetracycline, recovery of adsorbent material and their desorption for reuse still have technical difficulties, leading to increased costs and their practicability.

8.2.6.3 Adsorption of Other Materials

Some soils and natural ores own promising capabilities to enrich tetracyclines. In particular, montmorillonite and metal ion-rich ores could effectively remove tetracycline through effective enrichment for removal of tetracycline. This deciphering of adsorption mechanism could help us understand the characteristics of tetracycline. Zou et al. [48] used other montmorillonite adsorbents for the static adsorption of tetracycline and oxytetracycline antibiotics. They investigated factors such as pH and cationic strength of two promising montmorillonite for treatment of antibiotics. At strong acidic pH conditions (pH 2 ~ 3), the adsorption capacity was observed to be strongest whereas in alkaline conditions, the adsorption capacity was observed to be weakest. With the increase of Ca ion concentration, the adsorption capacity of montmorillonite gradually weakened. Therefore, from various studies, it was concluded that the changes in salt concentration almost did not affect the adsorption capability of tetracyclines toward both adsorbents. It was also observed that the presence of Cu could augment adsorption by complexation with montmorillonite. As mentioned afterwards, some example models for adsorption/desorption could be postulated for practical assessments.

8.2.6.4 Adsorption/Desorption Model

Regarding the mathematical model for antibiotic adsorption/desorption, the reversible reaction can be written as follows:

$$\text{Adsorption: } C \xrightarrow{K_1(Q_m - Q)} Q$$

$$\text{Desorption: } Q \xrightarrow{K_2} C,$$

where C is the concentration of adsorbate (i.e. antibiotic) in solution and Q is the concentration of adsorbed adsorbate. The kinetic parameters K_1 and K_2 are the forward and reverse rate coefficients, respectively, and Q_m is the maximum capacity of the adsorbent. Thus, Langmuir isotherms at equilibrium may be described as:

$$Q_e = \frac{Q_m C_e}{K_d + C_e} \tag{8.1}$$

where Q_e and C_e refer to Q and C at equilibrium, respectively. In addition, specific adsorption capacity q can be defined as (Q/X), where X is adsorbent concentration (g L^{-1}). Transient dynamics of antibiotic adsorption may be expressed as follows:

$$\frac{dc}{dt} = -K_1(Q_m - Q)C + K_2 Q \quad C(t = 0) = C_0 \tag{8.2}$$

which satisfies mass balance equation $C_0 = C(t) + Q(t)$. Introducing the relation $K_2 = K_d K_1$, one may modify Equation (8.2) as the form of Riccati differential equation as follows:

$$\frac{dC}{dt} + P \cdot C + S \cdot C^2 = R \tag{8.3}$$

where, P, S, and R are $K_1(K_d + Q_m - C_0)$, K_1 and $K_1 K_d C_0$, respectively. Equation (8.3) becomes:

$$\frac{d^2M}{dt^2} + P\frac{dM}{dt} - RSM = 0 \tag{8.4}$$

upon making the substitution $C = \frac{1}{S}\frac{dM/dt}{M}$.

Thus, the analytical solution of Equation (8.2) is shown as follows:

$$C(t) = \frac{C_\infty(\alpha^- - C_0)\exp(C_\infty K_1 t) + \alpha^-(C_0 - C_\infty)\exp(\alpha^- K_1 t)}{(\alpha^- - C_0)\exp(C_\infty K_1 t) + (C_0 - C_\infty)\exp(\alpha^- K_1 t)} \tag{8.5}$$

where C_∞ denote $C(t = \infty)$ and $\alpha^- K_1$ and $\alpha^+ K_1$ ($\alpha^- < \alpha^+ = C_\infty$) are two distinct real roots of the characteristic equation:

$$\Gamma^2 - K_1(C_0 - Q_m - K_d)\Gamma - K_1^2 K_d C_0 = 0 \tag{8.6}$$

8.2.6.5 Model for Two-Step Degradation [49]

According to Julson and Ollis [50], the kinetic model for the degradation of antibiotic could be summarized to two steps as shown below:

$$\textit{Antibiotic} \xrightarrow{k_1} \textit{Intermediates} \xrightarrow{k_2} \textit{Product}(s)$$

where k_1 and k_2 is the first order rate constants of the two serial steps of proposed mechanism, respectively. As the proposed kinetics revealed, one could obtain analytical solution for time-series profiles of antibiotic degradation as follows:

$$Ab_t = Ab_0((k_3 \frac{k_1}{k_2 - k_1})(\exp(-k_1 t) - \exp(-k_2 t) + \exp(-k_1 t)) \tag{8.7}$$

where Ab_t was the antibiotic concentration at time t, and Ab_0 denoted the antibiotic concentration of initial sample. The kinetic constant k_3 is the constant related to the ratio of adsorption of antibiotic and intermediate. The constants of k_1, k_2, and k_3 can be calculated through the equations at initial rate (t small) and asymptotically stable rate (t sufficiently large) determination as shown below:

$$\frac{Ab_t}{Ab_0} = (k_3 - 1)k_1 t + 1 \tag{8.8}$$

$$\ln\left(\frac{Ab_t}{Ab_0}\right) = -k_2 t + \ln\left(\frac{-k_3 k_1}{k_2 - k_1}\right) \tag{8.9}$$

8.3 Biotoxicity Assessment after Treatment

For biotoxicity assessment upon treated and untreated antibiotics, dose-response assessment popularly used in toxicology can be adopted. A typical dose-response curve includes four parameters: the threshold (baseline dose; EC_0), the maximum treatment concentration (top dose; EC_{100}), the slope of response curve, and the concentration to provoke a 50% response (median effective dose; EC_{50}). The EC_0 and EC_0 can be defined as the maximum concentration to have a detectable response (i.e. 0 + %) and minimum concentration to have 100% response, respectively. In addition, EC_0, EC_{20}, and EC_{50} are tabulated to produce an obvious diagram of metallic toxicity series in terms of response-curve slope of various metals. For example, if values of effective concentrations (e.g. EC_{20}, EC_{50}) for $material_A$ are much less than those for $material_B$, $material_A$ is more toxic than $material_B$. Note that in this study, only interpolated values of EC (e.g. EC_{20}, EC_{50}) were chosen for toxicity comparison, since EC_0, EC_{100} are usually extrapolated from dose-response curves. The postulated concept of response variables in metal toxicity can be illustrated as indicated in Figure 8.2. When the presence of a toxicant has become a serious threat to the population, the cells must induce some biochemical or physiological mechanism to handle this threat in a manner to prevent toxic responses (e.g. loss in cellular viability or metabolic capability) from occurring. As toxicant concentration C increases (i.e. $+\infty > \ldots > C_n > \ldots > C_3 > C_2 > C_1 \geq 0$), the growth rate of bacterial cells (i.e. $0 \cong \mu_\infty|C_\infty < \ldots < \mu_n|C_n < \ldots < \mu_3|C_3 < \mu_2|C_2 < \mu_1|C_1 < +\infty$) tends to drop. Note that maximum specific growth rate μ (1/h) (i.e. the growth rate in the exponential phase) represents the rate of maximum growth per unit amount of biomass (i.e. $(1/X)$ (dX/dt)) [51]. In addition, cells living in an environment with a higher toxicant concentration require a longer acclimatization time (i.e. $+\infty = L_\infty|C_\infty < \ldots < L_n|C_n < \ldots < L_3|C_3 < L_2|C_2 < L_1|C_1 \leq 0$) to seek a metabolically-effective mechanism for survival

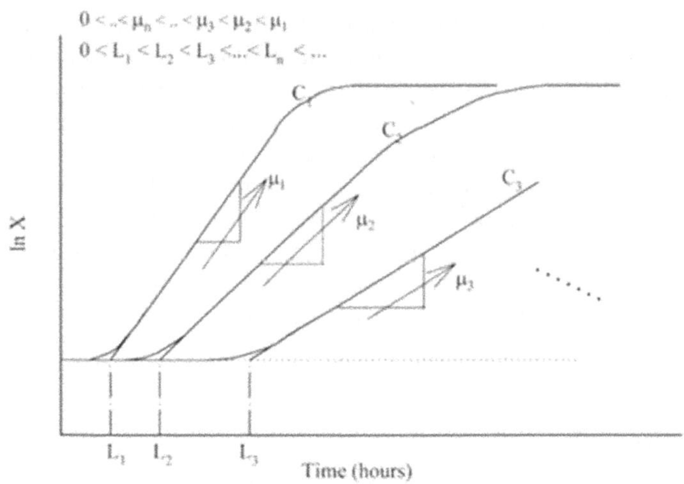

FIGURE 8.2
Typical growth curves for bacterial cultures laden with various toxicants. Note that living in a higher toxicant-laden environment (i.e. $+\infty > \cdots > C\ n > \cdots > C\ 3 > C\ 2 > C\ 1 \geq 0$) may result in a significantly larger decrease in growth rate (i.e. $0 \cong \mu_\infty|C_\infty < \cdots < \mu_n|C_n < \cdots < \mu_3|C_3 < \mu_2|C_2 < \mu_1|C_1 < +\infty$), and even a longer lag time (i.e. $+\infty = L_\infty|C_\infty < \cdots < L_n|C_n < \cdots < L_3|C_3 < L_2|C_2 < L_1|C_1 \leq 0$) to adapt it. (Data from Chen, B.-Y. et al., *Process Biochemistry*, 39: 737–748, 2004.)

and growth. Note that $\mu_j|C_j$ and $L_j|C_j$ denote the maximum specific growth rate and lag time of cultures in response to toxicant concentration C_j, respectively. Here, the measure of "chronic" response (i.e. $(\mu_0 - \mu)/\mu_0$) is defined as a decrease of maximum specific growth-rate (i.e. $(\mu_0 - \mu)$) in the presence of toxicant divided by maximum specific growth rate in the absence of this toxicant (i.e. μ_0), where subscript 0 indicates the culture in the absence of toxicant. This growth rate (μ) is determined from the slope of the semilogarithmic plot of cell density versus culture time in the exponential growth phase. The time (τ_d) required to double the microbial mass can be determined by $\tau_d = \ln 2/\mu$, where τ_d is the doubling time of cells. In the exponential growth phase, the cells have adjusted to their new environment to multiply in a maximum rate as they can. Therefore, the growth rate can be used as an equilibrium outcome of metabolic status in the cells in response to "chronic toxicity" of toxicant. As a dose (i.e. toxicant concentration) increases over EC_0 at which toxic effects begin to exert cells, the seriousness of those effects $(\mu_0 - \mu)/\mu_0$ also increases. Theoretically, the EC_0 can be explained as the maximum concentration to have non-detectable response (i.e. $\lim_{Z \to EC0} P = \lim_{Z \to EC0} \mu = \mu_0$). The maximum response occurs at a dose (i.e. EC_{100}) that the population can no longer tolerate and death intervenes (i.e. $(\mu_0 - \mu)/\mu_0 \cong 1$ or $\mu \cong 0$). When the toxicant concentration exceeds EC_{100}, 100% toxicity response (i.e. complete loss of cellular viability) will be inevitable (i.e. $\lim_{Z \to EC100} P = \lim_{Z \to EC100} \mu = 0$). In addition, we selected $1\text{-}L_0/L$ to present "acute toxicity" of toxicants, as the time lag (L) indicates a delay of cellular response to conduct an immediate adaptation mechanism in response to toxic or/and inhibitory stress for microbial growth. The lag time is a quantitative variable to show cellular adaptation to a new environment. In particular, in the presence of toxicants microbial cells are required to reorganize their metabolic status in molecular levels for survival. Depending upon toxicant species to be administered, the intracellular "defense mechanism" (e.g. the operon for specific toxicant) will be significantly expressed for survival to the fittest. Simultaneously, cells might temporarily repress molecular capabilities not directly related to growth or defense. All changes reflect an immediate mechanism in intracellular level for metabolic adjustment to the environment and thus are termed "acute responses."

8.4 Amended Perspectives of Abiotic Treatment

Compared to aforementioned aspects, the treatment of materials could interactively combine with adsorption and photo-degradation, and the synergistic effect will be much more promising. The following is a practical example of tetracycline antibiotics treated with Ti-MCM-41. The details of experimental methods could be shown as follows:

8.4.1 Photocatalytic Degradation

First, 1 L of simulated wastewater solution was prepared using different water samples and poured into a reactor. The effective antibiotic concentration was 20 μmol/L. The unregulated pH was 5.8 ± 0.3. The cycle water between the inner and outer tube of the reactor was controlled at a constant temperature of 25°C. The solution in the reactor was homogenized by an electromagnetic stirring system. Low pressure mercury lamp (APUV-12F, Sparxic, 254 nm, light intensity 0.97 Wm^{-2}) was used as the single light source, located in a quartz tube of the reactor centre. The reactor was wrapped with tin foil paper to inhibit external light. When 0.1 g catalyst was added into the solution, the lamp was turned on and the initial reaction time was noted. Samples were taken at a predetermined time interval, filtered with 0.2 μm filter, stored at 4°C, and analyzed within 24 h.

8.4.2 Case Study of Adsorption

To reveal kinetic modeling of adsorption, approx. 100 mL of OTC at different concentrations were prepared (e.g., 5, 10, 20 and 40 mg/L) in shake flasks. Unless specifically mentioned, pH of the solution was adjusted to 7.0, which was optimal pH for adsorption. These samples were incubated at different temperatures (e.g., 15°C, 25°C, 35°C, and 45°C). The adsorption of OTC continued for 5 h to obtain equilibrium. In addition, different weight percentages of TiO_2 (e.g., 2, 5, 10, and 20 % of composite-TiO_2/MCM) was used. Four different dosages of catalyst, such as 1, 2, 4, and 8 g/L, were also used. Since adsorption efficiency varies based on above-mentioned parameters, the effect of these parameters on adsorption was also studied. The effect of pH, including 3, 5, 7, 9, and 11, on adsorption performance was also assessed.

8.4.2.1 Effect of Initial Concentration of Pollutant and Temperature

In this study, MCM41 material was used for adsorption study. Approximately 10% TiO_2 doped MCM41 was examined with the above-mentioned different initial concentrations at 25°C and adsorption capacity was calculated. The experiments were also repeated for other three temperatures and adsorption capacity was also calculated. As Figure 8.3(a)–(d) indicate below, time courses of adsorption efficiency of four initial concentrations of OTC as aforementioned at 15°C, 25°C, 35°C, and 45°C were shown. Furthermore, Langmuir adsorption isotherms of the above experiments were determined. According to **Adsorption/Desorption Model**, the maximum adsorption capacity

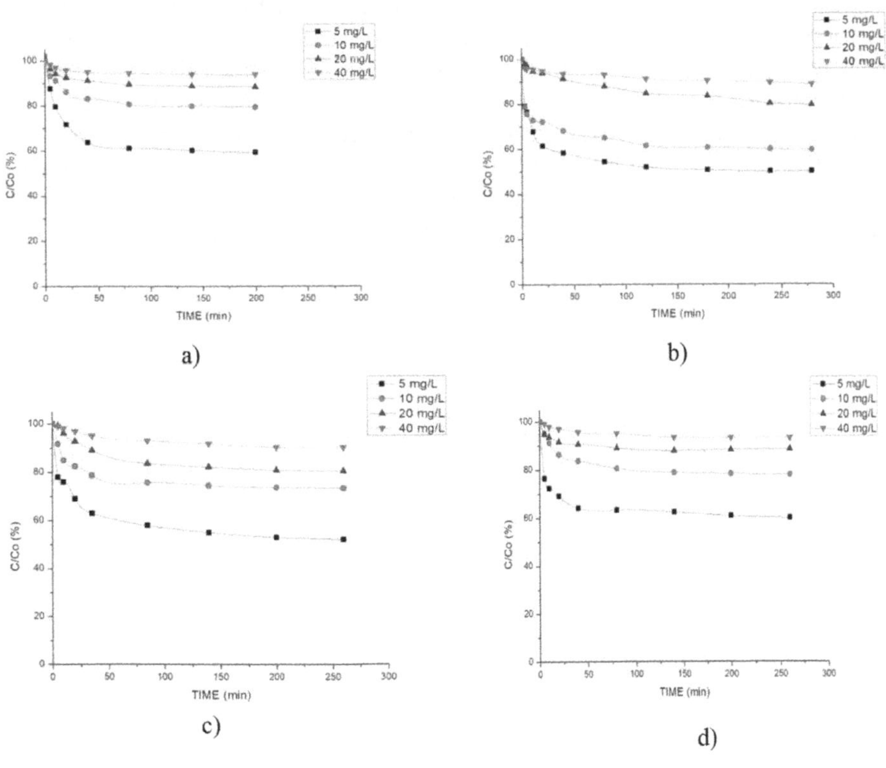

FIGURE 8.3
Effect of different temperature on OTC degradation of different concentration. (a) 15°C (b) 25°C (c) 35°C and (d) 45°C.

(Q_m) and kinetic parameters can be determined from inverse polynomial regression (Ce/Qe vs. Ce) to minimize least squares of deviations as best-fit Langmuir isotherm. The results are explained in Figure 8.4. At each temperature, Qe and Ce data can be calculated for different initial concentrations of OTC, where Ce concentration of OTC at equilibrium and Qe is adsorption capacity at equilibrium.

Kinetic parameters K_d and Q_m in Langmuir isotherm were determined by plotting graph between Ce and Ce/Qe for different temperature, where Q_m is maximum adsorption capacity and K_m is rate constant. The slope equals to $1/Q_m$ and intercept equals to K_d/Q_m as evaluated in Figure 8.5 for the plot between Ce and Ce/Qe. Thus from the above graph the K_d and Q_m values were determined for different temperatures. Using the equation A5, the K_1 and K_2 values were determined by fitting the concentration values determined by the equation A5 to experimental data curve where K_1 and K_2 are adsorption and desorption rate constants, respectively. Thus, K_1 and K_2 values are determined for four different temperatures as shown in the Figure 8.6. Similar to the above-mentioned graph, some other fitting was also performed for the other three temperatures to determine adsorption and desorption rate constants. The values of rate constants for four different temperatures were tabulated as below. As indicated in Tables 8.6 and 8.7, the best adsorption efficiency could be achieved at 25°C. From the above result, we concluded that the adsorption efficiency of test material decreased with increasing initial concentration of OTC. This reason may be that increasing concentration will lead to equilibrium within a short time since the number of adsorption active sites is constant. Therefore, OTC with higher concentration could not adsorb effectively. Thus, 5 mg/L initial concentration of OTC

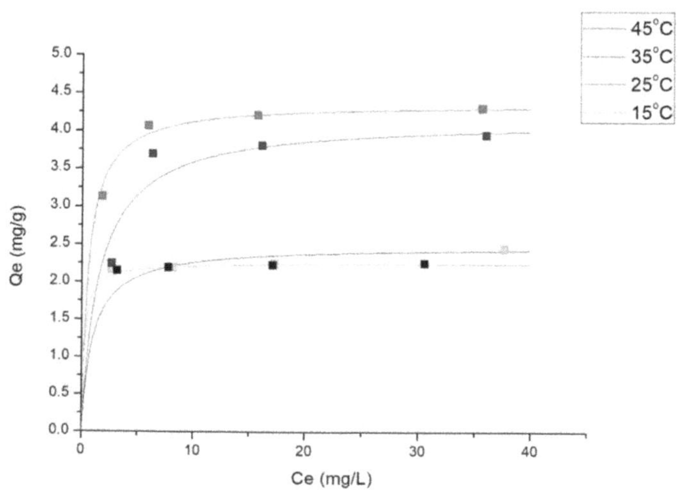

FIGURE 8.4
Langmuir Isotherm of OTC adsorption at different Temperature.

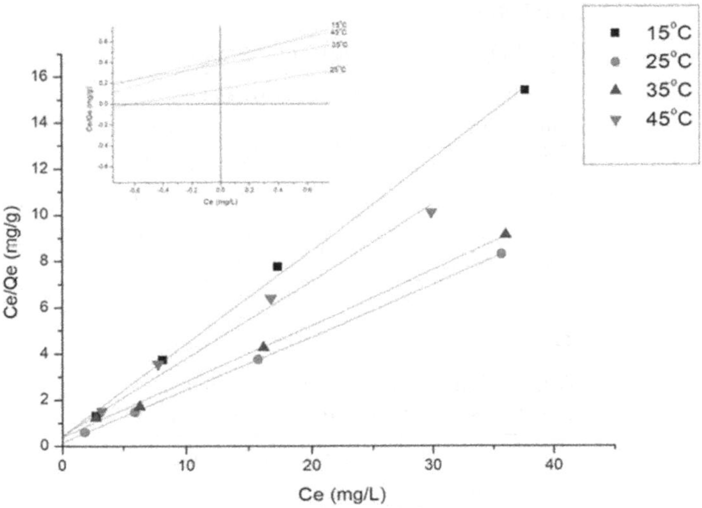

FIGURE 8.5
Langmuir plot for determination of K_d and Q_m value.

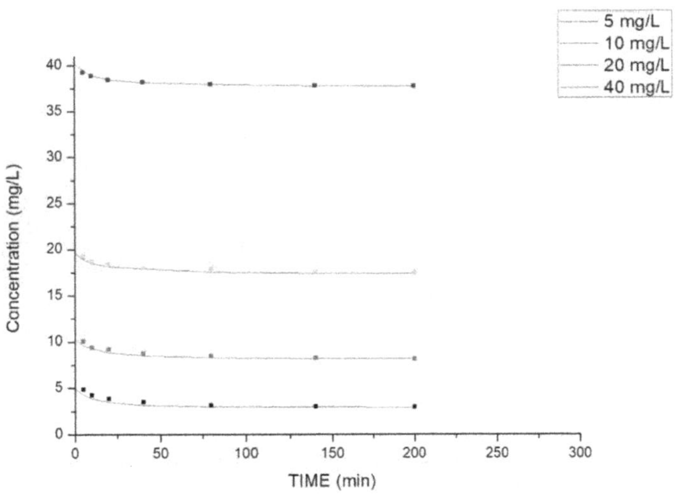

FIGURE 8.6
Fitting curve for determination of adsorption and desorption rate constants for 15°C.

adsorption capacity was studied at different temperatures. Figure 8.7(a) and (b) represent the curve graph and bar graph of adsorption for the mentioned concentration at various temperatures. Thus, from the above results it was concluded that the best adsorption is achieved at 25°C and 5 ppm of initial concentration of OTC. Further studies will be performed only in above temperature and initial concentration of OTC.

TABLE 8.6

K_d and Q_m Values for Different Temperature

	Kinetic Parameters		
Temperature	**Q_m mg/g**	**K_d**	**R^2**
15° C	2.49	1.0358	0.9981
25° C	4.37	0.6225	0.9999
35° C	4.14	1.5736	0.9981
45° C	2.70	1.1186	0.9977

TABLE 8.7

Adsorption (K_1) and Desorption (K_2) Rate Constants for Different Temperature

	Kinetic Parameters	
Temperature	**K_1**	**K_2**
15° C	0.0086	0.0089
25° C	0.0098	0.0061
35° C	0.0048	0.0076
45° C	0.0072	0.0081

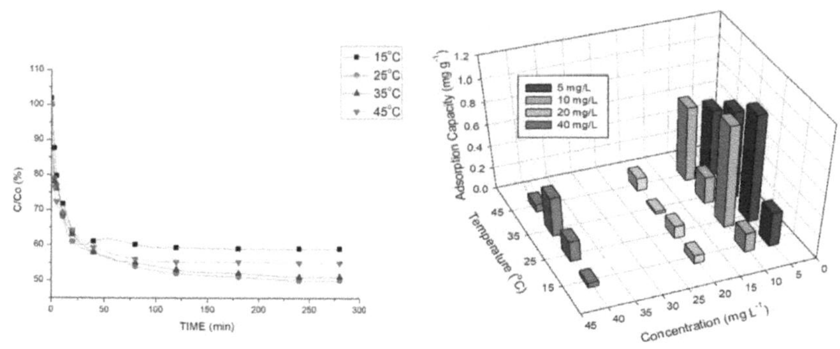

FIGURE 8.7
Effect of different temperature on OTC degradation for OTC initial concentration of 5 mg/L. (a) Plot graph and (b) Bar graph.

8.4.2.2 Effect of Different Dosage of TiO_2 in Doping and Dosage of Adsorbent Used in Solution

In a further study, only 5 mg/L initial OTC concentration was used and all reaction was examined in 25°C. Different weight percept of TiO_2, such as 2, 5, 10, and 20%, was used and their adsorption study was performed as same before. The results are shown in Figure 8.8. The adsorption capacity in equilibrium (Qe) was calculated for composite with different dosage of TiO_2. The values are listed in below table. Thus, from the above results the best adsorption

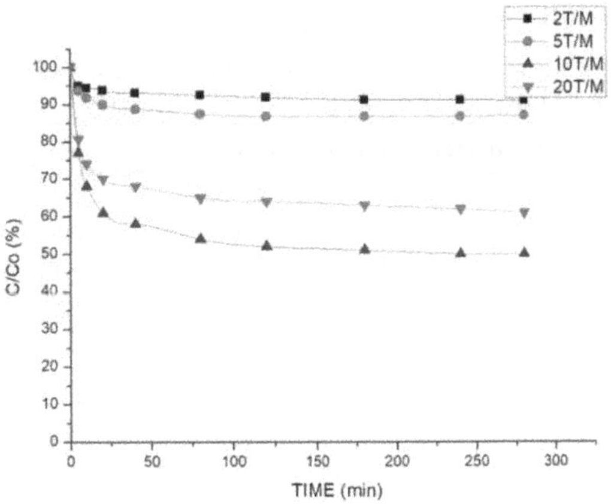

FIGURE 8.8
Effect of different dosage of TiO_2 in adsorption material on OTC degradation.

takes place only when there is 10% of TiO_2 in TiO_2-MCM41 composite. This result may come from the fact that increasing the dosage of composite may damage the geometry of adsorbent material and lead to poor adsorption efficiency. In contrast, decreasing the dosage of composite may lead to insufficient development of pore in material and cause poor adsorption efficiency. Now different dosage of adsorbent was used in solution, such as 1, 2, 4, 8 g/L. From this experiment, we can conclude that increase in dosage of adsorbent leads to increase in adsorption efficiency due to increasing surface area of adsorbent material for adsorption. Result of this experiment was described in Figure 8.9. The maximum adsorption capacity (Q_m) was calculated for different dosage of adsorbent. Since we are using a standard temperature and initial concentration of OTC for this experiment, the K_d value will be constant

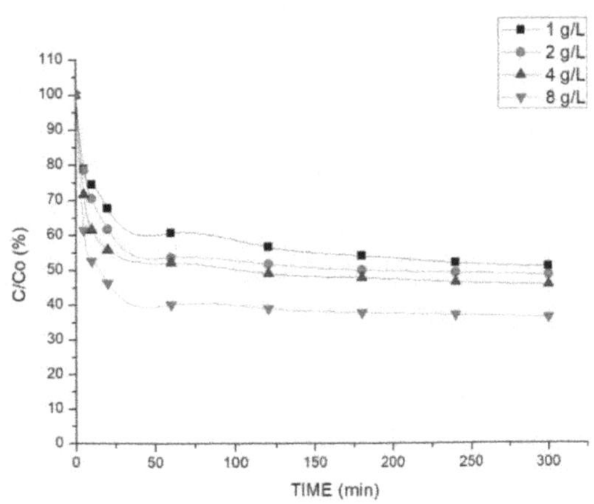

FIGURE 8.9
Effect of different dosage of catalyst used in solution on OTC degradation.

TABLE 8.8

Adsorption Capacity at Equilibrium (Qe) Value for Different Dosage of TiO_2 in Composite

Dosage of TiO_2 in Our Composite	Qe (mg/g)
2%	0.50
5%	0.75
10%	2.84
20%	2.81

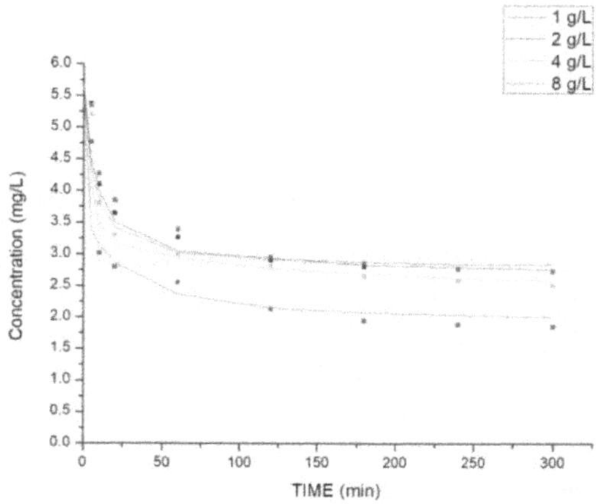

FIGURE 8.10
Fitting curve for determination of maximum adsorption capacity (Q_m) for different dosage of adsorbent.

for all dosages. By fitting the data to experimental curve, the Q_m value was determined. The plot was presented in Figure 8.10. The values of Q_m are listed in Table 8.8. Thus, from the above results the best adsorption will take part when there is increase in dosage of adsorbent. We can conclude that increasing dosage of adsorbent used will increase the adsorption efficiency. From the above study, we can conclude that adsorption plays main role in adsorption of OTC. Thus, it is essential to study the adsorption property of adsorption material before it is using along with catalyst for degradation.

8.4.3 Analysis

Antibiotics and their by-products were analyzed using a HPLC-MS/MS system, with a positive ion electrospray ionization source and ACQUITY CSH C18 reverse phase column (5 μm, 4.6 × 250 mm). The flow rate was 1.2 mL/min. The value of capillary voltage, ion source temperature, desolvation temperature, desolvation gas flow rate, and cone voltage were 2.5 kV, 150°C, 500°C, 1000 L/hr, and 16 V, respectively. Detection mode was multiple reaction monitoring. The ion concentration was determined by an ion chromatography system (811/761, Metrohm).

8.4.4 Toxicity Tests

The toxicity of photocatalytic degradation solution was evaluated with indicator bacterium *Escherichia coli* DH5α. The strain was activated in 50 mL of Luria-Bertani (LB) broth (the composition of the media used is as follows: 10 g/L peptone, 5 g/L yeast, 10 g L-1 NaCl) at 37°C for 12 h in a constant-temperature shaking incubator at 125 rpm. Optical density at 600 nm (OD_{600}) of activated bacterial suspension was measured and later was adjusted to 0.020 using double concentrated LB broth. The treated TCs solution of 2 mL at each sampling time point was added into 2 mL of the activated and diluted bacterial solution in a culturing tube. A series of untreated TCs solution was prepared at concentrations of 20, 10, 5, 1, 0.5, 0.1, 0.05, 0.01 μmol/L without degraded products. The TCs solution of 2 mL was also added into 2 mL of the activated and diluted bacterial solution with the same bacterial concentration. After homogenisation, the concentration of antibiotics was 10 to 0.005 μmol/L, and the corresponding logC was 1 to –2.30. Meanwhile, 2 mL of sterile distilled water was used instead of TCs solution as control. The initial bacterial concentration of all mixed solution was ca. 1×10^6 cfu/mL. All mixed solutions were cultured in a constant-temperature shaking incubator at 3°C for 8 h. Samples of 0.5 mL were taken at terminal of 8 h from each culturing tube to measure their OD_{600} recorded as A and A_0 for TCs solution and sterile distilled water, respectively. The growth inhibition (I) of each solution was calculated from Equations (8.9–8.10). The growth inhibition I was plotted against logC, and the curve was fitted according to Equations (8.9–8.11). All biotic experiments were repeated three times.

$$I = 1 - \frac{A}{A_0} \times 100\% \tag{8.10}$$

$$y = y_0 + \frac{a}{1 + e^{-\frac{x - x_0}{b}}} \tag{8.11}$$

8.4.5 Results of Photocatalytic Degradation

OTC solution was prepared using various water samples to simulate the antibiotic-bearing wastewater from hennery. The OTC solution contained 15.82 mg of veterinary OTC, because the veterinary OTC with low purity contained a large amount of organic impurities. The effective OTC concentration in the simulated wastewater was 20 μmol/L as much as in the pure OTC solution. The photocatalytic degradation of OTC in different water samples has

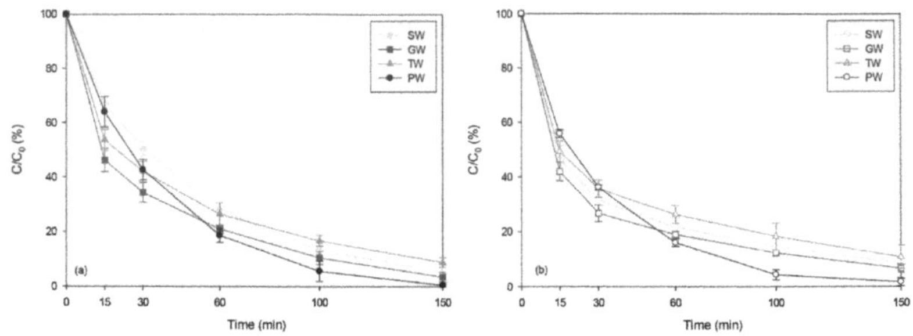

FIGURE 8.11
Structural elements in lignocellulose molecules. (Data from Zhou, K. et al., *Applied Surface Science*, 416: 248–258, 2017.)

been shown in Figure 8.11, in which (a) is for pure OTC and (b) is for veterinary OTC. In the three natural water samples, the OTC removal efficiency in 150 min was lower than that of deionized water. The possible reasons include that the ions and soluble organic impurities in the natural water samples have effects on the degradation reaction, as well as the lower light transmittance of the natural water samples at 254 nm. The removal of veterinary OTC in natural water solution is similar to that of pure OTC, indicating the ignorable effect of organic impurities in veterinary OTC. It can be further speculated that the effect of trace amount of organic matters in natural water samples on degradation also can be neglected. Although the pH value of each water sample was different, the pH value became 5.8 ± 0.2 after dissolving OTC, thus pH is not the main factor to reduce the removal efficiency. It can be deduced that the ions in the natural water samples affected the degradation.

The added ion concentration was the same as that of ground water in order to explore the effect of each ion, and the results are shown in Figure 8.12. The effect of the cations, Mg^{2+}, K^+, and Ca^{2+} on the photocatalytic degradation of OTC was not significant. Similar results were observed for the anion NO_3^-. The other two anions show significant influence. The inhibition of SO_4^{2-} on

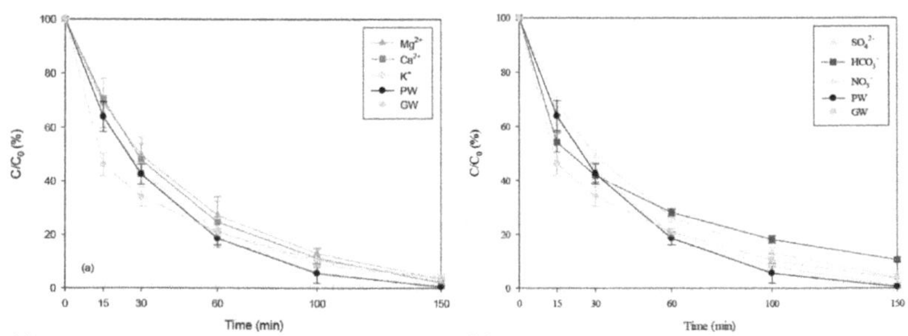

(a) (b)

FIGURE 8.12
Effect of cations (a) and anions (b) on OTC degradation. (Data from Zhou, K. et al., *Applied Surface Science*, 416: 248–258, 2017.)

the photocatalytic degradation of OTC was slight, probably due to the concentration being too low. Many studies have shown that the presence of SO_4^{2-} leads to a decrease in degradation efficiency because it reacts with HO• free radicals, depleting HO• to produce $SO_4{\bullet}^-$, which is less reactive than HO• free radicals and reacts with water molecules to turn back to SO_4^{2-}, as shown in Equations (8.13–8.14):

$$SO_4^{2-} + HO\bullet \rightarrow SO_4\bullet^- + OH^- \tag{8.12}$$

$$SO_4\bullet^- + H_2O \rightarrow HO\bullet + SO_4^{2-} + H^+ \tag{8.13}$$

The HCO_3^- changed the degradation efficiency greatly in this experiment. HCO_3^- ion can remove HO• free radicals, resulting in reduced degradation rate and efficiency, according to the following Equation (8.14).

$$HCO_3^- + HO\bullet \rightarrow CO_3\bullet^- + H_2O \tag{8.14}$$

Obviously, the synergistic effect of ion on the degradation of OTC was more than single operation. Since it was difficult to accurately simulate the proportions of various ions in the natural water samples, the interaction of the ions has not been explored in this study.

8.4.6 Adsorption Capability

As shown in Table 8.9, all three materials reached their maximum saturated adsorption capacity at pH 5, while the saturated adsorption capacity reached lowest value at pH 3 or 11. The maximum adsorption capacity of MCM-41 and T/M-25 were determined to be 221.2 and 203.3 μmol/g. The maximum saturated adsorption capacity of T/M-25 is smaller than that of MCM-41 because the decrease in specific surface area and pore size leads to decrease in adsorption capacity. The adsorption capacity of TiO_2 was poor, and its maximum saturated adsorption capacity was only 76.2 μmol/g.

TABLE 8.9

Langmuir Adsorption Isotherm of OTC on MCM-41, T/M-25 and TiO_2

	MCM-41		T/M-25		TiO_2	
pH	q_m (μmol/g)	R^2	q_m (μmol/g)	R^2	q_m (μmol/g)	R^2
3	5.9	0.971	5.9	0.994	5.9	0.942
5	221.2	0.997	203.3	0.954	76.2	0.995
7	113.1	0.998	128.2	0.984	35.4	0.966
9	50.4	0.995	46.7	0.966	8.3	0.980
11	19.1	0.958	18.7	0.972	0.5	0.940

The pH value also has a great impact on the adsorption capacity, as it determines the charging properties of adsorbent materials and adsorbate. The pH_{PZC} of TiO_2, MCM-41, and T/M-25 were measured to be 6.2, 3.8, and 4.5, respectively. When the pH is above the pH_{PZC}, the material surface is positively charged, otherwise negatively charged. At pH 3, the main form of OTC is cation OTC^+, while the surface of the material is positively charged. The same charge rejection leads to poor adsorption performance. At pH 5, the main form of OTC is zwitterion H_2OTC, thus the charge repulsive force is small and the adsorption capacity is large. There are zwitterionic H_2OTC and a small amount of anion $HOTC^-$ at pH 7, which is slightly repulsive to the negatively charged material. The anion $HOTC^-$ is the predominant form at pH 9 and a small amount of double anions OTC^{2-} also exist. The negative charge of OTC and the negative charge on the surface of the material produce strong charge repulsion. When the pH increased to 11, the double anion OTC^{2-} expressed the tendency to be the predominant form, the repulsive force would be stronger than that of pH 9 and the saturated adsorption capacity significantly reduced to a limited small value.

8.4.7 Residual Biotoxicity

The photodegradation and photocatalytic degradation of TCs has been shown in Figure 8.13. Photodegradation of 1 L of 20 μmol/L TCs solution by only UV light was time-consuming, which spent 280 min, 1200 min, and 570 min for TC, OTC, and CTC to the concentration of the detection limit. The log Kow values of TC, OTC, and CTC are −3.4, −1.1, and −1.6, respectively, supporting that TC is the most hydrophilic and OTC is the most hydrophobic species. The high water solubility leads to a greater mobility, easier bioavailability, and the shorter half-life via aqueous phase. The photocatalytic degradation with T/M-25 greatly shortened the treatment time, which required 220 min, 160 min, and 300 min for TC, OTC, and CTC, respectively. The degree of balance between the positive and negative surface potentials also describe the surface electrostatic potentials, and the electrostatic balance values of TC, OTC, and CTC are 0.19, 0.13, and 0.22, respectively. Chlortetracycline owned the highest electrostatic balance because of the presence of a single Cl atom;

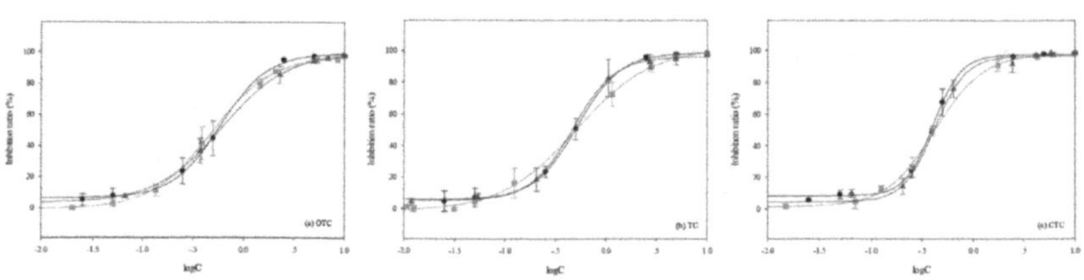

FIGURE 8.13
Inhibition of *E. coli* in TCs solution after photocatalytic degradation. (Data from Zhou, K. et al., *Applied Surface Science*, 416: 248–258, 2017.)

although it is an electron-withdrawing substituent, it distributes its charge over a large volume. A lower electrostatic balance contributes to the greater molecular contact on the molecular surface and further, a greater reactivity.

As indicated in dose response curves, the half effect concentration (EC_{50}) was calculated for the comparative assessment. The corresponding pseudo-half effect concentration of the photolysis and photocatalysis solution was denoted as EC_{50}-UV and EC_{50}-T/M. As revealed in Figure 8.14, three fitting curves for each TCs solution are almost overlapped. For the three antibiotics, the ranking of biotoxicity potency was EC_{50} < EC50-UV < EC50-T/M. Many studies reported a slight increase in the toxicity of TCs solution after degradation because of the higher toxicity of intermediate than other pollutants [53–56]. Therefore, EC_{50}-UV value should be less than EC_{50} if the toxicity significantly augmented after the degradation, while the result of this study seemed to be opposite. That is to say, there was no intermediate with higher toxicity potency detected after treatment. The inconsistent result with previous studies was likely because of the different methods of testing toxicity. The experimental time of luminous bacteria used in previous studies might be short. Therefore, this toxicity potency should be measured in terms of acute toxicity rather than chronic toxicity prone to short-term inhibition increase toxicity, while long-term inspection of bacterial culture may show no inhibition [57]. The results indicate that sufficient defense mechanism to overcome inhibitory characteristics of such toxic intermediates could be effectively induced for resistance of long-term toxicity. At present, conventional treatment techniques, membrane treatment, oxidation, and adsorption have their own disadvantages. T/M catalysts exhibited a certain potential in the treatment of antibiotic bearing wastewater due to its excellent adsorption and photocatalytic performance. To have conclusive data for economic feasibility,

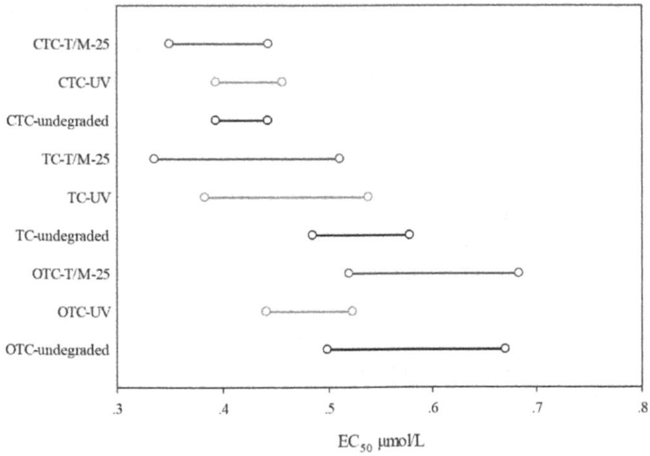

FIGURE 8.14
EC_{50} to TCs of *Escherichia coli* after photocatalytic degradation. (Data from Zhou, K. et al., *Applied Surface Science*, 416: 248–258, 2017.)

more detailed exploration of treatment mechanism should be implemented prior to further scale-up for industrial applications. Tetracycline antibiotics was just as a typical example for the comparative assessments. Further studies on a myriad of antibiotics should be carried out with consideration of practicability for not only environmental friendliness, but also for ecological viability.

References

1. Hicks, L., Taylor, T., Hunkler, R. (2013) More on US outpatient antibiotic prescribing, 2010 REPLY. *New England Journal of Medicine* 369: 1175–1176.
2. Carvalho, I. T., Santos, L. (2016) Antibiotics in the aquatic environments: A review of the European scenario. *Environment International* 94: 736–757.
3. Silbergeld, E. K., Graham, J., Price, L. B. (2008) Industrial food animal production, antimicrobial resistance, and human health. *Annu. Rev. Public Health* 29: 151–169.
4. Su, H. C., Ying, G. G., He, L. Y., Liu, Y. S., Zhang, R. Q., Tao, R. (2014) Antibiotic resistance, plasmid-mediated quinolone resistance (PMQR) genes and ampC gene in two typical municipal wastewater treatment plants. *Environmental Science: Processes & Impacts* 16: 324–332.
5. Marshall, B. M., Levy, S. B. (2011) Food animals and antimicrobials: Impacts on human health. *Clinical Microbiology Reviews* 24: 718–733.
6. Looft, T., Johnson, T. A., Allen, H. K., Bayles, D. O., Alt, D. P., Stedtfeld, R. D., Sul, W. J., Stedtfeld, T. M., Chai, B., Cole, J. R. Hashsham, S. A. (2012) In-feed antibiotic effects on the swine intestinal microbiome. *Proceedings of the National Academy of Sciences* 109: 1691–1696.
7. Zhu, Y. G., Johnson, T. A., Su, J. Q., Qiao, M., Guo, G. X., Stedtfeld, R. D., Hashsham, S. A. Tiedje, J. M., (2013). Diverse and abundant antibiotic resistance genes in Chinese swine farms. *Proceedings of the National Academy of Sciences* 110: 3435–3440.
8. Novais, C., Freitas, A. R., Silveira, E., Antunes, P., Silva, R., Coque, T. M., Peixe, L. (2013) Spread of multidrug-resistant Enterococcus to animals and humans: An underestimated role for the pig farm environment. *Journal of Antimicrobial Chemotherapy* 68: 2746–2754.
9. Commission E. (2005) Ban on antibiotics as growth promoters in animal feed enters into effect [EB/OL]. Brussels, Belgium: European Commission. http://europa.eu/rapid/press-release_IP-05-1687_en.htm.
10. FDA (2005) Final decision of the FDA commissioner: Withdrawal of approval of the new animal drug application for enrofloxacin in poultry [EB/OL]. Rockville, MD: Center for Veterinary Medicine, U.S. Food and Drug Administration. http://www.fda.gov/NewsEvents/Newsroom/PressAnnouncements/2005/ucm108467.htm.
11. Wang, S., Wang, H. (2015) Adsorption behavior of antibiotic in soil environment: A critical review. *Frontiers of Environmental Science & Engineering* 9: 565–574.

12. Chen, H., Peng, Y. P., Chen, K. F., Lai, C. H., Lin, Y. C. (2016) Rapid synthesis of Ti-MCM-41 by microwave-assisted hydrothermal method towards photocatalytic degradation of oxytetracycline. *Journal of Environmental Sciences* 44: 76–87.
13. Berlini, C., Guidotti, M., Moretti, G., Psaro, R., Ravasio, N. (2000) Catalytic epoxidation of unsaturated alcohols on Ti-MCM-41. *Catalysis Today* 60: 219–225.
14. Peng, R., Zhao, D., Dimitrijevic, N. M., Rajh, T., Koodali, R. T. (2012) Room temperature synthesis of Ti–MCM-48 and Ti–MCM-41 mesoporous materials and their performance on photocatalytic splitting of water. *The Journal of Physical Chemistry C* 116: 1605–1613.
15. Zhou, K., Xie, X. D., Chang, C. T. (2017) Photocatalytic degradation of tetracycline by Ti-MCM-41 prepared at room temperature and biotoxicity of degradation products. *Applied Surface Science* 416: 248–258.
16. Boxall, A. B. A., Fogg, L. A., Blackwell, P. A., Blackwell, P., Kay, P., Pemberton, E. J., Croxford, A. (2004) Veterinary medicines in the environment. In Reviews of environmental contamination and toxicology (1-91). Springer, New York.
17. Gao, L. H., Shi, Y. L., Li, W., Liu, J., Cai, Y. (2013) Environmental behavior and impacts of antibiotics. *Environmental Chemistry* 32: 1619–1633.
18. Kemper, N. (2008) Veterinary antibiotics in the aquatic and terrestrial environment. *Ecological Indicators* 8: 1–13.
19. Homem, V., Santos, L. (2011) Degradation and removal methods of antibiotics from aqueous matrices—A review. *Journal of Environmental Management* 92: 2304–2347.
20. Blackwell, P. A., Kay, P., Ashauer, R., Boxall, A. B. (2009) Effects of agricultural conditions on the leaching behaviour of veterinary antibiotics in soils. *Chemosphere* 75: 13–19.
21. Chee-Sanford, J. C., Mackie, R. I., Koike, S., Krapac, I. G., Lin, Y. F., Yannarell, A. C., Maxwell, S., Aminov, R. I. (2009) Fate and transport of antibiotic residues and antibiotic resistance genes following land application of manure waste. *Journal of Environmental Wuality* 38: 1086–1108.
22. Gothwal, R., Shashidhar, T. (2015) Antibiotic pollution in the environment: A review. *Clean—Soil, Air, Water* 43: 479–489.
23. Manzetti, S., Ghisi, R. (2014) The environmental release and fate of antibiotics. *Marine Pollution Bulletin* 79: 7–15.
24. Michael, I., Rizzo, L., McArdell, C. S., Manaia, C. M., Merlin, C., Schwartz, T., Dagot, C., Fatta-Kassinos, D. (2013) Urban wastewater treatment plants as hotspots for the release of antibiotics in the environment: A review. *Water Research* 47: 957–995.
25. Daghrir, R., Drogui, P. (2013) Tetracycline antibiotics in the environment: A review. *Environmental Chemistry Letters* 11: 209–227.
26. Johnson, A. C., Jürgens, M. D., Nakada, N., Hanamoto, S., Singer, A. C., Tanaka, H. (2017) Linking changes in antibiotic effluent concentrations to flow, removal and consumption in four different UK sewage treatment plants over four years. *Environmental Pollution* 220: 919–926.
27. Li, B., Zhang, T. (2011) Mass flows and removal of antibiotics in two municipal wastewater treatment plants. *Chemosphere*, 83: 1284–1289.
28. Xu, J., Xu, Y., Wang, H., Guo, C., Qiu, H., He, Y., Zhang, Y., Li, X., Meng, W. (2015) Occurrence of antibiotics and antibiotic resistance genes in a sewage treatment plant and its effluent-receiving river. *Chemosphere* 119: 1379–1385.

29. Prado, N., Ochoa, J., Amrane, A. (2009) Biodegradation by activated sludge and toxicity of tetracycline into a semi-industrial membrane bioreactor. *Bioresource Technology* 100: 3769–3774.
30. Batt, A. L., Kim, S., Aga, D. S. (2007) Comparison of the occurrence of antibiotics in four full-scale wastewater treatment plants with varying designs and operations. *Chemosphere* 68: 428–435.
32. Fan, Y., Qi, P., Zhao, R. (2002) Hydrolysisacidrogenesis pretreated penicillin, oxytetracycline wastewater experimental. *Environmental Sciences* 28: 19–21.
32. Kanakaraju, D., Glass, B. D., Oelgemöller, M. (2014) Titanium dioxide photocatalysis for pharmaceutical wastewater treatment. *Environmental Chemistry Letters* 12: 27–47.
33. Giraldo, A. L., Erazo-Erazo, E. D., Flórez-Acosta, O. A., Serna-Galvis, E. A., Torres-Palma, R. A. (2015) Degradation of the antibiotic oxacillin in water by anodic oxidation with Ti/IrO2 anodes: Evaluation of degradation routes, organic by-products and effects of water matrix components. *Chemical Engineering Journal* 279: 103–114.
34. Nunes, M. J., Monteiro, N., Pacheco, M. J., Lopes, A., Ciríaco, L. (2016). Ti/β-PbO2 versus Ti/Pt/β-PbO2: Influence of the platinum interlayer on the electrodegradation of tetracyclines. *Journal of Environmental Science and Health, Part A* 51: 839–846.
35. Auguste, A. F. T., Quand-Meme, G. C., Ollo, K., Mohamed, B., Placide, S. S., Ibrahima, S., Lassine, O. (2016) Electrochemical oxidation of amoxicillin in Its commercial formulation on thermally prepared RuO2/Ti. *Journal of Electrochemical Science and Technology* 7: 82–89.
36. Palma-Goyes, R. E., Vazquez-Arenas, J., Ostos, C., Ferraro, F., Torres-Palma, R. A., Gonzalez, I. (2016) Microstructural and electrochemical analysis of Sb2O5 doped-Ti/RuO2-ZrO2 to yield active chlorine species for ciprofloxacin degradation. *Electrochimica Acta* 213: 740–751.
37. Sun, Y., Li, P., Zheng, H., Zhao, C., Xiao, X., Xu, Y., Sun, W., Wu, H., Ren, M. (2017) Electrochemical treatment of chloramphenicol using Ti-Sn/γ-Al2O3 particle electrodes with a three-dimensional reactor. *Chemical Engineering Journal* 308: 1233–1242.
38. Fadhli, M., Khedher, I., Fraile, J. M. (2016) Modified Ti/MCM-41 catalysts for enantioselective epoxidation of styrene. *Journal of Molecular Catalysis A: Chemical* 420: 282–289.
39. Silvestre-Alberó, J., Domine, M. E., Jordá, J. L., Navarro, M. T., Rey, F., Rodríguez-Reinoso, F., Corma, A. (2015) Spectroscopic, calorimetric, and catalytic evidences of hydrophobicity on Ti-MCM-41 silylated materials for olefin epoxidations. *Applied Catalysis A: General* 507: 14–25.
40. Nguyen, V. H., Lin, S. D., Wu, J. C. S. (2015) Synergetic photo-epoxidation of propylene over VTi/MCM-41 mesoporous photocatalysts. *Journal of Catalysis* 331: 217–227.
41. Wu, H. Y., Bai, H., Wu, J. C. (2014) Photocatalytic reduction of CO2 using Ti–MCM-41 photocatalysts in monoethanolamine solution for methane production. *Industrial & Engineering Chemistry Research* 53: 11221–11227.
42. Aboul-Gheit, A. K., Abdel-Hamid, S. M., Mahmoud, S. A., El-Salamony, R. A., Valyon, J., Mihályi, M. R., Szegedi, A. (2011) Mesoporous Ti-MCM-41 materials as photodegradation catalysts of 2, 4, 6-trichlorophenol in water. *Journal of Materials Science* 46: 3319–3329.

43. Wang, S., Shi, Y., Ma, X. (2012) Microwave synthesis, characterization and transesterification activities of Ti-MCM-41. *Microporous and Mesoporous Materials* 156: 22–28.
44. Chen, H., Peng, Y. P., Chen, K. F., Lai, C. H., Lin, Y. C. (2016) Rapid synthesis of Ti-MCM-41 by microwave-assisted hydrothermal method towards photocatalytic degradation of oxytetracycline. *Journal of Environmental Sciences* 44: 76–87.
45. Ji, L., Chen, W., Bi, J., Zheng, S., Xu, Z., Zhu, D., Alvarez, P. J. (2010) Adsorption of tetracycline on single-walled and multi-walled carbon nanotubes as affected by aqueous solution chemistry. *Environmental Toxicology and Chemistry* 29: 2713–2719.
46. Ji, L., Wan, Y., Zheng, S., Zhu, D. (2011) Adsorption of tetracycline and sulfamethoxazole on crop residue-derived ashes: Implication for the relative importance of black carbon to soil sorption. *Environmental Science Technology* 45: 5580–5586.
47. Ji, L., Liu, F., Xu, Z., Zheng, S., Zhu, D. (2010) Adsorption of pharmaceutical antibiotics on template-synthesized ordered micro-and mesoporous carbons. *Environmental Science & Technology* 44: 3116–3122.
48. Zou, S., Xu, W., Zhang, R., Tang, J., Chen, Y., Zhang, G. (2011) Occurrence and distribution of antibiotics in coastal water of the Bohai Bay, China: Impacts of river discharge and aquaculture activities. *Environmental Pollution* 159: 2913–2920.
49. Zhang, Q., Jing, Y. H., Shiue, A., Chang, C. T., Chen, B. Y., Hsueh, C. C. (2012) Deciphering effects of chemical structure on azo dye decolorization/degradation characteristics: Bacterial vs. photocatalytic method. *Journal of the Taiwan Institute of Chemical Engineers* 43: 760–766.
50. Julson, A. J., Ollis, D. F. (2006) Kinetics of dye decolorization in an air–solid system. *Applied Catalysis B: Environmental* 65: 315–325.
51. Atkinson, B., Mavituna, F. (1991) Biochemical engineering and biotechnology handbook. 2nd ed. New York: M. Stockton Press.
52. Chen, B.-Y., Liu, H.-L., Liu, Chen, Y.-W., Cheng, Y.-C. (2004) Dose–response assessment of metal toxicity upon indigenous Thiobacillus thiooxidans BC1. *Process Biochemistry* 39:737–748.
53. Zhao, C., Deng, H., Li, Y., Liu, Z. (2010) Photodegradation of oxytetracycline in aqueous by 5A and 13X loaded with TiO2 under UV irradiation. *Journal of Hazardous Materials* 176: 884–892
54. Gómez-Pacheco, C. V., Sánchez-Polo, M., Rivera-Utrilla, J., López-Peñalver, J. J. (2012) Tetracycline degradation in aqueous phase by ultraviolet radiation. *Chemical Engineering Journal* 187: 89–95.
55. López-Peñalver, J. J., Sánchez-Polo, M., Gómez-Pacheco, C. V., Rivera-Utrilla, J. (2010) Photodegradation of tetracyclines in aqueous solution by using UV and UV/H2O2 oxidation processes. *Journal of Chemical Technology and Biotechnology* 85: 1325–1333.
56. Yuan, F., Hu, C., Hu, X., Wei, D., Chen, Y., Qu, J. (2011) Photodegradation and toxicity changes of antibiotics in UV and UV/H2O2 process. *Journal of Hazardous Materials* 185: 1256–1263.
57. Wammer, K. H., Slattery, M. T., Stemig, A. M., Ditty, J. L. (2011) Tetracycline photolysis in natural waters: Loss of antibacterial activity. *Chemosphere* 85: 1505–1510.

9

Revalorization of Agro-Food Residues as Bioadsorbents for Wastewater Treatment

Gassan Hodaifa, Cristina Agabo García, and Santiago Rodriguez-Perez

CONTENTS

9.1 Introduction

In agricultural and agro-industrial activities it is common to see the accumulation of by-products during the raw material collection and transformation (husks, hulls, seeds, peel, shell, skin, etc.) in the process (washing, processing, preparing, and/or packaging food, etc. [1]). Agro-food waste is directly linked with the environmental, economic, and social impacts [2].

The agro-food processing industry sector generates substantial quantities of residue and several authors have predicted an increase up to 44% by 2025 [3–5]. Previous authors have indicated 19–29% of global greenhouse gases emissions to come from the agro-food sector [6]. In fact, Nanda et al. [1] determined the CO_2 equivalent greenhouse gases emissions derivate from the agro-food waste in 20 million tonnes yearly. According to the Food and Agricultural Organization [7], every year more than one-third of the food produced is lost, which generate loses for the value of $750 billion [1].

Waste from agro-food is mainly disposed into landfilling, which may produce environmental problems by means of fermentative processes due to high organic fraction and moisture [8]. Incineration is another option occasionally used for the agro-food waste, but the poor calorific value and high moisture show energy inefficient processes [8]. The traditional valorization consists of the use of wastes as raw material for feeding livestock and the direct return to agriculture as a soil amendment [1]. Therefore, the underutilization of the agro-food waste is contributing to disposal problems, aggravating environmental pollution, and increasing the operating costs [3].

According to previous studies, the European Union (EU-28) generates around 100 million tons of food waste per year [2], China 600 millon tons [5], and USA 251 million tons [9–11]. The produced agro-food waste accounts around 20% of the total food waste [2,8]. In EU-28 the agro-food waste is the second sector contributing to the food waste production [2].

Currently, international policies such as Directive 2008/98/EC promote the industrial ecology, circular economy approach as well as "zero waste" process intensification [12]. In this way, research and development are intended to use of agro-food waste, by-products or side streams, produced in the agro-food processes as raw material for new products and applications. New cross-sector interconnections are being created due to this industrial symbiosis trend. In recent years, the research studies focused on the development of environmental eco-friendly technologies, which try to maintain materials. Agro-food waste-based on materials have been increased intensively in the last decades [8,12].

An increasing number of researchers have been studying different ways of agro-food residues revalorization in different fields such as: energy production (as biofuels), bioactive compounds recovery (biopolymers), dietary supplements (fibers, enzymes, organic acids, nutraceuticals), pollutant adsorption, natural fertilizer, additive in animal nutrition, etc. [12–15]. Among them, bioadsorption is an eco-friendly operation, simple and economic method of pollutant elimination and it has been studied in the removal of new pollutants from wastewater [16].

The aims of this work are to determine the state of the art of bioadsorbents and its application on the wastewater treatment processes as well as the evaluation of the potential of these bioadsorbents on the reduction of contamination level and volume of polluted wastewater.

9.2 Wastewater Treatment

There are many technologies involved in wastewater treatment management. However, most of these technologies usually show high operational or maintenance costs, complicated flow charts, waste or toxic generation, and scalability problems [17]. According to literature, the adsorption process is one of the best technologies to be applied in the wastewater treatment due to its easy operation, design, and convenience [17,18]. Figures 9.1 and 9.2 show that bioadsorbents have been growing as a research field since 1980, but it has become a key research field in recent years. The most commonly used adsorbent is activated carbon [19].

A huge range of pollutants such as heavy metals, dyes, organic compounds, and others can be removed from the wastewater by adsorption processes [20]. However, the high operational costs, regeneration capacity, and disposal at the end-of-life are significant constrains [21]. Therefore, the cost-effective and eco-friendly solution for this issue is the substitution of the activated carbon for low-cost and eco-friendly adsorbents. Many authors have proved the adsorbent capacity of some agro-food waste such as those derivates from rice processing, groundnut shell, sawdust, soybean hull, apricot, coffee, citrus,

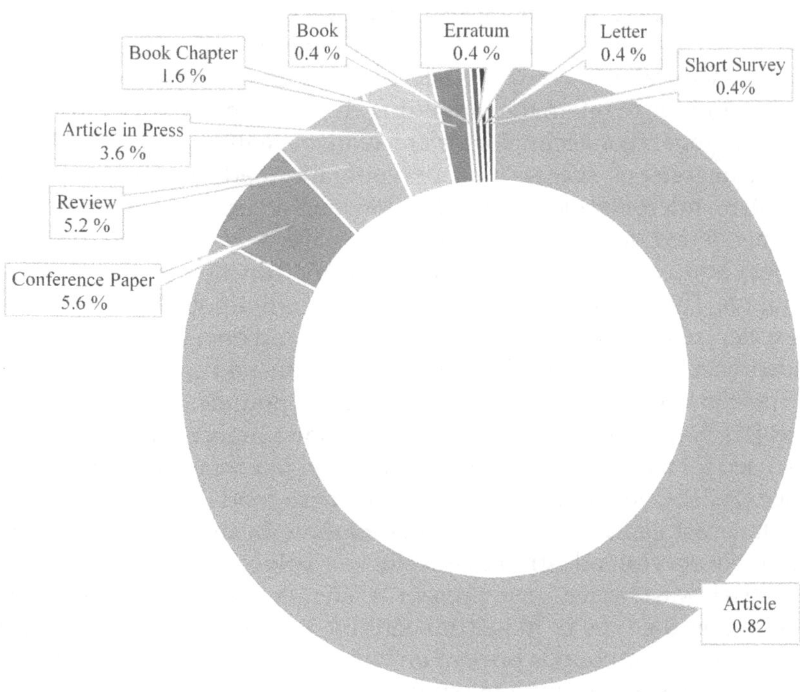

FIGURE 9.1
Type of published document on biosorbent study in the last 28 years.

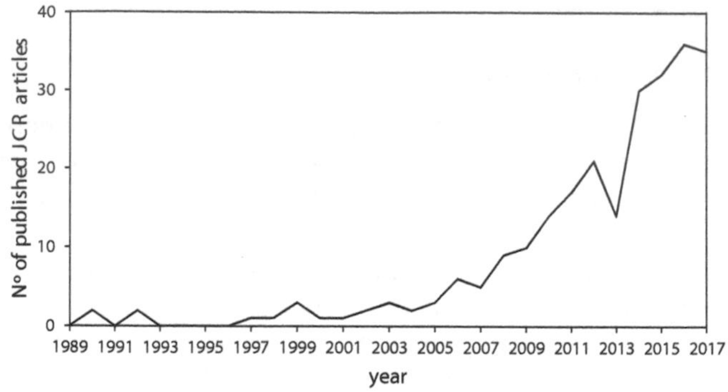

FIGURE 9.2
Published documents number about the biosorbents study in the last 28 years.

olive oil production, and many other different natural products [22–30]. Not only would the use of agro-food wastes as adsorbents be a sustainable solution in the agro-food waste management but also a sustainable solution for the wastewater treatment.

In the wastewater treatment plants (WWTPs), the following sequence of treatments are commonly used: preliminary (pre-treatment), primary, secondary or biological, and in some depuration plants can be seen a tertiary treatment. After main operations for organic matter reduction (normally biological or chemical oxidation), it is necessary to apply a physicochemical separation technique to retain the remaining dissolved pollutants. In this sense, a variety of separation technologies are available with specific functions and different degrees of success like pervaporation, adsorption, coagulation-flocculation, microfiltration, nanofiltration, and reverse osmosis [31].

All mentioned separation technologies have a key role in the tertiary treatment and each technique seeks a specific function in the wastewater depuration. The adsorption process is considered a potential alternative in wastewater treatment due to the simple design and operation [18]. Currently, the adsorbent most industrially employed is activated carbon because of its high specific surface, its high affinity with compounds, and its easy regeneration [19]. Nevertheless, activated carbon shows higher cost because it cannot be used for a long period of time in agitated reactors due to the mechanic damage [18,21]. Comparatively, wastes from agro-food industries are appearing as low-cost alternatives to activated carbon. In this way, bioabsorption capability of several industrial by-products is being examined.

Industrial wastewater may present a complex and toxic composition, depending on the type of industrial activity. For this reason, the identification of the toxic compounds present on wastewater is crucial in order to seek the adequate absorbents with maximum adsorbent capacity. Different agro-food residues are used as bioadsorbents with the aim of pollutant removals such as dyes, aromatic compounds, or heavy metals. This is possible because agro-food residues are composed by different parts of the plant that are rich in cellulose, hemicellulose, and lignin. These compounds have in common their chemical structures with functional groups as hydroxyl, acetamide, or amino. These groups have excellent selectivity toward aromatic structures and heavy metals. Moreover, these biopolymers show chemical stability and high reactivity [17].

Among others, heavy metals are highly toxic due to their long-term presence in the polluted wastewater. Agro-food adsorbents have generally high-efficiency in metal removal and they also allow the possibility of recovery metals. However, the adsorption capacity would vary depending on the type of agro-food industry residue, an element of metal, structural modification of bioadsorbents, and operational conditions of bioadsorption. The layer of metals attached to the adsorbents is usually monomolecular due to the weak bounds which bind both compounds. For this reason, the adsorbents must be extremely porous in order to increase the area of adsorption [16].

Different researchers have modified bioadsorbents by physical and chemical pre-treatments, leading to better surface areas, surface chemistry, and pore size distribution. All these changes favor adsorption of targeted species by the mechanisms involved (Figure 9.3) in heavy metal bioadsorption [16].

1. Physical bioadsorption: This kind of adsorption occurs in the bioadsorbents surface by precipitation and diffusion through cell wall and membrane and no exchange of electrons done. The adsorbate is attracted to the bioadsorbent surface by van der Waals forces and multiple layers may be formed [28].

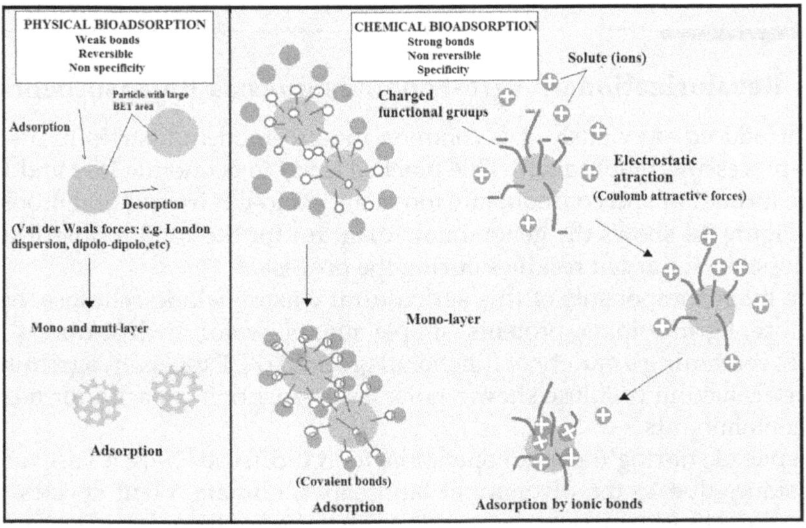

FIGURE 9.3
Schematic representation of removal contaminants via different bioadsorption mechanisms.

2. Chemical bioadsorption: An exchange of electrons between active sites of bioadsorbent surface and solute molecules on the solution occurs. The result of this interaction is the formation of chemical bonds. This kind of adsorption is stronger and more stable than physical bioadsorption. Usually, in chemisorption a monolayer forms [32].
3. Electrostatic bioadsorption (Ion exchange). The attraction between bioadsorbent (charged functional groups) and the solute (ions) on the solution is governed by Coulomb attractive forces [29].

According to researchers, in the election of bioadsorbent the following criteria must be taken into account:

- Cost-effectiveness.
- Availability.
- Adsorption and desorption capacity.
- Regeneration.

All these criteria are directly related to the optimal operating conditions of adsorption, such as contaminant concentration, modification of bioadsorbent, contact time, agitation, pressure, temperature, and pH. In addition, it is important to study the adsorbent regeneration capacity with the aim of reusing it more than once. For this reason, the majority of researchers normally vary the operational conditions.

9.3 Revalorization of Agro-Food Residues as Bioadsorbents

As introduced previously, it is common in agro-food industries to discard food-processing by-products. This practice leads to economic loss and food waste formation and consequently too many socio-environmental problems [33]. Figure 9.4 shows the general flow diagram for the food industries and the possible discarded residues during the process.

The main components of this agricultural waste include cellulose, hemicellulose, lignin, lipids, proteins, simple sugars, water, hydrocarbons, and starch, containing a variety of functional groups [17]. Especially, agricultural materials having cellulose show a potential adsorption capacity for numerous contaminants.

In spite of sharing the main operation, food industries have a vast variety of residues due to the diversity of landscapes, climate, plant species, and economic activities all over the world. A great deal of them have been examined for the removal of different contaminants from aqueous solutions and wastewater.

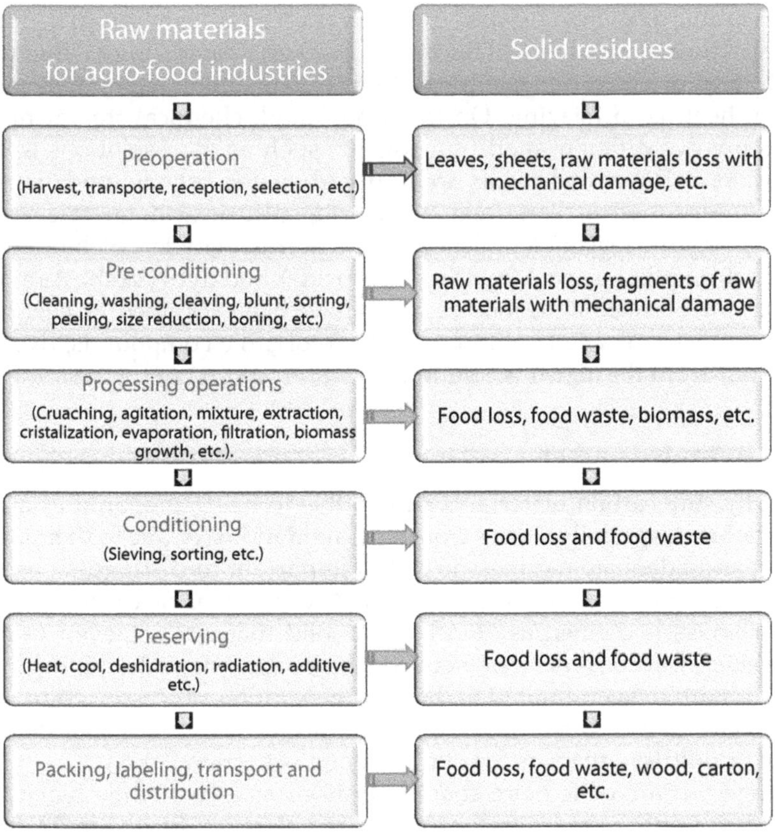

FIGURE 9.4
General flow diagram for the processes in the agro-food industry and residues generation during the process.

In addition, the following agricultural residues were shown as low-priced and readily accessible resources of bioadsorbent in the reviewed bibliography: agave bagasse [34], almond shell [35], apricot shell [35], barley straw [36], cashew nut shell [37,38], citric acid [36], corncob [39], cotton and gingelly seed shell [40], depectinated pomelo peel [41], Egyptian mandarin peel [42], fruit juice residue [43], garden grass [44], garlic peel [45], grapefruit peel [46], hazelnut shell [35], lentil shell [47], mango peel waste [48], Mosambi (Citrus limetta) peel [49], muskmelon peel [50], pine sawdust [51], pongam seed shell [40], groundnut shell [22], olive stone [52], plum kernel [53], pomegranate peel [53], pomelo peel [41], potato peel [54], rice shell [47], rice straw [55], sugarcane bagasse [56], walnut shell [35], banana peel [57], cane pith [58], coir pith [59], yellow passion fruit [60], orange peel [61], rice husk [43], sawdust carbon [62], soymeal hull [63], sunflower stalk [64], white ash [65], white rice husk ash [66], wood derived biochar [67], and coconut shells [68] have been examined for the removal of many dyes from aqueous solutions.

While some agro-food wastes have been used by researchers in their natural form, the majority of them have been used as a bioadsorbent after different physical or chemical treatments. Physical treatment techniques include cutting, heating, or sieving. On the other hand, chemical treatment techniques involve different modifying agents such as base solutions (sodium hydroxide, calcium hydroxide, sodium carbonate), organic and inorganic acid solutions (hydrochloric acid, nitric acid, sulfuric acid, tartaric acid, citric acid), organic reagent (ethylenediamine, formaldehyde, epichlorohydrin, methanol), oxidizing agent (hydrogen peroxide), and dyes [23,47]. As a result, different revalorized agro-food residues have been obtained as new bioadsorbent which are able to eliminate soluble organic compounds, dyes, and metal ions from the aqueous solutions.

9.3.1 Activated Carbon

Generally, any carbon material is susceptible to being prepared as an activated carbon, especially wastes from agro-food industry due to their low ash content and adequate hardness [69]. It is important to distinguish between activated carbon, biochar, and charcoal. The charcoal or carbon activated from biomass is defined as "Biochar," a solid material achieved from the thermochemical conversion of biomass (lignocellulosic material) in an environment with oxygen-limitation. Biochar generation processes go from traditional ovens and earth mounds to engineered processes founded on a flatbed or fluidised reactors for pyrolysis or gasification [70,71].

However, charcoal is more commonly used as a fuel while biochar can be used, not only as an energy producer but also as a fertilizer [72] or bioadsorbent [73]. Basically, the difference lies in the objective of the carbonic product. Therefore, due to the use of biochar as an adsorbent, it will be manufactured to a different set of adsorption properties that increase the internal microporosity of the original carbon-rich source material. There are a set of activation techniques that promote the removal of individual carbon atoms and creates nooks and crannies, increasing the adsorption sites. Properties of matrix, pyrolysis conditions (e.g. heating temperature and residence time), and posterior activation treatment will determine the adsorptive properties of the final activated carbon-based bioadsorbent [74]. They may be used to treat the total flow, or directed to a specific contaminant as part of a multi-stage approach. In this sense, biochar shares adsorption properties with activated carbon, but also exhibits a significant cation exchange capacity derived from side functional groups joined to graphitic backbones such as carboxylic, phenolic, hydroxyl, and carbonyl [72]. However, functionalization techniques avoid any residual side chain aliphatic groups in activated carbons, reducing the ionic interactions. Nevertheless, activated carbon has major bulk density and mechanical hardness favouring their regeneration and recycling, and their efficiency in purification. For these reasons, it is normally used as a packed bed adsorbent in the purification treatment of organic compounds from air and water streams and the demand of activated carbon has increased 10% in the last 5 years [74].

Conversely, biochar has a low-density bulk and mechanical hardness. However, an advanced technology in the development and functionalized of biochar could be crucial to replace commercial activated carbon with low-cost bioadsorbent [70]. As an example, Mohan et al. [75] showed similar removal efficiencies of lead and cadmium by oak bark char and a commercial activated carbon (Calgon F-400).

There is a widespread range of potential raw materials for biochar generation and production as wood and agricultural wastes, rice husks, manure, straw, leaves, food waste, paper sludge, bagasse and many other residues [73,76]. Different researchers have studied agricultural wastes derived from biochar for adsorbing heavy metals such as Pb, Cu, Zn, Cd, Cr, As, and Hg [77,78]. Additionally, forest residues may offer a good resource of biomass, although care must be taken to avoid any soil nutrient degradation by over removal [79]. Other types of waste such as manure or sewage sludge can be potentially applied for pyrolysis provided the excess process heat is used to dry the feedstock [80].

Activated carbon is used for other value-added applications, as filler of filters and for composite materials, catalysts, electronic material, and bioenergy applications [76,81]. The different properties of biochar make it a versatile material with several uses in the chemical and pharmaceutical industries [76]. Additionally, activated carbon has multiple applications as an adsorbent material for pollutant removal such as dyes. As an example, Juang et al. [39] used corncob-AC to remove acid blue 25 obtaining a maximum capacity, q_{max} = 1060 mg/g. Tseng et al., [82] used pine-AC to reduce acid blue 264 and basic blue 69 obtaining q_{max} = 1176 mg/g and 1119 mg/g, respectively. Other biorecalcitrant compounds studied as a target of AC-agro-food wastes were phenolic compounds or herbicides [76,81]. Furthermore, as it was remarked in the introduction section, activated carbon is one of the most effective bioadsorbent for removing a wide range of wastewater pollutants. The goal of tertiary treatment of wastewater is to complete the elimination of some compounds (perhaps non-biodegradables) that could not be removed during the previous treatments (preliminary, primary, and secondary) in urban wastewater treatment plants. Among others, filtration and adsorption are feasible methods used in order to refine the final water quality at the outlet of the treatment plants. It would be interesting to use low-cost bioadsorbent for removal of some compounds such as nitrogen or phosphate. In this sense, Pillai et al. [68] used AC-coconut shells to reduce urea obtaining a maximum adsorption capacity q_{max} = 60 mg/g. On the other hand, Kizito et al. [67] used AC-rice to reduce ammonium nitrogen registering a maximum adsorption capacity equal to 44.6 mg/g.

9.3.2 Fruit Residues

Fruits wastes, peels, and other by-products are formed in great amounts during industrial processing. Since last decades, efforts have been focused on reusing fruit residues. They are normally collected in the landfill or used as feed to livestock, fertilizer, or production of biochemicals [1].

They are usually rich on valuable biocompounds such as essential and edible oils, polyphenolic compounds, pigments, food additives, anti-carcinogenic compounds, dietary fibres, enzymes, bio-ethanol, bio-degradable plastic, and other miscellaneous products which can be recovered. In addition, fruit waste peels can be used in the removal of multiple water pollutants. Being a natural and a renewable resource, fruit residues are considered a new promising resource for environmental technology wastewaters treatments.

9.3.2.1 Orange (Citrus) Residue

Among different fruit residues, citrus production reaches 135×10^6 tons in 2013 [7] and their residues constitute around 15×10^6 tons of food waste annually [83]. Orange peel wastes are the most studied citrus residues as a bioadsorbent by the researchers due to their worldwide production and their potential capacity of adsorption [7].

Furthermore, high market demand leads to vast amounts of orange derived products and consequently a wide availability of orange peel wastes. This waste is disposed of in landfill sites where some greenhouse gases are released due to the microbiota digestion as well as the accumulation of polluting substances such as phenolic compounds [17,84]. For this reason, orange peel has been used as an example of fruit residue used as a bioadsorbent.

The composition of orange peel is known: moisture, 40.7%, ash, 7.39%, fat, 1.85%, pectin, 7.0%, lignin, 6.4%, crude fiber, 7.8%, total sugar, 14.08%, reducing sugars, 10.7% and non-reducing sugar, 3.70% [85]. These compounds contain OH functional groups that facilitate the adsorption of pollutants. Different researchers have modified orange peel by physical pre-treatments such as cutting, drying and grinding. However, the most elevated adsorption capacity of metals was obtained after chemical pre-treatment of orange peel. One of the most common modification comprises alkali saponification using NaOH, NH_4OH, $Ca(OH)_2$, MgCl, or KCl that facilitate the ion exchange mechanism between Na^+, K^+, NH_4^+, Ca^{2+}, Mg^{+2}, and the bivalent metal ions such as Ni, Cu, Zn, Cd, and Pb [86,87]. Different acid oxidation after saponification has been also studied, as a method of introducing a carboxyl group into cellulose structure that enhances the adsorption process [86,88].

In general, the structure of cellulose can be modified by oxidation agents and consequently, the properties of modified celluloses differ widely. It seems that the attack of oxidizing agents on cellulose structure is mainly limited to three positions: (i) the aldehyde end-groups can be reacted to carboxyl groups, (ii) the primary alcohol groups, which can be reacted to the aldehyde or carboxyl groups, and (iii) the glycol group (the 2,3-dihydroxy group), which can be reacted to the ketone, aldehyde, or carboxyl stage. The behaviour of the ion exchange of carboxylate groups in 2,2,6,6-tetramethylpiperidine-1-oxy radical (TEMPO)-oxidized fibrous cellulose prepared from cotton linters was compared with that of fibrous carboxymethyl cellulose (F-CMC) with almost the same carboxylate content as that of the TEMPO-oxidized cellulose [89].

Moreover, orange peel has been also modified by sulfuration or co-precipitation of Fe_3O_4 nanoparticles used as a potential adsorbent of Pb^{+2}, Zn^{+2}, or Cd^{+2} due to metallic complexation or ion exchange [90]. Additionally, it was also chemically modified and loaded with Zr(IV), La(III), Ce(III), and/or Fe(III) in order to eliminate Mo(VI).

On the other hand, it is possible to modify the operational conditions in order to establish the most favourable ones that lead to major adsorption capacity. In Table 9.1, the adsorption capacity in terms of maximum adsorption capacity (q_{max}) of the same absorbent (orange peel) for different metal ions at different operational conditions is shown. Table 9.2 shows the maximum adsorption capacity of orange peel for different organic substances and dyes, and Table 9.3 shows the maximum adsorption capacity of metal ions onto other fruit residues registered.

9.3.2.2 *Other Fruit Residues*

Many fruit residues were reported as good adsorbent as pomegranate peel [91], Litchi seeds [27], watermelon rind [92], coconut coir pith [93], Durian Shell waste [94], Mosambi peel [49], Muskmelon peel [50], Ponkan peel [60], Egyptian mandarin [95], yellow passion fruit [60], pomegranate peel [91], and citrus limetta residues [43]. Some of them can be observed in Table 9.3, where the maximum adsorption capacity is q_{max} = 140 mg/g.

9.3.3 Olive (Olea Europaea) Residues

Olive oil industry is one of the most important commercial sector in the agro-food industries, above all in Mediterranean countries [96]. Hence, there is desired a feasible method for management of olive and olive oil residues produced, such as olive stones or olive tree pruning.

Olive stone residue is normally incinerated or dumped without control in the olive oil industry. However, nowadays the olive stone is being used as fuel [97]. It has been also investigated as a bioadsorbent in a raw state or after different treatments. Fiol et al. [52] tested olive stone dried and sieved (particle size of 0.75–1.5 mm) to Cu(II), Pb(II), Ni(II), and Cd(II) adsorption obtaining 3.19×10^{-5}, 4.47×10^{-5}, 3.63×10^{-5}, and 6.88×10^{-5} mol/g at 20°C. Ronda et al. [30] tested natural olive stones as bioadsorbent of Pb(II) and with three treatments: nitric acid (HNO_3), sulfuric acid (H_2SO_4), and sodium hydroxide (NaOH). Different operational conditions were also tested.

TABLE 9.1

Maximum Adsorption Capacity Determined (q_{max}) for Different Metal Ions Using Natural and Modified Orange Peel Residue as Bioadsorbent

Metal Ion	Treatment	Operational Condition	q_{max} (mg/g)	Reference
As(III)	Washed and grounded	$As(III)_0$ = 50 mg/L, 30°C, 90 min, 4 g, pH = 7 and 180 rpm.	3.43	Kamsonlain et al., (2012)
Sb(IV)	Fe(III)-loaded SOW	$Sb(III)_0$ =15 (mg/L) batch-adsorption 25 mg Adsorbent/ 15 mL, 30°C, 140 rpm, 24 h and pH = 10.	14.8	Biswas et al., (2009)
Cd(II)	Mercapto-acetic acid or grafted copolymerization	Batch-adsorption; 25°C; 50 mg/L single metal ion (Cd_2^+) 5 g/L adsorbents.	136.1 or 293.3	Liang et al., (2009)
Cd(II)	Citric acid after alkali saponification	Batch-adsorption, 0.025 g adsorbent/15 mL aqueous solution; pH = 5.5-6.0; 120 min; 80°C.	0.101	Li et al., (2007)
Cu (II)	Mercapto-acetic acid	Batch-adsorption; 25°C; 50 mg/L single metal ion (Cu_2^+)/5 g/L adsorbents.	70.7	Sha et al., (2009)
Ni (II)	Unmodified	0.2 g/100 mL of Ni(II), room temperature, pH = 5, 14 min and 200 rpm.	62.3	Gonen and Serin (2012)
Ni (II)	Grafted copolymerization	30°C, 0.050 g/25 mL of metal ion, pH = 5.5; 120 rpm and 3 h.	70.7	Feng et al., (2011)
Pb (II)	NaOH y $CaCl_2$	25°C, 0.10 g/25 mL of metal ions solution, 120 rpm, pH 5.5 and 10 min.	209.8	Feng and Guo (2012)
Pb (II)	Grafted copolymerization	30°C, 0.050 g/25 mL of metal ion, pH = 5.5, 120 rpm and 3 h.	476.1	Feng et al., (2011)
Zn (II)	NaOH y $CaCl_2$	25°C, 0.10 g/25 mL of metal ions solution, 120 rpm, pH = 5.5 and 10 min.	56.2	Feng and Guo (2012)
Cu (II)	Biomass acidification (acetic acid 98%) and copolymer grafting.	0.12 g/12 mL of adsorbate solution; pH = 5, agitation intermittently 0.5 min to 1 h. Time = 1 h, Adsorbate = 30 mg/L.	289.0	Lugo-Lugo et al., (2009)

At mild operational conditions (T = 50°C, pH = 5, chemical solution concentration = 0.1 M) results showed 3.15, 2.79, and 4.59 mg/g, respectively.

Olive-oil factories generate variable quantities of wastewaters, which need a treatment for their disposal or reuse. In this sense, Nieto et al. [98] proposed Fenton's oxidation process as a solution to get irrigation water. In this process, olive stone residue was used as a filler of the filters to retained particles and iron ions. Nowadays, for olive oil mill wastewater reuse regulations are becoming gradually stricter regarding the parameters determined in these

TABLE 9.2
Maximum Adsorption Capacity of Orange Peel for Different Organic Substances and Dyes

Organic Substance	q_{max}, mg/g	Reference
Congo red	22.4	Namasivayam et al., (1996)
Procion orange	1.30	Namasivayam et al., (1996)
Rhodamine-B	3.22	Namasivayam et al., (1996)
Acid violet 17	19.9	Sivaraj et al., (2001)
Direct Red 23	10.7	Arami et al., (2005)
Direct Red 80	21.1	Arami et al., (2005)
Direct blue-86	33.8	Nemr et al., (2009)
Direct Yellow 12	75.8	Khaled et al., (2009a)
Direct Navy Blue 106	107.5	Khaled et al., (2009b)
Toluidine blue	314.3	Lafi et al., (2014c)
Carbofuran	84.5	Chen et al., (2012)
Furadan	161.3	Choi et al., (2013)

TABLE 9.3
Maximum Adsorption Capacity of Fruits Used as Bioadsorbents for Metals

Residue	Residue Treatment	Metal	q_{max}, mg/g	Reference
Lemon residues	$FeCl_3$	As(V)	0.47	Marin-Rangel et al., (2012)
Pineapple peel fiber	Succinic anhydride	Cd (II)	34.2	Hu et al., (2011)
Cortex banana wastes	Natural	Cu (II)	36.0	Kelly-Vargas et al., (2012)
Pomelo	Cutted. Washed. dried 70°C and sieved (particle size ≤ 0.42 mm or < 35 mesh)	Cu(II)	19.7	Tasaso (2014)
Grape fruit peel	Washed. dried 70°C and sieved (particle size = 20–40 mesh)	U(VI)	140.8	Zou et al., (2012)
Mango peel waste	Washed. dried 70°C and sieved (particle size = 0.85–1.00 mm)	Cu(II)	46.1	Iqbal et al., (2009)

effluents. In Spain, the Resolution by the President of the Hydrographical Confederation of the Guadalquivir on water use [99] set parameter limits as follows: pH = 6.0–9.0, total suspended solid = 500 mg/L, and maximum values for COD and BOD_5 were fixed at 1 g O_2/L.

In addition, table olive and olive-oil industries produce olive stones as a by-product. In Spain, more than 3.7×10^5 tons of triturated olive stones are generated each year [100], most of which are been destined for combustion and for production of activated carbon [101]. In spite of the environmental benefits of using this biomass as a fuel or a raw material for activated carbon, some problems persists such as air pollution. Additionally, an economically and environmentally interesting alternative would be its use as a bioadsorbent for heavy metals ions. In this sense, many studies have shown that olive stones could be revalorized as biomass sources to eliminate pollutants such as phenols [102], dyes [103], or heavy metals like Cd (II), Pb (II), Ni (II) [52,104].

The characterization of a bioadsorbent is an important task for understand the behavior or the mechanism of iron elimination on the surface of bioadsorbent. Pore sizes are classified in accordance with the classification accepted by the International Union of Pure and Applied Chemistry (IUPAC), that is, micropores (diameter < 2 nm), mesopores (2 nm < diameter < 50 nm), and macropores (diameter > 50 nm). In Table 9.4, it can be seen that the porosity of agro-food materials (olive stones) have been classified into two different ranges, pores with diameter from 3.8 to 50 nm (represented 12 and 38% of total specific volume of pores), and pores with diameter higher than 50 nm (represented 66 and 88% of total specific volume pores), and therefore olive stone can be identified as macroporous bioadsorbent.

Bioadsorption of iron on olive stones is governed by the structural and morphological features of the support. The olive stones which were measured in this study were mesopores (pore width ca.2 to ca.50 nm) and macropores (pore width > 50 nm) (Table 9.4). This structure permits its use in iron bioadsorption. The pore and particle sizes are a key to control the bioadsorption process. Firstly, the influence of particle size in iron bioadsorption onto

TABLE 9.4

Characteristics of Crude Olive Stones

Particle Size (mm)	Pore Diameter (nm)	Total Cumulative Volume ($cm^3\,g^{-1}$)	Pore Diameter Ranges (nm)	Specific Volume ($cm^3\,g^{-1}$)
< 4.8[a]	13.8	0.17	3.8–50	0.02
			>50	0.15
2–3	8.8	0.14	3.8–50	0.03
			>50	0.11
2–1.4	8.8	0.09	3.8–50	0.03
			>50	0.06
1.4–1	8.8	0.13	3.8–50	0.03
			>50	0.10
<1	8.8	0.12	3.8–50	0.03
			>50	0.09
< 4.8[b]	4.4	0.11	3.8–50	0.02
			>50	0.09

[a] Original crude olive stones before the adsorption.
[b] Original crude olive stone after the adsorption. Adsorption conditions: Fe (III): 20 mg dm-3; olive stone: 37.5 g dm-3; stirring rate: 117 rpm; pH = 2.9; temperature: 293 K.

olive stones was analysed. The total specific surface area, as well as the total cumulative specific volume, had no alteration and it may be considered the average values 23 $m^2 g^{-1}$ (Standard deviation = 2.4) and 0.12 $cm^3 g^{-1}$ (Standard deviation = 0.019), respectively. The low total cumulative volume indicates that the surface of olive stone had no porous structure, but the surface had a roughness which mainly manifests as macropores when the porosity was determined.

The average diameter value of the pores in the original raw olive stone is equal to 13.8 nm, and this value is shifted to lower values with the crushing and separation of particles. It occurs the same when data are obtained after bioadsorption. In summary, no significant variations in the distribution of the pores were registered (Table 9.4).

As an industrial application of olive stone as bioadsorbent, Nieto et al. [28,99] confirmed the efficiency of the use of raw olive stone to remove residual iron ions registered in the effluent after real olive oil mill wastewater treated by Fenton process in real conditions. Bioadsorbent is installed at the end of the wastewater treatment plant located in Baeza (Spain) operating using Fenton's reagent, consisting of a system composed of three packed filters (sand, olive stones, and gravel), showed in Figure 9.5.

Wastewater treatment processes for metals removal include operations such as precipitation-flocculation and filtration. The clarified water still

FIGURE 9.5
Three packed filters (sand, olive stone, and sand) operated in the Industrial Wastewaters Treatment Plant of the company of S.A.T. Olea Andaluza situated in Baeza (Spain).

contains toxic amounts of heavy metal ions. Chromium is one of the most important trace metals in water effluents due to its ability to form very stable complexes in solution [105]. With the aim to remove very low concentrations of chromium, an adsorption operation can be added to the wastewater treatment processes. Bioadsorption of Cr (III) in low-cost carbons has acquired importance through the years. The main parameters influence in the bioadsorption are pH, temperature, ion concentration, and the presence of other inorganic ions [106].

On the other hand, olive tree pruning is one of the most important lignocellulosic residues from Mediterranean countries with an average of 3000 kg/ha per year worldwide [107]. Nowadays, over 10×10^6 ha area worldwide is cultivated by about 900×10^6 olive trees, 98% of this area are located in the Mediterranean coast and covering an area of 5.163×10^6 ha [108].

Olive tree pruning as a residue is usually eliminated by burning, grinding, or scattering. As a revalorized by-product, it can be used as domestic firewood [109], as a source of sugars for biorefinery of valuable compounds [110], or as a source of cellulosic matrix for cellulosic derived products [111].

According to the data of the International Olive Council, the worldwide consumption of olive oil rose 78% between 1990 and 2010. The rise in olive oil production implies a proportional growth in olive mill wastes. The annual world olive oil production yields has valued a huge quantities of wastes 8.1×10^6 m^3 of OMW, 3.2×10^6 tons of olive press cake, and 0.3×10^6 tons of leaves [112]. Furthermore, the total amount of twigs and leaves resulting from pruning each year reaches 15×10^6 tons [113]. As a consequence of this growth trend, olive mills are fronting severe environmental impacts due to the absence of feasible and/or cost-effective solutions to olive oil mill waste management.

9.3.4 Garlic (Allium Sativum L.) Residue

Garlic is an essential vegetable that has been widely consumed as a culinary product as well as an herbal remedies with high effectiveness, few side effects, and relatively low cost [114]. It is known that garlic peel has adsorbent properties. In this sense, garlic peel has been investigated as a bioadsorbent for dyes [115] and phenols [116]. Results showed major adsorption affinity by the dye than phenol with amounts absorbed of 38.0 and 14.5 mg/g, respectively. Garlic also has been studied as a bioadsorbent of several metals ions such as Pb^{+2}, Cu^{+2}, and Ni^{+2} [117]. Results showed that maximum adsorption capacity registered to Pb^{+2} was 209 mg/g, but the adsorption capacity was influenced by the presence of other metals ions. This occurred because of the ion exchange mechanism between Ca^{+2} from garlic peel and heavy metal ions in the solutions. Bioadsorption of garlic peels to Pb^{+2} after mercerization was also investigated by this group. Obtaining an adsorption capacity increased 2.1 times from native garlic peel.

9.3.5 Cassava (Manihot Esculenta Crantz) Residue

Cassava, also called manioc, yuca, or tapioca, is commonly cultivated in many countries across Africa, Asia, and Latin America [118]. Cassava plant is the most important staple root crop in the world and the cheapest source of starch used in more than 300 industrial products [119]. The residue obtained from starch extraction is called cassava bagasse and preserve 40–50% of starch and 10% of cellulose in the dry matter [120]. Cassava bagasse has been proved to have great potential as an adsorbent for metal ions and dyes in aqueous solutions [121,122] used Cassava bagasse treated by 1M of thioglycolic acid to Cd^{+2}, Cu^{+2}, and Zn^{+2} bioadsorption. Results showed adsorption capacities of 26.3 mg/g, 90.9 mg/g, and 83.3 mg/g, respectively. In another study conducted by Wang et al. [123], cassava bagasse pyrolyzised at different temperatures were used to adsorb the antibiotic ofloxacin. In that study, also analysed was the effect of pH and water metal ions interactions. In the case of pH, only at extreme conditions was ofloxacin adsorption influenced, and the operation range of 3–9 had a minor effect on ofloxacin adsorption to biochars. Finally, it was concluded that ofloxacin sorption capacities of biochars were enhanced by supplementing with Zn^{2+} and Al^{3+}. Cu^{2+} reduced the bioadsorption of ofloxacin when cassava bagasse was pyrolyzed at 450°C and 550°C. On the other hand, cations K^{+} and Ca^{2+} had no significant effect on the ofloxacin sorption.

Cassava bagasse was also used to prepare microspheres, spherical carbon, and activated carbon as bioadsorbents [124,125]. As an example, Xie et al. [125] prepared microspheres as a Cu(II) bioadsorbents. The resulted magnetic particle had a uniform size about 206 nm and good reusability. Maximum Cu (II) bioadsorption capacity "q_{max} = 110.5 mg/g" was achieved when the following operation conditions were used: pH = 6.4, time = 105 min, temperature = 25°C, and initial Cu(II) concentration = 200 mg/L. In addition, recycling microsphere was tested and after the fifth cycle, the adsorption capacity was 92.1 mg/g. Activated carbon adsorption based on cassava peel also has been studied in copper and other metals adsorption by Moreno-Piraján and Giraldo [126] and Owamah [78]. Also, Alcantara et al. [127] demonstrated that cassava plants can capture Hg and Au from the soil.

9.3.6 Cucumber (Cucumis Sativus L.) Residue

Cucumber is the fourth most important vegetable worldwide [128]. The origin of cucumber is in India but its cultivation has been extended to Western Asia, where China is the main producer country followed by Russian Federation, Iran, Turkey, and Spain in the top five producers countries (Figure 9.4). The cucumber is worldwide consumed in salads, fermented (pickles), or as a cooked vegetable due to the nutrients content. In addition, the seeds of cucumber served as a source of proteins and amino acids and table oils [129]. The vast demand provokes a massive production of wastes (cucumber peel) rich in lignocellulosic material disposed of as garbage as a source of pollution. For this reason, the reuse of cucumber peel has been investigated as a new low-cost bioadsorbent to metals such as lead (Pb) or cadmium (Cd). Basu et al. [130] reported that optimal adsorption conditions for Pb adsorption using dried and heated cucumber waste were pH = 5 where 90% of the metal was removed in 15 min. Cucumber backbone account with carboxyl and phosphate groups that played crucial roles in adsorption because of ion exchange mechanism. Moreover, the presence of cadmium does not affect the lead bioadsorption. All this data shows the capability of decontaminating lead in industrial effluents as an approach to practical application. Cadmium also was intended to be adsorbed by acid-modified cucumber peel waste [131]. The optimal Cd sorption was 58.1 mg/g at pH = 5, temperature = 20°C, and time = 90 min.

The majority of cucumber peel adsorption experiments were applied to remove dyes in the literature. Methylene Blue was the target for Akkaya et al. [132]. They determine optimal conditions with a q_{max} = 111.1 mg/g at pH = 7–10, bioadsorbent dose = 0.1 g/50 mL, temperature = 20°C and desorption eluent = 0.1 M-HCl with a recovery of dye 93.7%. Shakoor and Nasar [133] also studied cucumber peel waste as bioadsorbent of Methylene blue. In this case, q_{max} = 21.5 mg/g were achieved at optimal operation conditions: adsorbent dose = 4 g/L, pH = 8, T = 20°C, contact time = 1 h and using HCl as desorption eluent with a recovery of dye 63.6%.

As it has been observed before, it is possible to make activated carbon from any kind of agro-food residue. In this sense, there are some researchers using activated carbon based on cucumber in order to remove dyes from water [134,135]. Santhi and Manonmani [134] used a batch experiment for the adsorption of Malachite Green using 0.2 g/50 mL of activated carbon based on cucumber peel. Results showed that at room temperature = 27–28°C, agitation speed = 150 rpm, and contact time = 80 min, and the amount of Malachite Green bioadsorbed was 36.2 mg/g. This result was obtained using a synthetic sample of 100 mg/L of Malachite Green. Furthermore, this group tested this experimental conditions in the adsorption treatment of a real wastewater. This wastewater belongs to a local dyeing and contained four dyes namely Yellow MR, Red M5B, Black B, and Malachite Green. In this case, the percentage removal of wastewater was 59.0%. In spite of incomplete adsorption and lack of more information, this investigation was a great step forward to using cucumber peel waste for a real adsorption treatment process.

9.3.7 Parsley (Petroselinum Crispum) Residue

The demand for spices and other natural preservatives is increasing due to their radical scavenging ability and the pathogenic protection capacity [136]. Among others, parsley is the most consumed aromatic herb in Europe and it is often used as kitchen spice and as nitrate additive in meat products because its antioxidant, antibacterial, and antifungal activities [136–138].

Parsley is usual in Mediterranean cuisine as a flavouring and aromatic food additive and as a folk medicinal plant because of its pharmacological properties [137]. Adsorptive properties on arsenite (As(V)) and arsenite (As(III)) of parsley biomass have been examined by Jiménez-Cedillo et al. [139]. In this research, 50 g of parsley non-living biomass was treated with a $FeCl_3$ solution with the aim of increasing iron content. It is known that iron has been used to modify biomass (e.g. tea fungal mat, *Staphylococcus xylosus, Aspergillus niger,* or maracuya biomass) due to its natural affinity toward arsenic species [140]. Therefore, both Fe-modified non-pyrolyzed and pyrolyzed biomass from parsley were tested in order to remove As(III) and As(V). Best results occurred when using iron-modified carbonaceous material obtained after pyrolysis of Parsley (q_{max} = 18.2 mg/g).

9.3.8 Lentils (Lens Culinaris) Residue

Many authors have studied the use of agro-food waste from lentils processing as bioadsorbent [47,141]. Aydin et al. [47] studied the removal of copper (II) from aqueous solution by adsorption onto low-cost adsorbents showing 8.98 mg Cu (II)/g adsorbent at 20°C. Basu et al. [141] evaluated the adsorption behaviour of lead and cadmium on the husk of lentils, finding a correlation between pH and the adsorption process. Basu et al. [141] showed that the optimum pH was 5.0, leading to an adsorption of 81.4 mg/g and 107.3 mg/g of cadmium on lentil husk, respectively. The authors have investigated the chemical modification of functional groups, which revealed that both hydroxyl and carboxyl groups played a crucial role in the binding process [19].

9.3.9 Apple (Malus Domestica) Residue

Waste derived from apple processing have been studied by Ozbay and Yargic [142] and Chand et al. [143]. Ozbay and Yargic [142] evaluated the bioadsorption of Cu (II) and Co (II) by carbonized apple pulp at 550°C as a low-cost bioadsorbent and achieved, in optimal conditions, the removal percentages of Cu (II) (90.5%) and Co(II) (65.1%). Additionally, the same authors showed that removal processes were pH-dependent, adsorbent dosage, temperature, and contact time. A Xanthate-Modified Apple Pomace (XMAP) has been tested as a bioadsorbent for the removal of Cd (II), Ni (II), and Pb (II) by Chand et al. [143]. The maximum bioadsorption capacities observed were 112 mg Cd (II)/g XMAP, 51 mg Ni (II)/g XMAP, and 178.5 mg Pb (II)/g XMAP. This study also showed the regeneration capacity of the XMAP, which could be re-used up to 8 times with a removal efficiency above 50%.

9.3.10 Corn (Zea Mays) Residue

Corn-based bioadsorbents have been evaluated by Alves et al. [144–146]. The platinum recovery using corn husk-based bioadsorbent was tested by Tavassolirizi et al. [145], which compared different corn husk-based bioadsorbents (acid and anime treatments). The bioadsorption capacity was 16.2 mg Pt/g bioadsorbent, hydroxyl, and anime functional groups showed as a key in the bioadsorption process and aluminates showed interference with the bioadsorption of platinum ions [145]. The bioadsorbent capacity of corn stalk-based adsorbent has also been tested by Xiong et al. [146], which synthesized the bioadsorbent based on the carbonize corn stalk (OCS) modified by ammonia (NH_3)–thiosemicarbazide (TSC) to the selective recovery of Ag(I). The maximum bioadsorption capacity in the Ag(I) single and Ag(I)–Cu(II)–Ni(II) ternary system were notified as 153.5 and 46.7 mg/g of bioadsorbent, respectively. Solid residues from corn processing such as corn cobs have been evaluated as bioadsorbent by Alves et al. [144], showing a maximum capacity for phenylalanine removal equal to 109 mg phenylalanine/g bioadsorbent. The bioadsorbent was prepared by means 3 min impregnation with H_3PO_4 followed by 1 h activation in a muffle furnace [144].

9.3.11 Wheat (Triticum) Residue

The wheat processing residues have been studied as bioadsorbent by Aydin et al. [47]. These authors evaluated shells of wheat as bioadsorbent to remove copper (II) from aqueous solution showing 16.1 mg Cu (II)/g bioadsorbent at 40°C. Effectively removal of cationic and anionic dyes by bioadsorbents derived from waste-wheat straw have been studied [147,148]. Bioadsorbents were prepared by acid modification and the removal efficiency was verified. The maximum bioadsorption capacities were 205.4 mg/g at 25°C [147] and 506 mg/g at pH = 2 [148]. You et al. [147] and Lin et al. [148] suggested that the bioadsorption was chemisorption. Additionally, Lin et al., [148] concluded that the bioadsorption process involved valence forces through sharing or exchange of electrons. Atrazine removal has also been studied by the biochars derived from wheat straws by Yang et al. [149]. N-doped porous carbon sheets resulted from wheat straws are fabricated by using molten salts via the carbonization-functionalization progress, showing a specific structure. This structure can bring about the desired performance of bioadsorption capacity of 82.8 mg/g [149].

9.3.12 Rice (Oryza) Residue

The annual production of rice processing waste comprises one-fifth of the annual gross rice production worldwide [150]. This agro-food waste has many functional groups such as hydroxyl that can bind metal ions [151]. In literature, the agro-food waste derivate from rice processing has been deeply studied [55,151–156]. Authors have investigated agro-food waste derivate from rice processing as bioadsorbent to removal many compounds as phenolic compounds, heavy metals (Cr, Ni, Zn, Cd, Mn, Co, Cu, Hg, Pb, and Au), and dyes (such as malachite green and Methylene blue). The removal of heavy metal ions using agro-food waste derivate from rice processing is the main goal in the studies revised. Singh et al. [152] used a bioadsorbent rice bran achieving 99.4% Cr (VI) removal at pH = 2.0 and temperature = 20°C. Gao et al. [153] indicated that carboxyl groups present on the biomass played an important role in chromium remediation, showing 3.15 mg/g of sorption capacity of rice straw. Krishnani et al. [154] prepared a biomatrix from rice husk for the removal of several heavy metals and to study the role of calcium and magnesium present in the biomatrix in ion exchange mechanism. The maximum bioadsorption capacity values obtained by Krishnani et al. [154] were Ni (0.094), Zn (0.124), Cd (0.149), Mn (0.151), Co (0.162), Cu (0.172), Hg (0.18), Pb (0.28), and Cr(III) (1.0) mmol/g. Krishnani et al. [154] concluded that enough calcium, magnesium in the biomatrix means the high presence of -OH and -COOH groups, which help the ion exchange with metal cations. Rice straw has been notified to be an excellent bioadsorbent for aquatic Cd removal by Ding et al. [55], whose determined ion exchange as the main sorption mechanism and functional groups such as C-C, C–O, O–H or carboxylic acids were involved in chelation step. Ding et al. [55] showed 13.9 mg/g as the high bioadsorption capacity. Xu et al. [151] determined the maximum bioadsorption capacity to remove Au(III) by rice husks-based bioadsorbent developed through fast and simple esterification reaction with hydroxylethylidenediphosphonic acid, which under the optimum process conditions could reach 3.25 ± 0.07 mmol/g. The effectiveness of rice straw-derived char, rice husk, rice husk ash to remove dyes has been evaluated in the literature [23,155,156]. Hameed and El-Khaiary [23] calculated the maximum rice straw-derivated char sorption capacity 148.7 mg malachite green/L at 30°C. Sharma et al. [155] used pretreated rice husk and rice husk ash for the removal of methylene blue from wastewater calculating the maximum bioadsorption capacity i.e. 1,347.7 mg methylene blue/g of rice husk and 1,455.6 mg methylene blue/g of rice husk ash at 50°C. Tolba et al. [156] found that the amorphous nano silica separated from rice-husk could be introduced as effective and reusable bioadsorbent for dyes. Tolba et al. [156] calculated thermodynamic parameters indicating that the bioadsorption of methylene blue is spontaneous and endothermic. Furthermore, the bioadsorbent could be recycled by a simple heat treatment retaining the high removal efficiency in four successive cycles [156].

9.3.13 Coffee Residue

Coffee waste is produced in huge quantities worldwide and recently researchers are showing its applicability in the wastewater treatment [157]. Dyes and heavy metals removal are among the most typical processes studied in literature using coffee waste as bioadsorbent [24,158–162]. Kyzas [158] studied the removal of Cu(II) and Cr(VI) from aqueous solutions with commercial coffee wastes, and observed that the maximum bioadsorption capacity of the coffee residues can reach to 70 mg/g and 45 mg/g for Cu(II) and Cr(VI) removal, respectively. Confirming the strong potential of reuse, this low-cost bioadsorbent achieved ten cycles of adsorption-desorption. Liu et al. [160] investigated the role of exhausted coffee on Cr(VI), Cu(II), and Ni(II) bioadsorption determining that an alkaline hydrolysis destroyed in part the structural compounds of the bioadsorbent which resulted in an insignificant decrease of Cr(VI) removal while a significant increase of Cu(II) and Ni(II) bioadsorption. Alhogbi [163], who evaluated coffee husk biomass waste as bioadsorbent to remove Pb(II), achieved a bioadsorption capacity of 19.0 mg/g. Baek et al. [24] studied the feasibility of employing degreased coffee beans as bioadsorbent for malachite green removal in dyeing wastewater, finding that degreased coffee beans have higher bioadsorption efficiency than raw coffee beans. Baek et al. [24] showed that the bioadsorption process is spontaneous and endothermic nature. Lafi et al. [159] evaluated the bioadsorption performance of coffee waste for the removal of two basic dyes, toluidine blue and crystal violet, achieving a maximum bioadsorption capacity equal to 142.5 mg/g for toluidine blue and 125 mg/g for crystal violet. In these cases, the bioadsorption processes are exothermic. Shen and Gondal [162] proved exhausted coffee ground powder as bioadsorbent for the removal of Rhodamine dyes founding maximum bioadsorption capacities value equal to 17.4 μmol/g, which may be caused through the hydrophobic interaction forces from the ester group (–$COOCH_3$). Ronix et al. [161] used coffee husk as precursor of hydrochar and applied in sorption studies of methylene blue dye. The authors estimated the maximum bioadsorption capacity equal to 34.9 mg/g. The thermodynamic study showed that bioadsorption occurs via spontaneous and endothermic reaction.

9.3.14 Other Residues

De Gisi et al. [19] analysed other low-cost adsorbent coming from other industries, sludge, sea materials, soil, and ore materials, as well as agricultural and household residues. In this study, the affinity of each agro-food-based bioadsorbent was compared with four pollutants studied in the literature: (i) Dyes, (ii) Heavy metals, (iii) Biorecalcitrant as herbicides and fungicides, and (iv) Nutrients as phosphates and nitrates. From agricultural and household residues, it was concluded that banana and orange peel were compounds with less selectivity, able to adsorb acid and basic dyes and heavy metals. In addition, agro-food residues studied were the only low-cost residues with affinity for nutrients such as phosphorous and nitrogen compounds. Furthermore, depending on the metal or organic compounds, the matrix from bioadsorbent would be more or less efficient. As an example, according to Bilal et al. [164], copper bioadsorption capacity of different bioadsorbent reviewed can be order as follows: activated carbon>algal> bacterial>agriculture and forest>fungal> yeast biomass.

9.4 Conclusions and Future of Bioadsorption Application

According to the bibliography, most of the researchers studied agro-food residues as bioadsorbents with synthetic wastewater. The physicochemical compositions of real wastewater are different and variant. For this reason, experimental results obtained from synthetic wastewater cannot be applied to wastewater treatment plants, indicating that the experimentation on a real wastewater is the key to accelerating the scale-up of this technology. Therefore, it is necessary and recommendable to develop novel adsorbents from biowastes. In this context, it is important to characterize the biowastes to determine the specific properties of such materials (porosity, surface functional groups, etc.) which make the materials interesting for wastewater purification by adsorption, without ignoring the evaluation of the adsorption capacity of the novel materials for real wastewater contaminants.

A growing interest in biochar production has been seen in literature. However, the lack of commercial suppliers prevents a cost-effective analysis in full-scale applications. In addition, biochar production process seems to be a cutting edge level-knowledge in which researchers are testing different production conditions and material sources. Therefore, such efforts seem to be an interesting future research trend about bioadsorbents production.

It has been demonstrated that the presence of a second metal ion may have an inverse effect on the sorption of the target metal [165]. In the case of decontaminating real wastewaters, a combination of several biowastes should be studied as adsorbents, which have metal adsorption selectivity. If different bioadsorbents show different adsorption mechanisms, no competition phenomenon takes place and consequently, it results in a synergic effect [17]. Therefore, not only the efficiency of bioadsorbent depends on the adsorption rate and capacity, but also on its selectivity. Further development of studies which evaluate different biowaste combinations for different wastewater seems to be the key to scale-up of the bioadsorption processes.

Results show that physical or chemical activation of bioadsorbents has been successfully proved in metals and other organic biorecalcitrant compounds uptake. Additionally, most of the reviewed studies have focused on bioadsorption in bed columns at batch experimental models. Consequently, further investigation based on continuous or semi-continuous processes seems to be required to help the scaling up of the biosorption processes. Moreover, despite many studies of bioadsorbent production and/or activation from agro-food waste, there is a lack of studies of regeneration and reuse of the bioadsorbents. This approach is important, because the experimental costs of the activation methods continue to be unbearable.

Acknowledgments

The activity was performed within the Project Ref.: AGR-7092 "Application of advanced oxidation technologies for treating of washing wastewaters of olive oil and olives" founded by the Junta of Andalusia and Ministry of Economy and Competitiveness (Spain).

References

1. Nanda, S., Isen, J., Dalai, A. K., Kozinski, J. A. (2016) Gasification of fruit wastes and agro-food residues in supercritical water. *Ener Conver Manage* 110: 296–306.
2. Stenmarck, A., Jensen C., Quested T., Moates G. (2016) Estimates of European food waste levels. Ed. Swedish Environmental Research Institute, Stockholm, Sweden.
3. Rossini, G., Toscano, G., Duca, D. et al. (2013) Analysis of the characteristics of the tomato manufacturing residues finalized to the energy recovery. *Biomass Bioenergy* 51: 177–182.
4. Ariunbaatar, J., Panico, A., Frunzo, L., Esposito, G., Lens, P. N. L., Pirozzi, F. (2014) Enhanced anaerobic digestion of food waste by thermal and zonation pretreatment methods. *J Environ Manage* 146: 142–149.
5. Algapani, D. E., Qiao, W., Su M. et al. (2016) Bio-hydrolysis and bio-hydrogen production from food waste by thermophilic and hyperthermophilic anaerobic process. *Bioresource Technol* 216: 768–777.
6. Miranda-Ackerman, M. A., Azzaro Pantel, C., Aguilar Lasserre, A. A. (2017) A green supply chain network design framework for the processed food industry: Application to the orange juice agrofood cluster. *Comput Ind Eng* 109: 369–389.
7. FAO, 2015. Food and Agricultural Organization of the United Nations. Food and Agricultural Organization of the United Nations Statistics Division. http://www.fao.org/faostat/en/#home. (accessed September 05 2017).
8. Van Dyk, J. S., Gama, R., Morrison, D., Swart, S., Pletschke B. I. (2013). Food processing waste: Problems, current management and prospects for utilisation of the lignocellulose component through enzyme synergistic degradation. *Renew Sust Energ Rev* 26: 521–531.
9. Gullón, B., Falqué, E., Alonso, J. L., Parajó, J. C. (2007) Evaluation of apple pomace as a raw material for alternative applications in food industries. *Food Technol Biotech* 45(4): 426–433.

10. Wilkins, M. R., Widmer, W. W., Grohmann, K., Cameron, R. G. (2007) Hydrolysis of grapefruit peel waste with cellulase and pectinase enzymes. *Bioresource Technol* 98: 1596–1601.
11. Emaga, T. H., Robert, C., Ronkart, S. N., Wathelet, B., Paquot, M. (2008) Dietary fibre components and pectin chemical features of peels during ripening in banana and plantain varieties. *Bioresource Technol* 99: 4346–4354.
12. Ravindran, R., Jaiswal, A. K. (2016) Exploitation of food industry waste for high-value products. *Trends Biotechnol* 34: 58–69.
13. Galanakis, C. M. (2012) Recovery of high added-value components from food wastes: Conventional, emerging technologies and commercialized applications. *Trends Food Sci Tech* 26(2): 68–87.
14. Yu, H., Wang, T., Yu, L. (2016) Remarkable adsorption capacity of Ni-doped magnolia-leaf-derived bioadsorbent for congo red. *J Taiwan Institute Chem Engineers* 64: 279–284.
15. Arevalo Gallegos, A., Ahmad, Z., Asgher, M., Parra Saldivar, R., Iqbal, H. M. (2017) Lignocellulose: A sustainable material to produce value-added products with a zero waste approach—A review. *Int J Biol Macromol* 99: 308-318.
16. Nguyen, T. A. H., Ngo, H. H., Guo, W. S. et al. (2013) Applicability of agricultural waste and by-products for adsorptive removal of heavy metals from wastewater. *Bioresource Technol* 148: 574–585.
17. Bhatnagar, A., Sillanpää, M., Witek Krowiak, A. (2015) Agricultural waste peels as versatile biomass for water purification—A review. *Chem Eng J* 270: 244–271.
18. Gautam, R. K., Mudhoo, A., Lofrano, G., Chattopadhyaya, M. C. (2014) Biomass-derived biosorbents for metal ions sequestration: Adsorbent modification and activation methods and adsorbent regeneration. *J Environ Chem Eng* 2(1): 239–259.
19. De Gisi, S., Lofrano, G., Grassi, M., Notarnicola, M. (2016) Characteristics and adsorption capacities of low-cost sorbents for wastewater treatment: A review. *Sust Mater Technol* 9: 10–40.
20. Lofrano, G. (2012) Emerging compounds removal from wastewater. Natural and solar based treatments 15–37. ed. Springer Netherlands, Dordrecht.
21. De Feo, G., De Gisi, S. (2014) Using MCDA and GIS for hazardous waste landfill siting considering land scarcity for waste disposal. *Waste Manag* 34(11): 2225–2238.
22. Malik, R., Ramteke, D. S., Wate, S. R. (2007) Adsorption of malachite green on groundnut shell waste based powdered activated carbon. *Waste Manag* 27: 1129–1138.
23. Hameed, B. H., El Khaiary M. I. (2008) Kinetics and equilibrium studies of malachite green adsorption on rice straw-derived char. *J Hazard Mater* 153: 701–708.
24. Baek, M. H., Ijagbemi, C. O., Jin, O. S., Kim, D. S. (2010) Removal of malachite green from aqueous solution using degreased coffee bean. *J Hazard Mater* 176: 820–828.
25. Chowdhury, S., Mishra, R., Saha, P., Kushwaha, P. (2011) Adsorption thermodynamics, kinetics and isosteric heat of adsorption of malachite green onto chemically modified rice husk. *Desalination* 265: 159–168.
26. Feng, N., Guo, X., Liang, S., Zhu, Y., Liu, J. (2011) Biosorption of heavy metals from aqueous solutions by chemically modified orange peel. *J Hazard Mater* 185: 49–54.
27. Flores-Garnica, J. G., Morales Barrera, L., Pineda Camacho, G., Cristiani Urbina, E. (2013) Biosorption of Ni(II) from aqueous solutions by Litchi chinensis seeds. *Bioresour Technol* 136: 635–643.

28. Nieto, L. M., Alami, S. B. D., Hodaifa, G. et al. (2010) Adsorption of iron on crude olive stones. *Ind Crop* Prod 32(3): 467–471.
29. Hodaifa, G., Alami, S. B. D., Ochando Pulido, J. M., Víctor Ortega, M. D. (2014) Iron removal from liquid effluents by olive stones on adsorption column: Breakthrough curves. *Ecol Eng* 73: 270–275.
30. Ronda, A., Martín Lara, M. A., Calero, M., Blázquez, G. (2015) Complete use of an agricultural waste: Application of untreated and chemically treated olive stone as biosorbent of lead ions and reuse as fuel. *Chem Eng Res Des* 104: 740–751.
31. Jiang, S., Li, Y., Ladewig, B. P. (2017) A review of reverse osmosis membrane fouling and control strategies. *Sci Total Environ* 595: 567–583.
32. Chen, C. Y., Chen, S. Y. (2004) Adsorption properties of a chelating resin containing hydroxy group and iminodiacetic acid for copper ions. *J Appl Poly Sci* 94 (5): 2123–2130.
33. Scialabba, N. E. (2014) Food wastage footprint: Full-cost accounting. http://www.fao.org/3/a-i3991e.pdf Accessed 01.01.16. (accesed September 20, 2017).
34. Velazquez-Jimenez, L. H., Pavlick, A., Rangel Mendez, J. R. (2013) Chemical characterization of raw and treated agave bagasse and its potential as adsorbent of metal cations from water. *Ind Crop Prod* 43: 200–206.
35. Aygun, A., Yenisoy Karakas, S., Duman, I. (2003) Production of granular activated carbon from fruit stones and nutshells and evaluation of their physical, chemical and adsorption properties. *Microporous Mesoporous Mater* 66: 189–195.
36. Pehlivan, E., Altun, T., Parlayici, S. (2012) Modified barley straw as a potential biosorbent for removal of copper ions from aqueous solution. *Food Chem* 135: 2229–2234.
37. Kumar P. S., Ramalingam, S., Kirupha, S. D., Murugesan, A., Vidhyadevi, T., Sivanesan, S. (2011) Adsorption behavior of nickel(II) onto cashew nut shell: Equilibrium, thermodynamics, kinetics, mechanism and process design. *Chem Eng J* 167: 122–131.
38. Kumar P. S., Ramakrishnan K., Kirupha S. D., Sivanesan S. (2011) Thermodynamic, kinetic, and equilibrium studies on phenol removal by use of cashew nut shell. *The Canadian J Chem Eng* 89: 284–291.
39. Juang, R. S., Wu, F. C., Tseng, R. L. (2002) Characterization and use of activated carbons prepared from bagasses for liquid-phase adsorption. *Colloids Surf A Physicochem Eng Asp* 201: 191–199.
40. Thinakaran N., Panneerselvam, P., Baskaralingam, P., Elango, D., Sivanesan, S. (2008) Equilibrium and kinetic studies on the removal of acid red 114 from aqueous solutions using activated carbons prepared from seed shells. *J Hazard Mater* 158: 142–150.
41. Tasaso, P. (2014) Adsorption of copper using pomelo peel and depectinated pomelo peel. *J Clean Energy Technol* 2: 154–157.
42. Husein, D. Z. (2013) Adsorption and removal of mercury ions from aqueous solution using raw and chemically modified Egyptian mandarin peel. *Desalin Water Treat* 51: 6761–6769.
43. Yadav, D., Kapur, M., Kumar, P., Mondal, M. K. (2015) Adsorptive removal of phosphate from aqueous solution using rice husk and fruit juice residue. *Process Saf Environ* 94: 402–409.
44. Hossain, M. A., Ngo, H. H., Guo, W. S., Setiadi, T. (2012) Adsorption and desorption of copper(II) ions onto garden grass. *Bioresour Technol* 121: 386–395.

45. Liu, W., Liu, Y., Tao, Y., Yu, Y., Jiang, H., Lian, H. (2014) Comparative study of adsorption of Pb(II) on native garlic peel and mercerized garlic peel. *Environ Sci Pollut Res* 21: 2054–2063.
46. Zou, W. H., Zhao, L., Zhu, L. (2012) Efficient uranium (VI) biosorption on grapefruit peel: Kinetic study and thermodynamic parameters. *J Radioanal Nucl Ch* 292(3): 1303-1315.
47. Aydin, H., Bulut, Y., Yerlikaya, Ç. (2008) Removal of copper (II) from aqueous solution by adsorption onto low-cost adsorbents. *J Environ Manage* 87: 37–45.
48. Iqbal, M., Saeed, A., Kalim, I. (2009) Characterization of adsorptive capacity and investigation of mechanism of Cu2+, Ni2+ and Zn2+ adsorption on mango peel waste from constituted metal solution and genuine electroplating effluent. *Sep Sci Technol* 44: 3770–3791.
49. Saha, R., Mukherjee, K., Saha, I., Ghosh, A., Ghosh, S., Saha, B. (2013) Removal of hexavalent chromium from water by adsorption on mosambi (Citrus limetta) peel. *Res Chem Intermed* 39: 2245–2257.
50. Huang, K., Zhu, H. (2013) Removal of Pb^{2+} from aqueous solution by adsorption on chemically modified muskmelon peel. *Environ Sci Pollut Res* 20: 4424–4434.
51. Akmil-Basar, C., Onal, Y., Kilicer, T., Eren, D. (2005) Adsorptions of high concentration malachite green by two activated carbons having different porous structures. *J Hazard Mater* 127: 73–80.
52. Fiol, N., Villaescusa, I., Martínez, M., Miralles, N., Poch, J., Serarols, J. (2006) Sorption of Pb(II), Ni(II), Cu(II) and Cd(II) from aqueous solution by olive stone waste. *Sep Purif Technol* 50: 132–140.
53. Moghadam, M., Nasirizadeh, N., Dashti, Z., Babanezhad, E. (2013) Removal of Fe(II) from aqueous solution using pomegranate peel carbon: Equilibrium and kinetic studies. *Int J Ind Chem* 4: 1–6.
54. Aman, T., Kazi, A. A., Sabri, M. U., Bano, Q. (2008) Potato peels as solid waste for the removal of heavy metal copper (II) from waste water/industrial effluent. *Colloids Surf B: Biointerfaces* 63: 116–121.
55. Ding, Y., Jing, D., Gong, H., Zhou, L., Yang, X. (2012) Biosorption of aquatic cadmium (II) by unmodified rice straw. *Bioresour Technol* 114: 20–25.
56. Khoramzadeh, E., Nasernejad, B., Halladj, R. (2013) Mercury biosorption from aqueous solutions by sugarcane bagasse. *J Taiwan Inst Chem Eng* 44: 266–269.
57. Annadurai, G., Juang, R. S., Lee, D. J. (2002) Use of cellulose-based wastes for adsorption of dyes from aqueous solutions. *J Hazard Mater* 92: 263–274.
58. Juang, R. S., Tseng, R. L., Wu, F. C. (2001) Role of microporosity of activated carbons on their adsorption abilities for phenols and dyes. *Adsorption* 7: 65–72.
59. Namasivayam, C., Radhika, R., Suba, S. (2001) Uptake of dyes by a promising locally available agricultural solid waste: Coir pith. *Waste Manag* 21: 381–387.
60. Pavan, F. A., Lima, E. C., Dias, S. L. P., Mazzocato, A. C. (2008) Methylene blue biosorption from aqueous solutions by yellow passion fruit waste. *J Hazard Mater* 150: 703–712.
61. Feng, N., Guo, X., Liang, S. (2009) Adsorption study of copper (II) by chemically modified orange peel. *J Hazard Mater* 164: 1286–1292.
62. Malik, P. K. (2003) Use of activated carbons prepared from sawdust and rice-husk for adsorption of acid dyes: A case study of acid yellow 36. *Dyes Pigments* 56(3): 239–249.

63. Arami, M., Limaee, N. Y., Mahmoodi, N. M., Tabrizi, N. S. (2006) Equilibrium and kinetics studies for the adsorption of direct and acid dyes from aqueous solution by soy meal hull. *J Hazard Mater* 135: 171–179.
64. Sun, G., Xu, X. (1997) Sunflower stalk as adsorbents for color removal from textile wastewater. *Ind Eng Chem Res* 36: 808–812.
65. Chou, K. S., Tsai, J. C., Lo, C. T. (2001) The adsorption of Congo red and vacuum pump oil by rice hull ash. *Bioresour Technol* 78: 217–219.
66. Tavlieva, M. P., Genieva, S. V., Georgieva, V. G., Vlaev, L. T. (2013) Kinetic study of brilliant green adsorption from aqueous solution onto white rice husk ash. *J Colloid Interface Sci* 409: 112–122.
67. Kizito, S., Wu, S., Kipkemoi Kirui W. et al. (2015) Evaluation of slow pyrolyzed wood and rice husks biochar for adsorption of ammonium nitrogen from piggery manure anaerobic digestate slurry. *Sci Total Environ* 505: 102–112.
68. Pillai, M. G., Simha, P., Gugalia, A. (2014) Recovering urea from human urine by bio-sorption ontomicrowave activated carbonized coconut shells: Equilibrium, kinetics, optimization and field studies. *J Environ Chem Eng* 2(1): 46–55.
69. Ahmedna, M., Marshall, W. E., Raw, R. M. (2000) Surface properties of granular activated carbons from agricultural by-products and their effects on raw sugar decolourization. *Bioresource Technol* 71: 103–112.
70. Mandu, I. Inyang, Bin Gao, Ying Yao, Yingwen Xue, Zimmerman, A., Mosa, A., Pullammanappallil, P., Ok, Y. S., Cao, X. (2016) A review of biochar as a low-cost adsorbent for aqueous heavy metal removal. *Crit Rev Environ Sci Technol* 46(4): 406-433.
71. Roy, P., Dias, G. (2017) Prospects for pyrolysis technologies in the bioenergy sector: A review. *Renew Sustain Energy Rev* 77: 59–69.
72. Novak, J. M., Busscher, W. J., Laird, D. L., Ahmedna, M., Watts, D. W., Niandou, M. A. S. (2009) Impact of biochar amendment on fertility of a south-eastern coastal plain soil. *Soil Sci* 174: 105–112.
73. Ahmad, M., Rajapaksha, A. U., Lim, J. E. et al. (2014) Biochar as a sorbent for contaminant management in soil and water: A review. *Chemosphere* 99: 19–33.
74. Park, J., Hung, I., Gan, Z., Rojas, O. J., Lim, K. H., Park, S. (2013) Activated carbon from biochar: Influence of its physicochemical properties on the sorption characteristics of phenanthrene. *Bioresource Technol* 149: 383–389.
75. Mohan, D., Pittman, C. U. Jr., Bricka, M. et al. (2007) Sorption of arsenic, cadmium, and lead by chars produced from fast pyrolysis of wood and bark during bio-oil production. *J Colloid Inter Sci* 310: 57–73.
76. Nanda, S., Dalai, A. K., Berruti, F., Kozinski, J. A. (2016) Biochar as an exceptional bioresource for energy, agronomy, carbon sequestration, activated carbon and specialty materials. *Waste Biomass Valor* 7: 201–235.
77. Tan, G., Sun, W., Xu, Y., Wang, H., Xu, N. (2016) Sorption of mercury (II) and atrazine by biochar, modified biochars and biochar based activated carbon in aqueous solution. *Bioresource Technol* 211: 727–735.
78. Owamah, H. I. (2014) Biosorptive removal of Pb(II) and Cu(II) from wastewater using activated carbon from cassava peels. *J Mater Cycles Waste Manage* 16: 347–358.
79. Gabrielle, B., Gagnaire, N. (2008) Life-cycle assessment of straw use in bio-ethanol production: A case study based on biophysical modelling. *Biomass Bioenergy* 32 (85): 431–441.

80. Huang, Y., Anderson, M., Mcllveen Wright, D. et al. (2015) Biochar and renewable energy generation from poultry litter waste: A technical and economic analysis based on computational simulations. *Appl Energy* 160: 656–663.
81. Qambrani, N. A., Rahman, M. M., Won, S., Shim, S., Ra, C. (2017) Biochar properties and eco-friendly applications for climate change mitigation, waste management, and wastewater treatment: A review. *Renew Sustain Energy Rev* 79: 255–273.
82. Tseng, R. L., Wu, F. C., Juang, R. S. (2003) Liquid-phase adsorption of dyes and phenols using pinewood-based activated carbons. *Carbon* 41: 487–495.
83. Marin, F. R., Soler Rivas, C., Benavente Garcia, O., Castillo, J., Perez Alvarez, J. A. (2007) Byproducts from different citrus processes as a source of customized functional fibres. *Food Chem* 100: 736–741.
84. Lam, S. S., Liew, R. K., Lim, X. Y., Ani, F. N., Jusoh, A. (2016) Fruit waste as feedstock for recovery by pyrolysis technique. *Int Biodeter Biodegr* 113: 325–333.
85. Ahmed, I., Zia, M. A., Hussain, M. A., Akram, Z., Naveed, M. T., Nowrouzi, A. (2016) Bioprocessing of citrus waste peel for induced pectinase production by Aspergillus niger; Its purification and characterization. *J Radiat Res Appl Sci* 9(2): 148–154.
86. Li, X., Tang, Y., Cao, X., Lu, D., Luo, F., Shao, W. (2008) Preparation and evaluation of orange peel cellulose adsorbents for effective removal of cadmium, zinc, cobalt and nickel. *Colloids Surface* A 317: 512–521.
87. Guo, X. Y., Liang, S., Tian, Q. H. (2011) Removal of heavy metal ions from aqueous solutions by adsorption using modified orange peel as adsorbent. *Adv Mater Res* 236–238: 237–240.
88. Sha, L., Xueyi, G., Ningchuan, F., Qinghua, T. (2009) Adsorption of Cu^{2+} and Cd^{2+} from aqueous solution by mercapto-acetic acid modified orange peel. *Colloid Surface B* 73(1): 10–14.
89. Saito, T., Yanagisawa, M., Isogai, A. (2005) TEMPO-mediated Oxidation of native cellulose: SEC–MALLS analysis of water-soluble and -insoluble fractions in the oxidized products. *Cellulose* 12(3): 305–315.
90. Liang, S., Guo, X., Tian, Q. (2011) Adsorption of Pb^{2+} and Zn^{2+} from aqueous solutions by sulfured orange peel. *Desalination* 275: 212–216.
91. Bhatnagar, A., Minocha, A. K. (2010) Biosorption optimization of nickel removal from water using Punica granatum peel waste. *Colloids Surf B* 76: 544–548.
92. Bhatti, H. N., Khadim, R., Hanif, M. A. (2011) Biosorption of Pb(II) and Co(II) on red rose waste biomass. *Iran J Chem Eng* 30: 80–88.
93. Suksabye, P., Thiravetyan, P. (2012) Cr(VI) adsorption from electroplating plating wastewater by chemically modified coir pith. *J Environ Manage* 102: 1–8.
94. Kurniawan, A., Kosasih, A. N., Febriano, J., Ju, Y. H., Sunarso, J., Indraswati, N. (2011) Evaluation of cassava peel waste as lowcost biosorbent for Ni sorption: Equilibrium, kinetics, thermodynamics, and mechanisms. *Chem Eng J* 172: 158–166.
95. Anwar, J., Shafique, U., Zaman, W., Salman, M., Dar, A., Anwar, S. (2010) Removal of Pb(II) and Cd(II) from water by adsorption on peels of banana. *Bioresour Technol* 101: 1752–1755.
96. Agabo García, C., Hodaifa, G. (2017) Real olive oil mill wastewater treatment by photo-Fenton system using artificial ultraviolet light lamps. *J Clean Prod* 162: 743–753.

97. Martín-Lara, M. A., Blázquez, G., Ronda, A., Pérez, A., Calero, M. (2013) Development and characterization of biosorbents to remove heavy metals from aqueous solutions by chemical treatment of olive stone. *Ind Eng Chem Res* 52: 10809–10819.
98. Nieto, L. M., Hodaifa, G., Rodriguez, V. S., Gimenez Casares, J. A., Ben Driss, S., Grueso, R. (2009) Treatment of olive-mill wastewater from a two-phase process by chemical oxidation on an industrial scale. *Water Sci Technol* 59: 2017–2026.
99. Resolution of Guadalquivir River Basin president (2006) Scattering washing waters from olive-oil mill. Ministry of Environment, June 9, Seville, Spain.
100. Andalusian Energy Agency (2008) Status of biomass in Andalusia, http://www.agenciaandaluzadelaenergia.es/agenciadelaenergia/portal/com/bin/contenidos/proyectos/areas/energiasRenovables/biomasa/proybiomasa1/1202479474122la biomasaenandalucxaen08.pdf. (accessed September 10, 2017).
101. Rodriguez, G., Lama, A., Rodriguez, R., Jimenez, A., Guillen, R., Fernandez Bolanos, J. (2008) Olive stone an attractive source of bioactive and valuable compounds: Review. *Bioresour Technol* 99: 5261–5269.
102. Stasinakis, A. S., Elia, I., Petalas, A. V., Halvadakis, C. P. (2008) Removal of total phenols from olive-mill wastewater using an agricultural by-product, olive pomace. *J Hazard Mater* 160: 408–413.
103. Akar, T., Tosun, I., Kaynak, Z. et al. (2009) An attractive agro-industrial by-product in environmental cleanup: Bye biosorption potential of untreated olive pomace. *J. Hazard Mater* 166: 1217–1225.
104. Pagnanelli, F., Toro, L., Veglio, F. (2002) Olive mill solid residues as heavy metal sorbent material: A preliminary study. *Waste Manag* 22: 901–907.
105. Barros, M. A. S. D. (2003) Evaluation of the Chromium Exchange Mechanism in Zeolitic Systems. PhD diss., Universidade Estadual de Maringá, Maringá, Brazil.
106. Arriagada, R., Bello, G., Cid, R., García, R. (2001) "Retención de cromo en carbonos activados" en Barros M.A.S.D. "Problemas ambientales en soluciones catalíticas I. El cromo en el curtido de pieles". Ciencia y Tecnología para el Desarrollo, Madrid, España.
107. Romero-García, J. M., Niño, L., Martínez-Patiño, C. (2014) Biorefinery based on olive biomass. State of the art and future trends. *Bioresource Technol* 159: 421–432.
108. Sesli, M., Yeğenoğlu E. D. (2009) RAPD-PCR analysis of cultured type olives in Turkey. *Afr J Biotechnol* 8: 3418–3423.
109. Cara, C., Romero, I., Oliva, J. M., Sáez, F., Castro, E. (2007) Liquid hot water pretreatment of olive tree pruning residues. *Appl Biochem and Biotech* 137: 379–394.
110. Negro, M. J., Manzanares, P., Ruiz, E., Castro, E., Ballesteros, M. (2017) Chapter 3. The biorefinery concept for the industrial valorization of residues from olive oil industry. Olive mill waste: Recent Advances for Sustainable management. ed. Academic Press, An imprint of Elsevier, Galanakis Laboratories, Chania, Greece.
111. Fillat, Ú., Wicklein, B., Martín Sampedro, R. et al. (2017) Assessing cellulose nanofiber production from olive tree pruning residue. *Carbohyd Polym* 179: 252–261.
112. Zervakis, G., Balis, C. (1996) Bioremediation of olive oil mill wastes through the production of fungal biomass In: Royse (Hrsg.), Mushroom biology and mushroom products. Proceedings of the 2nd International Conference. Penn State Univ., University Park.
113. Nefzaoui, A. (1995) Olive by-products recycling. Proc. of the International Symposium on olive oil processes and by-products recycling, Granada.

114. Suleria, H. A. R., Butt, M. S., Khalid, N. et al. (2015). Garlic (Allium sativum): Diet based therapy of 21st century—A review. *Asian Pac J Trop Disease* 5(4): 271–278.
115. Oliveira, D. Q. L., Gonçalves, M., Oliveira, L. C. A., Guilherme, L.R.G. (2008) Removal of As(V) and Cr(VI) from aqueous solutions using solid waste from leather industry. *J Hazard Mater* 151: 280–284.
116. Muthamilselvi, P., Karthikeyan, R., Kumar, B. S. M. (2014) Adsorption of phenol onto garlic peel: Optimization, kinetics, isotherm, and thermodynamic studies. *Desalin. Water Treat* 57(5): 1–15.
117. Srivastava, S. K., Singh, A. K., Sharma, A. (1994) Studies on the uptake of lead and zinc by lignin obtained from black liquor—A paper industry waste material. *Environ Technol* 15: 353–361.
118. El Sharkawy, M. A. (2012) Stress-Tolerant cassava: The role of integrative ecophysiology Breeding research in crop improvement. *Open J Soil Sci* 2(2): 162–186.
119. Zainuddin, I. M., Fathoni, A., Sudarmonowati, E., Beeching, J. R., Gruissem, W., & Vanderschuren, H. (2017) Cassava post-harvest physiological deterioration: From triggers to symptoms. *Postharvest Biology and Technology.*
120. Polachini, T. C., Betiol, L. F. L., Lopes Filho, J. F., Telis Romero, J. (2016) Water adsorption isotherms and thermodynamic properties of cassava bagasse. *Thermochim Acta* 632: 79–85.
121. Gupta, V. K., Srivastava, S. K., Mohan, D. (1997) Equilibrium uptake, sorption dynamics, process optimization, and column operations for the removal and recovery of malachite green from wastewater using activated carbon and activated slag. *Ind Eng Chem Res* 36(6): 2207–2218.
122. Horsfall, M., Abia A. A., Spiff A. I. (2006) Kinetic studies on the absorption of cadmium, copper and zinc ions from aqueous solutions by cassava tuber bark waste. *Bioresour Tech* 97: 283–291.
123. Wang, B., Jiang, Y. S., Li, F. Y., Yang, D. Y. (2017) Preparation of biochar by simultaneous carbonization, magnetization and activation for norfloxacin removal in water. *Bioresource Technol* 233: 159–165.
124. Jin, D., Yang, X., Zhang, M. et al. (2015) A novel high surface area spherical carbon from cassava starch. *Mater Lett* 139: 262–264.
125. Xie, X., Xiong, H., Zhang, Y., Tong, Z., Liao, A., Qin, Z. (2017) Preparation magnetic cassava residue microspheres and its application for Cu (II) adsorption. *J Environ Chem Eng* 5(3): 2800–2806.
126. Moreno-Piraján, J. C., Giraldo, L. (2010) Adsorption of copper from aqueous solution by activated carbons obtained by pyrolysis of cassava peel. *J Anal Appl Pyrol* 87(2): 188–193.
127. Alcantara, H. J. P., Doronila, A. I., Kolev, S. D. (2017) Phytoextraction potential of Manihot esculenta Crantz. (cassava) grown in mercury-and gold-containing biosolids and mine tailings. *Miner Eng* 114: 57–63.
128. Tatlioglu, T. (1993) Cucumber Cucumis sativus L. Genetic improvement of vegetable crops. Tarrytown: Pergamon Press Ltd. 197–234.
129. Alnadif, A. A. M., Mirghani, M. E. S., Hussein, I. H. (2017) Unconventional oilseeds and oil sources. *Part A: Principless of oil extraction, Processing and Oils composition: Unconventional oils from Annual Herbs, and vegetables. 89-93 X* ed. Elsvier Inc Academic Press.
130. Basu, M., Guha, A. K., Ray, L. (2017). Adsorption of Lead on Cucumber Peel. *J Clean Prod* 151: 603–615.

131. Pandey, R., Ansari, N. G., Prasad, R. L., Murthy, R. C. (2014) Removal of Cd(II) ions from simulated wastewater by HCl modified Cucumis sativus peel: Equilibrium and kinetic study. *Air Soil Water Res* 7: 93–101.
132. Akkaya, G., Güzel, F. (2013) Application of some domestic wastes as new low-cost biosorbents for removal of methylene blue: Kinetic and equilibrium studies. *Chem Eng Commun* 201: 557–578.
133. Shakoor, S., Nasar, A. (2017) Adsorptive treatment of hazardous methylene blue dye from artificially contaminated water using cucumis sativus peel waste as a low-cost adsorbent. *Groundwater Sust Devel* 5: 152–159.
134. Santhi, T., Manonmani, S. (2011) Malachite green removal from aqueous solution by the peel of Cucumis sativa fruit. *Clean–Soil, Air, Water* 39(2): 162–170.
135. Smitha, T., Thirumalisamy, S., Manonmani S. (2012) Equilibrium and kinetics study of adsorption of crystal violet onto the peel of Cucumis sativa fruit from aqueous solution. *E J Chem* 9(3): 1091–1101.
136. Farah, H., Elbadrawy, E., Al Atoom, A. A. (2015) Evaluation of antioxidant and antimicrobial activities of ethanolic extracts of Parsley (Petroselinum crispum) and Coriander (Coriandrum sativum) plants grown in Saudi Arabia. *Int J Adv Res* 3(4): 1244–1255.
137. Farzaei, M. H., Abbasabadi, Z., Ardekani, M. R. S., Rahimi, R., Farzaei F. (2013) Parsley: A review of ethnopharmacology, phytochemistry and biological activities *J Tradit Chin Med* 33(6): 815–826.
138. Riel, G., Boulaaba, A., Popp, J., Klein, G. (2017) Effects of parsley extract powder as an alternative for the direct addition of sodium nitrite in the production of mortadella-type sausages—Impact on microbiological, physicochemical and sensory aspects. *Meat Science* 131: 166–175.
139. Jiménez Cedillo, M. J., Olguín, M. T., Fall, C., Colin Cruz, A. (2013) As(III) and As(V) sorption on iron-modified non-pyrolyzed and pyrolyzed biomass from Petroselinum crispum (parsley). *J Environ Manage* 117: 242–252.
140. Aryal, M., Ziagova, M., Liakopoulou Kyriakides, M. (2010) Study on arsenic biosorption using Fe(III)-treated biomass of Staphylococcus xylosus. *Chem Eng J* 162: 178–185.
141. Basu, M., Guha, A. K., Ray, L. (2017). Adsorption behaviour of cadmium on husk of lentil. *Process Saf Environ* 106: 11–22.
142. Ozbay, N., Yargic, A. S. (2017) Statistical analysis of Cu (II) and Co (II) sorption by apple pulp carbon using factorial design approach. *J Ind Eng Chem. In press.*
143. Chand, P., Bafana, A., Pakade, Y. B. (2015) Xanthate modified apple pomace as an adsorbent for removal of Cd (II), Ni (II) and Pb (II), and its application to real industrial wastewater. *Int Biodeter Biodegr* 97: 60–66.
144. Alves, C. C. O., Franca, A. S., Oliveira, L. S. (2013) Removal of phenylalanine from aqueous solutions with thermo-chemically modified corn cobs as adsorbents. LWT - *Food Sci Technol* 51: 1–8.
145. Tavassolirizi, Z., Shams, K., Omidkhah, M. R. (2015) Platinum recovery from model media and a Pt–Sn/alumina spent catalyst extract using corn husk-based adsorbent. *J Ind Eng Chem* 23: 119–127.
146. Xiong, Y., Wan, L., Xuan, J., Wang, Y., Xing, Z., Shan, W. (2016) Selective recovery of Ag(I) coordination anion from simulate nickel electrolyte using corn stalk based adsorbent modified by ammonia–thiosemicarbazide. *J Hazard Mater* 301: 277–285.

147. You, H., Chen, J., Yang, C., Xu, L. (2016) Selective removal of cationic dye from aqueous solution by low-cost adsorbent using phytic acid modified wheat straw. *Colloid Surface A* 509: 91–98.
148. Lin, Q., Wang, K., Gao, M., Bai, Y., Chen, L., Ma, H. (2017) Effectively removal of cationic and anionic dyes by pH-sensitive amphoteric adsorbent derived from agricultural waste-wheat straw. *J Taiwan Inst Chem Eng* 76: 65–72.
149. Yang, F., Sun, L., Xie, W. et al. (2017) Nitrogen-functionalization biochars derived from wheat straws via molten salt synthesis: An efficient adsorbent for atrazine removal. *Sci Total Environ* 607–608: 1391–1399.
150. Kenes, K., Yerdos, O., Zulkhair, M., Yerlan, D. (2012) Study on the effectiveness of thermally treated rice husks for petroleum adsorption. *J Non-Cryst Solids* 358: 2964–2969.
151. Xu, M., Yin, P., Liu, X., Tang, Q., Qu, R., Xu, Q. (2013) Utilization of rice husks modified by organomultiphosphonic acids as low-cost biosorbents for enhanced adsorption of heavy metal ions. *Bioresour Technol* 149: 420–424.
152. Singh, K. K., Rastogi, R., Hasan, S. H. (2005) Removal of Cr(VI) from wastewater using rice bran. *J Colloid Interface Sci* 290: 61–68.
153. Gao, H., Liu, Y., Zeng, G., Xu, W., Li, T., Xia, W. (2008) Characterization of Cr(VI) removal from aqueous solutions by a surplus agricultural waste-rice straw *J Hazard Mater* 150: 446–452.
154. Krishnani, K. K., Meng, X., Christodoulatos, C., Boddu, V. M. (2008) Biosorption mechanism of nine different heavy metals onto biomatrix from rice husk. *J Hazard Mater* 153: 1222–1234.
155. Sharma, P., Kaur, R., Baskar, C., Chung, W. J. (2010) Removal of methylene blue from aqueous waste using rice husk and rice husk ash. *Desalination* 259: 249–257.
156. Tolba, G. M. K., Barakat, N. A. M., Bastaweesy, A. M. et al. (2015) Effective and highly recyclable nanosilica produced from the rice husk for effective removal of organic dyes. *J Ind Eng Chem* 29: 134–145.
157. Anastopoulos, I., Karamesouti, M., Mitropoulos, A. C., Kyzas, G. Z. (2017) A review for coffee adsorbents. *J Mol Liq* 229: 555–565.
158. Kyzas, G. Z. (2012) Commercial coffee wastes as materials for adsorption of heavy metals from aqueous solutions. *Materials* 5: 1826–1840.
159. Lafi, R., Fradj, A. B., Hafiane, A., Hameed, B. H. (2014) Coffee waste as potential adsorbent for the removal of basic dyes from aqueous solution. *Korean J Chem Eng* 31: 2198–2206.
160. Liu C., Pujol, D., Olivella, M. A. et al. (2015) The role of exhausted coffee compounds on metal ions sorption. *Water Air Soil Pollut* 226:289.
161. Ronix, A., Pezoti, O., Souza, S. L., Souza, I. P. A. F., Bedin, K. C. (2017) Hydrothermal carbonization of coffee husk: Optimization of experimental parameters and adsorption of methylene blue dye. *J Environ Chem Eng* 5: 4841–4849.
162. Shen, K., Gondal, M. A. (2017) Removal of hazardous Rhodamine dye from water by adsorption onto exhausted coffee ground. *J Saudi Chem Soc* 21: S120–S127.
163. Alhogbi, B. G. (2017) Potential of coffee husk biomass waste for the adsorption of Pb(II) ion from aqueous solutions. *Sust Chem Pharm* 6: 21–25.
164. Bilal, M., Shah, J. A., Ashfaq, T. et al. (2013) Waste biomass adsorbents for copper removal from industrial wastewater—A review. *J Hazard Mater* 263: 322–333.
165. Kongsuwan, A., Patnukao, P., Pavasant P. (2009) Binary component sorption of Cu (II) and Pb (II) with activated carbon from Eucalyptus camaldulensis Dehn bark. *J Ind Eng Chem* 15: 465–470.

10

Adsorption on Activated Carbon: Role of Surface Chemistry in Water Purification

Vivekanand Gaur

CONTENTS

10.1 Introduction

Activated carbon is an exceptional carbonaceous material with a high level of micro porosity and extended internal surface area. In general, activated carbons contain more than 90% carbon element arranged in a hexagonal graphitic ring with six carbon atoms at each corner, called as graphitic layers randomly placed in basal planes separated from each other. Distance between the graphitic layers and the basal plane is the deciding factor for the shape and pore size distribution in the activated carbon. In other words, angular orientations of the planes in activated carbon are random to each other, making it different from those in graphene wherein they are well ordered. Activated carbon is invariably associated with appreciable amounts of hetero atoms such as oxygen, hydrogen, sulfur, and nitrogen, which exist in the form of functional groups and/or atoms chemically bonded to the structure [1,2].

Activated carbon is considered to be a versatile adsorbent material extensively used in various industrial applications based on liquid and gas-phase filtrations, such as air and water purification, metal recovery, and separation and modification of a variety of species. In this chapter, properties of activated carbon and its surface modification for water purification applications are discussed.

Carbon in various morphological forms such as soot, charcoal, granules, and powder have been used in multi functional roles for a long time in various chemical processes, for example, adsorption, bleaching, deodorization, and catalytic reactions. In the latter processes, carbon is used as a support to various metal catalysts. In the last couple of decades, relatively newer forms of carbon, such as activated carbon fabric or fiber (ACF), carbon molecular sieves (CMS), and nanostructured carbons, particularly carbon nanotubes (CNTs), have been developed for a variety of applications. The CNTs, on doping with suitable metal atoms inside the cylindrical central hollow of the tube have been used as electrical conducting nano-wires [3]. The average conductivity of nano tubes of 15–50 nm diameter and 20 μm length is reported to be as high as 1000–2000 S/cm [4]. The CMS finds application in molecular separation on the basis of adsorption rate rather than the amount of adsorption determined by adsorption equilibrium constants [5]. The composite of CMS with polyfurfuryl alcohol and activated carbon substrate has recently been developed and used as a catalyst support for Pt, exhibiting shape selectivity for the solutes [6]. Thus, 1-butene may be selectively hydrogenated, whereas 3-butene present in the same reactant stream is not. Carbon fabric or fiber is synthesized from various polymeric precursors such as polyacrylonitrile (PAN), coal tar pitches, cellulose derivatives, and phenolic resin [7]. Because of its hydrophobic characteristic, resistance to almost all types of organic solvents (aliphatic and aromatic), and specific surface interactions (e.g. π-π interactions) with aromatic molecules, ACF is perceived to be one of the most potential adsorbents for the removal of pollutant species. Developments of the advanced carbon materials for effective applications in different areas are still in progress. Furthermore, there are growing efforts to use different waste materials such as agricultural waste, municipal and industrial wastes, as a raw material for producing activated carbons. Specifically, these include hard nut shells, rice husks, bagasse, biomass residues, pulp industry wastes, bones, and waste tires [8].

In general, conventional activated carbon can be derived from different precursor materials of plants, animals, or petrochemical sources. Basic raw materials used for producing activated carbon include coal (anthracite or bituminous coal), peat, lignite, petroleum coke, wood, and nut shells, for example, coconut shell. Basic properties of activated carbon, for example, pore size distribution, surface area, hardness, and ash content, largely depend on the types of precursor for carbon, the manufacturing process and operating conditions. The manufacturing process consists of mainly two steps, carbonization and activation.

Carbonization, a process of pyrolytic decomposition in inert atmosphere at temperature between 500 and 1000°C, is essentially a de-volatilizing step to eliminate non-carbon element containing functional groups, such as oxygen, nitrogen, hydrogen and sulfur. The residual elementary carbon atoms group themselves into sheets and form the condensed aromatic ring systems having a certain degree of planar structure. There is also an increase in the C/H and C/O ratios as well as the randomly bonded hexagonal carbonaceous and aromatic contents with increasing carbonization temperatures up to a certain limit. The aromatic sheets have an irregular arrangement with free interstices between them, forming fine pores. In most cases, these pores are inter-connected paths through which the decomposing gases make their way out of the residual carbon.

In general, carbonization of the precursor materials is carried out in rotary furnaces, also called as rotary kilns. Interesting to learn is that large scale carbonization of various shells including the coconut shells is widely carried out in the stationary pits and drums in which the shells are packed and undergo pyrolysis. However, because of the extensive fume emissions and air pollution issues, industries have used closed reactors by recycling gases/fumes to generate thermal energy. At the end of the carbonization stage, adsorption capacity of the carbonized materials remains insignificant, because most of the mouths or entrances of the pores created during carbonization are blocked with some tarry substances which prevent adsorption at the sites. Further increase in the carbonization temperature does not remove tar from the pores, rather it seals the pores against any external sorbate. Sorption properties in these materials may be induced by subsequently carrying out activation.

Char produced after carbonization contains pore structure which is not yet completely developed, and therefore, the produced carbon has to be further treated via an oxidative step; this process is called activation. Activation of different types of precursor materials can be carried out using different processes, including thermal and chemical means.

10.1.1 Thermal Activation

During activation, the carbonized material is exposed to the oxidizing environment consisting of gases such as steam, air, carbon dioxide or their combinations, at high temperatures between 700 and 1000°C. This is widely used in the synthesis of coconut shell-based activated carbons.

10.1.2 Chemical Activation

Chemical activation process is used for the raw materials that are relatively inferior in hardness and contain high level of cellulose such as rice husk, nutshells, and also, wood. Chemical activation is brought about by the action of various chemical agents producing dehydration effect. The most commonly used reagents for chemical activation are H_3PO_4 [9], $ZnCl_2$ or other halides [10,11], KOH or NaOH [12,13], and H_2SO_4 or HNO_3 [14–16]. The pyrolytic temperature depends on the type of raw material, in particular its hardness and cellulose content, and it can vary between 600 and 800°C, which is much lower than that used for steam activation. Also, activation time is shorter for chemical activation as compared to the more conventional steam activation. At the end of the chemical activation step, the unreacted chemical agents are washed out from activated carbon. In some cases, the chemically treated char is activated in CO_2 or steam stream.

10.2 Forms of Activated Carbon for Filtration Application

Activated carbon is recognized as the most efficient and versatile sorbent material because of its unique properties and ready availability. From the early 1900s, large scale production of activated carbon was driven by various industrial and environmental applications. Starting with an appropriate precursor and manufacturing step, activated carbon can be produced in different forms such as granules, powder, cylindrical pellets, spherical beads, fibers, cloth, and many more. Amongst these, activated carbon fiber and cloth are known to contain high surface area and amenability to surface modification for the enhanced removal of volatile organic compounds and various contaminants in aqueous phase [17–23]. Furthermore, carbon powder can also be transformed into different shapes including carbon blocks, composites, films, and sheets. Based on the shapes, sizes, and structures, activated carbon is considered to be a family of microporous materials used in a myriad of commercial applications such as water purification and gaseous filtration.

Primary types of activated carbon are granular activated carbons (GAC) and powdered activated carbons (PAC), which are differentiated by particle sizes. The PAC have typical particle size less than 100 μm. Although the particles with larger than 100 μm are considered to be GAC, the mean particle size in GACs is 1–3 mm. The GACs are designated by mesh sizes, such as 8 × 16, 12/20, and 20/50. The carbonaceous materials coming out from the kiln after activation is called as ex-kiln material, which then undergo different stages of crushing, milling, and screening. Based on the top and bottom screening meshes, the GACs are accordingly identified. The sizing commonly starts with mesh 4/8 and 8/16 and so on.

Activated carbon can be produced in the other forms, among which the most commonly used in air and water filtration applications is extruded activated carbon (EAC), which is more commonly known as carbon pellets.

10.3 Properties of Activated Carbon

Activated carbon may be visualized as a solid porous material consisting of "pores" distributed in different sizes and shapes within the material.

According to the pore widths, the International Union of Pure and Applied Chemistry (IUPAC) has classified three types of pores: micropores which have pore width lower than 2 nm, meso pores which have pore width between 2 and 50 nm, and macropores which have pores width larger than 50 nm.

During the production of activated carbons, macro pores are first formed by surface oxidation. The macro pores are then extended and narrowed down into secondary channels, also known as capillaries, to form micro and meso pores. Figure 10.1 shows a schematic diagram of macro, meso and micro pores in the activated carbon.

Porous structure or network is developed when spaces between the layers of aromatic sheets become free from non-organized carbons or carbonaceous compounds during carbonization and activation steps. Porous carbon is a non-graphitic class of carbon, characterized by internal surface areas ranging from 800 to over 2500 m^2/g. Activated carbon, the best known example of porous carbon, consists of parallel layers of aromatic planes. These planes form crystallites, which are composed of the turbostratic layer stacks and are similar to those in graphite. Further, each carbon atom within a plane is bonded to three adjacent carbon atoms by σ-bonds, the C-C bond length being 0.142 nm, with the fourth electron participating in π-bond. The form and dimensions of the crystallites are generally characterized using X-ray diffraction technique, by which the parameters L_a, L_c, and d_{002} are measured. L_a and L_c represent crystallite dimensions in direction parallel and perpendicular to the basal planes, respectively (i.e. width and thickness of the stacks), whereas d_{002} represents the inter-layer spacing. The values of L_a and L_c generally range from tenths of a nanometer to 20 nm. The value of d_{002} in crystallites of different types of carbon fibers lies within 0.330–0.395 nm and is considerably greater than that of graphite (0.3358 nm) [24].

Based on the structure, carbons are reported to be divided into two categories, the graphitizable and non-graphitizable carbons. Different experimental techniques such as X-ray diffraction and high-resolution transmission

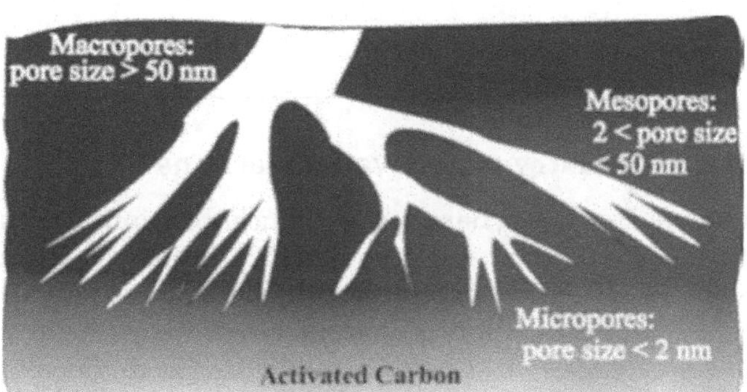

FIGURE 10.1
Schematic of pore size distribution in activated carbon. (From Cecen, F., Carbon, *Activated, Kirk-Othmer Encyclopedia of Chemical Technology*, John Wiley & Sons, 1, 2014.)

electron microscopy, coupled with adsorption experiments with molecules of various sizes and shapes, have confirmed that the structure of non-graphitic carbon consists of irregular aromatic sheets with micropores-spaces between them. Dresselhaus et al. [25] and Kaneko et al. [26] have proposed the micrograhitic models of both non-graphitizable and partially graphitizable carbon fibers. These models are found to be suited for activated carbons as well. The model indicated the presence of several kinds of micropores in carbon. The interstices between micrographitic walls, which are mutually connected by sp^3 linkages, may be classified into the open and partially closed slit or wedge shaped pores, or an interstitial edge surrounded by the neighboring micrographities.

Because of a wide distribution of pores in the structure, activated carbon has extensively large surface area per unit mass, which makes it to be the best adsorbent material. Depending on the raw carbon material and carbonization and activation conditions, surface area and pore volumes of activated carbons vary over a wide range. Amount of the material adsorbed can be extremely high because of the large internal surface area of the activated carbon. The internal surface area, known as Brunauer-Emmet-Teller (BET) area plays an exceptional role in the adsorption of a number of solutes. Activated carbons used in air and water purification generally have a BET area ranging between 1000 and 1500 m^2/g.

The other important parameters for adsorption are total pore volume and pore size distribution of the adsorbent material, which are represented by the plot of differential pore volume distribution (in cm^3/g) with respect to pore radius (in nm). As earlier stated, activated carbon exhibits a broad pore size distribution. It is interesting to note that activated carbon is such a complex material that even if two carbons are manufactured by the same process (say, steam activation) in the same reactor, the produced activated carbons may exhibit different properties, in particular different pore size distributions because of difference in the original pristine materials.

10.4 Surface Chemistry of Activated Carbons

In addition to major surface characteristics, including internal surface area, micro and total pore volumes, and pore size distribution, chemical compositions of the surface effect the adsorption behavior and capacity of the adsorbent material. Activated carbon is invariably associated with appreciable amounts of heteroatoms such as oxygen and hydrogen chemically bonded to the material structure, and with inorganic ash components. Oxygen surface groups are by far the most important in influencing the surface characteristics and adsorption behavior of activated carbon. Therefore, the structures pertaining to the surface properties have attracted much attention in the last few years [27].

At the edges of the basal planes of carbon atoms, where bonding in the plane is terminated, there are unsaturated carbon atoms. These sites are associated with high concentrations of unpaired electrons, and, therefore, they play a significant role in the chemisoption of oxygen, nitrogen, hydrogen, and halides, etc. to give rise to the non-stoichiometric stable surface compounds called surface complexes. Among these surface complexes, oxygen complexes are the most common and important because they are responsible for many physico-chemical and surface properties of carbon. The carbon-oxygen surface complex formation may also result from the reaction with many oxidizing gases (ozone, nitrogen oxides, oxygen, carbon dioxide, etc.), and with oxidizing solutions (nitric acid, sodium hypochlorite, hydrogen peroxide, etc.), allowing the modification of the types and amount of oxygen complexes of the carbon material. The precise characteristics of carbon-oxygen structures is not entirely established. However, findings of many studies using different experimental techniques conclude that there may be several types of oxygen functional groups and subsequently different acidic-basic characters of carbon [28]. These surface oxygen complexes have also been proposed to influence the micropores formed between graphitic platelets. The surface functional groups, existing on the edges of the graphene layers, are assumed to fill the micropores, resulting in decrease of the micropore volume.

10.5 Mechanism of Removal of Water Impurities

Intermolecular attractions in the pores result in adsorption. Molecules of the contaminants present in the water stream are adsorbed on the surface of the activated carbon via either physical attraction or chemical reaction. Physical attractions does not alter the adsorbate molecular structure, whereas chemical adsorption results in altering the adsorbate molecular structure. These phenomenon are referred as physisorption (physical adsorption) and chemisorption (chemical adsorption), respectively. In general, activated carbon has good adsorption capacity because of its inherent textural properties. However, it may not be sufficient to meet the desired specifications of the final end products (e.g. filters) produced from activated carbon. Adsorption on activated carbon proceeds through three basic steps after adsorbate molecules reach the active surface of carbon in the flowing stream:

1. Solutes adsorb on the exterior surface
2. Solutes diffuse into the carbon pores with the highest adsorption energy

3. Solutes adsorb/adsorb on the interior graphitic platelets of the material

In a typical carbon matrix, oxygen is the predominant heteroatom present in the form of surface functional groups, including carboxyl, carbonyl, phenols, lactones, quinones, and others. The unique adsorption properties of activated carbon can be significantly influenced by such surface functional groups. These groups facilitate surface reactions and the formation of surface oxygen complexes with several species, which is termed as chemisorption. However, the extent of these surface functionalities may be too low in the raw activated carbon to cause significant chemisorption. To design a water filter with desired filtration efficiencies to the specific contaminants, it is essential to modify the carbon surface for imparting the desired kinetics as well as adsorption capacity in the provided space.

10.6 Surface Modification of Activated Carbon

Based on the type and properties of the targeted impurities, the carbon surface may be modified in several ways to its enhance affinity towards the impurities. In general, activated carbon is post treated using wet process. However, some conventional carbon manufacturers treat carbon in the rotary kiln during activation process. In the kilns, activation is carried out at high temperatures (~ 1000°C) in the central zone of the kiln. The kilns have enough length beyond the central heating zone where temperature is relative lower, in the range of 500 to 700°C. In this zone reactive gases can be purged to modify the carbon surface by creating active surface functional groups. The drawback in this process is that it has limited controls because (1) carbon particle size at this stage is fairly large, and (2) temperature and residence time in this zone of the kiln operation are restricted.

To have increased flexibility in treating approximately the same particle size as that in filtration application, under controlled conditions, it is essential to perform surface modification after the activation step. Considering that filtration using carbon materials deal with chemical reactions, a majority of the surface modification techniques can be grouped under chemical treatment. As shown in Figure 10.2, the first modification step is to chemically treat carbon with different chemicals, divided into acidic and basic functional groups. Then, carbon can be impregnated with different metal or metal oxide crystallites. Carbon surface is immensely complex. Therefore, as soon as the material is impregnated with chemical reagents, it shows high affinity towards the solute by transforming several chemical changes in structure and subsequently acquiring the charged functional groups on its basal planes. Similarly, when carbon is impregnated with metal or metal

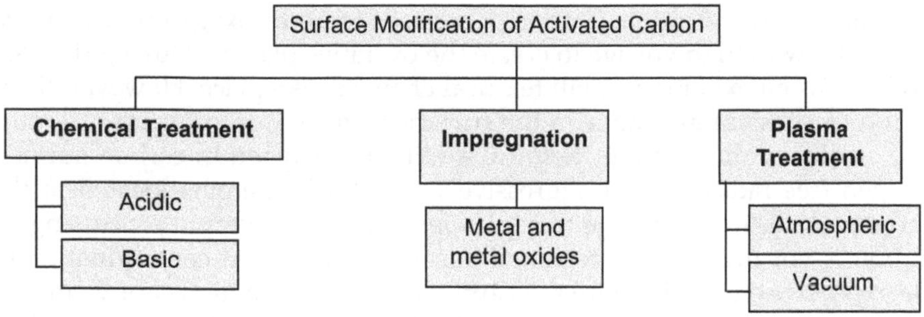

FIGURE 10.2
Surface modification techniques.

oxide suspension, the crystallites are dispersed in the pores. The functional groups and crystallites dispersed in carbon pores are said to be active sites, and when the density of these sites is sufficiently large to enhance the reaction rate during filtration, it's said to be catalytic carbon [29]. Figure 10.3 shows the schematic of various possible functional groups attached on a carbon surface.

The other class of surface modification methods is plasma treatment, wherein carbon is treated with air or oxygen plasma under vacuum or atmospheric plasma conditions. Likewise chemical impregnation, plasma treatment also creates different oxygen functional groups to render carbon relatively

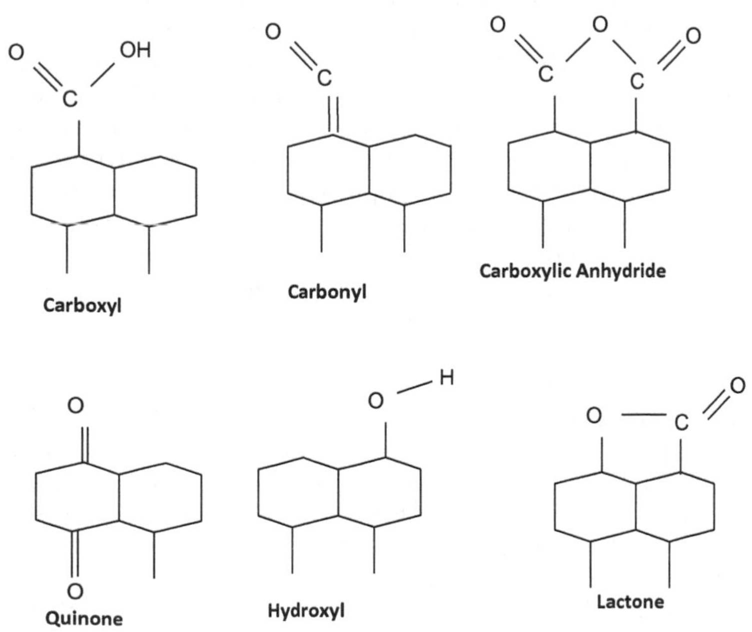

FIGURE 10.3
O-containing surface functional groups.

more active to adsorption. Carbon is exposed to air or oxygen between two electrodes with high voltage to create the oxidative plasma. During the controlled plasma oxidation, small textural changes take place. However, there would be substantial change in the surface chemistry, in particular the surface acidity because of the chemical addition of oxygen to carbon surface. Oxygen free radicals react aggressively with carbon atoms located at the peripheral surface of the graphitic platelets. Ironically, plasma processing on carbon is not fully commercialized because of potential contamination of the expensive electrodes with the carbon dusts, and the high captive investment on plasma reactor as compared to the wet process. However, the biggest advantage of plasma oxidation is that it is a dry process and no drying is involved. In future, the plasma treatment could be a source of specialty carbons to remove specific contaminants present in water. Researches are pursuing research to exploit the potential of plasma treatment for removing targeted chemical species, in particular metals such as lead, cadmium, arsenic, and cadmium for the water stream [30–32].

10.7 Removal of Water Impurities by Activated Carbon

Major toxic impurities present in water are inorganics or ionics such as fluoride, nitrate, perchlorate, and cyanide, heavy metals such as lead, cadmium, mercury, arsenic, and chromium, organics such as phenol and trichloroethylene, and microbiological such as cyst, bacteria, and virus. Different impurities emanate from different sources, and therefore, the concentration levels vary from season to season and region to region. That is why there is no universal solution to the removal of impurities. Broadly, water impurities are removed by different techniques including adsorption, ion-exchange, ultrafiltration (UF), reverse-osmosis (RO), and intensive processes such as chlorination and ozonation. However, filtration through activated carbon filters is the most common and cost effective.

Whereas activated carbon as such has good affinity and adsorption capacity for volatile organic compounds and heavy metals, it has a limited efficiency. Therefore, it is necessary to surface treat carbon to modify the adsorptive and catalytic properties and subsequently enhance its efficiency to remove such contaminants [33,34]. For the removal of heavy metals from drinking water, ion exchange and membrane filtration are effective processes. Surface modified activated carbon can be used for enhanced heavy metals removal from drinking water. A mechanism is shown in Figure 10.4.

Chlorination has been an effective and the most popular disinfection method for drinking water supply. It inhibits the growth of microorganisms such as bacteria and viruses, and in turn saves lives from illness caused by

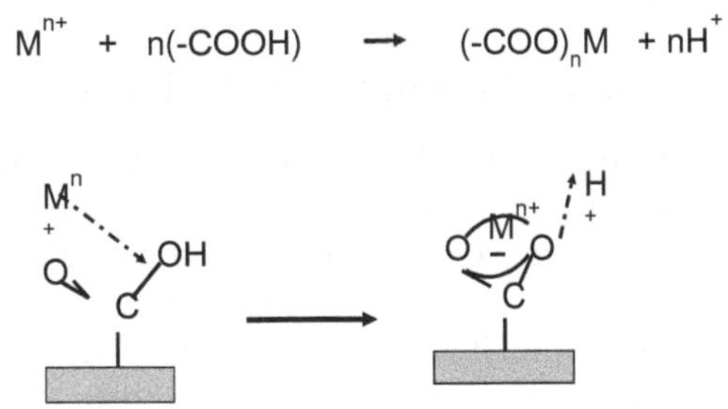

FIGURE 10.4
Representation of heavy metal reduction by surface acidic groups created on activated carbon.

such microbes in water. However, chorine reacts with naturally occurring matter (NOM) in water to form several disinfection by products (DBPs) such as trihalomethanes (THMs) and Haloacetic acids (HAAs), which are linked with the increased risks to human health. Activated carbon is considered to be the most effective adsorbent to remove free chlorine from drinking water. In fact, in any type of water purifier, including gravity filter system (GFS) without electricity or online filters running with pump and electricity, activated carbon is essentially used in different forms, more commonly granules or blocks/candles. The life of the filter is decided by the type of carbon and the size of filter. The mechanism for chlorine removal by activated carbon can be represented as:

$$\text{Step 1: } HOCl + H_2O + C^* \rightarrow 3H^+ + Cl^- + CO^* \rightarrow HCl + CO^*$$

where C^* represents the activated carbon adsorption site and CO^* is the intermediate complex formed during adsorption.

Ozone, chlorine dioxide, and chloramines are some of the disinfecting agents used as alternatives to replace chlorine or to be used as secondary disinfectant in order to minimize the chlorine's adverse effects on human. Chloramines, especially monochloramine, have been used as a disinfectant in public water supplies for more than 70 years. During the last decade, chloramines have shown increased popularity among most of the water supply utilities. The main reason for selecting chloramines is that they are stable for relatively longer period and take longer to decompose in the utility's distribution system. Also, they react less with organic compounds compared to chlorine.

It is important to understand the mechanism of monochloramine formation. The formation is ensured by adding ammonia to free chlorine (HOCl–) containing water at the later stage of chlorination by controlling the solution pH and the ratio of ammonia and chlorine concentrations. Hence, it has become easy for drinking water utilities to switch to chloramines, which subsequently reduces the disinfection by-products (DBP) formation.

10.8 Removal of Monochloramine by Activated Carbon

Challenges in enhancing removal of chloramines and several other contaminants led to attempts for developing effective surface modification of carbon. Typically, activated carbon is used to remove many contaminants including chlorine from water. However, standard activated carbon materials are not effective in reducing high volume of chloramines. It is also known that improvements in removal of chloramine can be realized by reducing the mean particle diameter of carbon and by increasing the carbon bed contact time. Whereas the effectiveness of ACs to act as adsorbents for a wide range of contaminants is well established, increasingly large numbers of research on the modification of activated carbon are gaining prominence because of the need to enable activated carbons to develop an affinity (reduction efficiency) for certain contaminants [31].

In general, activated carbon has no significant amount of surface groups and it is deficient in basic elements. To enhance the chloramine removal efficiency by activated carbon, it is surface treated nitrogen ("N") containing compounds such as ammonia, urea at high temperature, in order to created N containing functional groups. At the high temperature, the graphitic layers become loose and basal plane of activated carbon becomes susceptible towards any reactive species. Under such conditions, the carbon matrix can easily be doped with specific catalytic species or create catalytic site in the form of foreign element (N) or functional N-groups. In particular, for the catalytic reduction of chloramines and similar other acidic contaminants such as hydrogen sulfide, the catalytic carbon is produced by doping with N and creating enormous basic functionalities. Higher the N content, higher is the catalytic activity.

Chloramine may be significantly removed by using catalytic activated carbon. In fact, activated carbon does not adsorb chloramines, rather it removes them through its ability to act as a catalyst for the chemical decomposition or conversion of chloramines to innocuous chlorides in water. The theoretical reaction mechanism may be explained in the following two-steps:

$$\text{Step 1: } NH_2Cl + H_2O + C^* \rightarrow NH_3 + H^+ + Cl^- + CO^*$$

$$\text{Step 2: } NH_2Cl + CO^* \rightarrow N_2 + 2H^+ + 2Cl^- + H_2O + C^*$$

The mechanism shows that the catalytically active sites (C^*) on activated carbon decompose the chloramine molecules, resulting in the formation of the carbon oxide intermediate (CO^*) which further decomposes the molecules into chlorine.

Catalytic carbon has been commercialized by several carbon manufacturers for the removal of chloramine from water, using different processes and carbon precursors. A literature review reveals that the coconut shell-based catalytic carbon has significantly higher catalytic activity as compared to the other commercially available coal and coconut shell-based catalytic carbons

[29]. The study has attributed the high activity of the coconut shell-based carbon to the relatively higher level of N content in the catalytic carbon matrix. Figure 10.5 shows the comparative data for the chloramines reduction carried out using identical carbon block filters, one fabricated from a regular carbon and another from a catalytic carbon. Each filter is subjected to 2 liters per minute (LPM) flow rate of 3 ppm influent concentration of monochloramine in water. Figure 10.6 shows the comparative breakthrough data for the chloramines reduction carried out in a packed bed of different types of carbon. The data clearly show the significantly different performances of thc filters fabricated from different carbon materials.

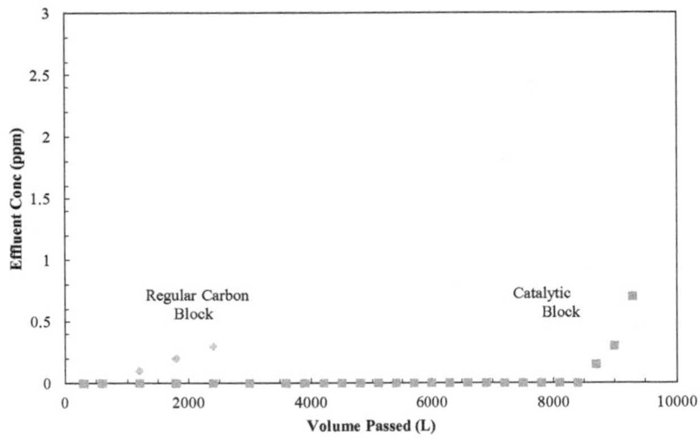

FIGURE 10.5
Comparative performance of chloramine removal regular and catalytic carbon blocks at 2 LPM flowrate of 3 ppm chloramine concentration. (From Gaur, V., Reduction of chloramine from drinking water using catalytic carbon, *WC&P International*, 2012.)

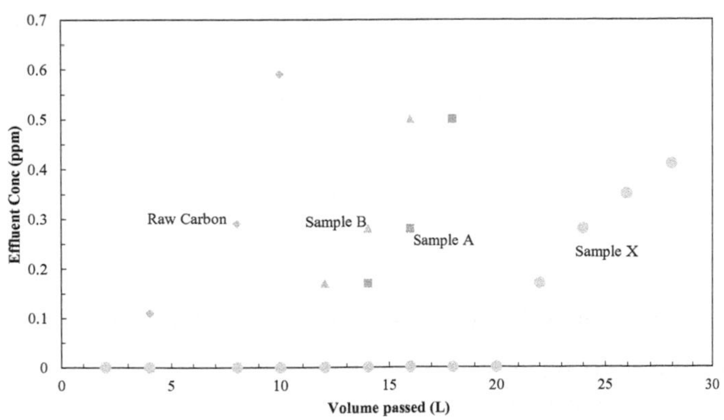

FIGURE 10.6
Comparative performance of catalytic carbons 20 × 40 (W = 5 g, Q = 50 ccpm, C_{in} = 3 ppm). (From Gaur, V., Reduction of chloramine from drinking water using catalytic carbon, *WC&P International*, 2012.)

TABLE 10.1

Comparative Data on the Characterization of Different Catalytic Carbon 20 × 40 Mesh

Parameter	Coconut Shell-Based Raw Carbon	Coconut Shell-Based Catalytic Carbon 20 × 40 (Sample X)	Coal-Based Catalytic Carbon 20 × 40 (Sample A)	Coconut Shell-Based Catalytic Carbon 20 × 40 (Sample B)
Ash (%)	< 4.0	< 3.0	3.0-5.0	< 3.0
pH	9.0–10.5	9.0–10.0	8.0–9.0	9.0–10.0
Peroxide Decomposition Capacity (%)[a]	< 15.0	46.76	22.66	19.06
Chloramine Activity (mg/g)[b] for 20 min	46.2	68.2	54.6	49.8
Chloramine Activity[b] (mg/g) for 30 min	54.1	90.8	56.2	56.8

Source: Gaur, V., Reduction of chloramine from drinking water using catalytic carbon, *WC&P International*, 2012.

[a] Shaking 0.5 g sample with 125 ml of 3% H_2O_2 for 10 min.

[b] Shaking 0.5 g sample with 100 ml of 500 ppm mono-chloramine solution for 20 min and 30 min.

Different parameters are used to characterize catalytic carbons, for example, elemental (N) content and the extent of catalytic activity for the decomposition of hydrogen peroxide, which is expressed as peroxide number. The peroxide number is the measure of catalytic activity for the reduction of chloramines, and expressed in term of time required to decompose a specific concentration of hydrogen peroxide or in terms of % decomposition of peroxide in a fixed time interval. Table 10.1 presents the comparative data for the performances of different carbon and precursor materials including catalytic carbon. As shown in the table, the coconut shell-based catalytic carbon performed superior to the other materials, with its greater peroxide number and less ash percentage.

10.9 Metal Impregnation for Microbial Contaminants

Silver is commonly used as an antibacterial agent to protect a wide variety of products against bacteria, fungus, mold, and other microbes. Colloidal silver, mostly as silver nitrate or silver acetate, is an effective bactericide for water treatment. Colloidal silver in drinking water has also been used with some limitations. Two issues facing surface activation or modification of carbon with colloidal silver are leach-level and cost. In order to overcome these issues, several research studies are being carried out to introduce nano silver impregnation to minimize leach and also to bring down the cost. Very few companies have developed a unique nano silver impregnation process to address both issues associated with the conventional colloidal silver impregnation.

Impregnation of coconut shell-based activated carbon has several advantages compared to conventional colloidal silver, including often non-detectable levels of silver in the treated water. Further, transition metals including copper, zinc, and iron in colloidal and specific phase of nanoparticles are also under development. Several methods are used to monitor these metals at very low (ppb level) concentration levels in water samples. Atomic spectrophotometric techniques such as flame and electro-thermal atomic absorption spectrophotometer (AAS), inductively coupled plasma atomic emission spectrometry (ICP-AES) and inductively coupled plasma mass spectrometry (ICP-MS) have been proposed for the determination of silver and other metals in water samples.

10.10 Summary

This chapter has briefly described the main features of activated carbon, which is the most commonly used adsorbent material, along with a wide spectrum of surface modification techniques used for the carbonaceous materials applied to water purification. Advantages and disadvantages of the techniques are discussed in the context of removing different types of water impurities. Chemical modifications, especially the acidic treatment, are most widely used techniques because of the simplicity and availability of the desired aqueous solutions. The oxidative treatments have been recommended for the removal of heavy metals from water, whereas the thermal treatment in general is preferred for removing organic contaminants from water. It is evident that selection of certain modification techniques for activated carbon should be carefully considered depending on the types of contaminant species in water. On the downside, chemical treatments to develop surface functionalities may adversely affect the physical properties of activated carbon. This opens new directions for modifications to simultaneously control chemical as well as physical characteristics of activated carbon.

References

1. Bansal, R. C., Goyal, M. (2005) *Activated carbon adsorption.* CRC press.
2. Gaur, V. (2005) *Ph.D Thesis.* Indian Institute of Technology Kanpur, India.

3. Smalley, R. E. (1993) From dopyballs to nanowires. *Materials Science and Engineering: B* 19(1–2): 1–7.
4. Kaneto, K., Tsuruta, M., Sakai, G., Cho, W. Y., Ando, Y. (1999) Electrical conductivities of multi-wall carbon nano tubes. *Synthetic Metals* 103(1–3): 2543–2546.
5. Schmiit, J. L. (1991) Carbon molecular sieves as selective catalyst supports—10 years later. *Carbon* 29(6): 743–745.
6. Foley, H. C. (1995) Carbogenic molecular sieves: Synthesis, properties and applications. *Microporous Materials* 4(6): 407–433.
7. Mochida, I., Korai, Y., Shirahama, M., Kawano, S., Hada, T., Seo, Y., Yoshikawa, M., Yasutake, A. (2000) Removal of SOx and NOx over activated carbon fibers. *Carbon* 38(2): 227–239.
8. Cecen, F. (2014) Carbon, *Activated. Kirk-Othmer Encyclopedia of Chemical Technology*. John Wiley & Sons, 1.
9. Dıaz-Dıez, M. A., Gómez-Serrano, V., González, C. F., Cuerda-Correa, E. M., Macıas-Garcıa, A. (2004) Porous texture of activated carbons prepared by phosphoric acid activation of woods. *Applied Surface Science* 238(1–4): 309–313.
10. Almansa, C., Molina-Sabio, M., Rodríguez-Reinoso, F. (2004) Adsorption of methane into ZnCl2-activated carbon derived discs. *Microporous and Mesoporous Materials* 76(1-3): 185–191.
11. Yue, Z., Mangun, C. L., Economy, J. (2002) Preparation of fibrous porous materials by chemical activation: 1. $ZnCl_2$ activation of polymer-coated fibers. *Carbon* 40(8): 1181–1191.
12. Lillo-Ródenas, M. A., Cazorla-Amorós, D., Linares-Solano, A. (2003) Understanding chemical reactions between carbons and NaOH and KOH: An insight into the chemical activation mechanism. *Carbon* 41(2): 267–275.
13. Guo, J., Lua, A. C. (1999). Textural and chemical characterisations of activated carbon prepared from oil-palm stone with H_2SO_4 and KOH impregnation. *Microporous and Mesoporous Materials* 32(1-2): 111–117.
14. Teng, H., Tu, Y. T., Lai, Y. C., and Lin, C. C. (2001) Reduction of NO with NH_3 over carbon catalysts: The effects of treating carbon with H_2SO_4 and HNO_3. *Carbon* 39(4): 575–582.
15. Gomez-Serrano, V., Acedo-Ramos, M., Lopez-Peinado, A. J., Valenzuela-Calahorro, C. (1997) Mass and surface changes of activated carbon treated with nitric acid. Thermal behavior of the samples. *Thermochimica Acta* 291(1–2): 109–115.
16. Cheng, P. Z., Teng, H. (2003) Electrochemical responses from surface oxides present on HNO3-treated carbons. *Carbon* 41(11): 2057–2063.
17. Gaur, V., Asthana, R., Verma, N. (2006) Removal of SO_2 by activated carbon fibers in the presence of O_2 and H_2O. *Carbon* 44(1): 46–60.
18. Das, D., Gaur, V., Verma, N. (2004) Removal of volatile organic compound by activated carbon fiber. *Carbon* 42(14): 2949–2962.
19. Gaur, V., Sharma, A., Verma, N. (2005) Catalytic oxidation of toluene and m-xylene by activated carbon fiber impregnated with transition metals. *Carbon* 43(15): 3041–3053.
20. Gaur, V., Asthana, R., Verma, N. (2006) Removal of SO2 by activated carbon fibers in the presence of O_2 and H_2O. *Carbon* 44(1): 46–60.
21. Adapa, S., Gaur, V., Verma, N. (2006) Catalytic oxidation of NO by activated carbon fiber (ACF). *Chemical Engineering Journal* 116(1): 25–37.

22. Gaur, V., Sharma, A., Verma, N. (2007) Removal of SO2 by activated carbon fibre impregnated with transition metals. *The Canadian Journal of Chemical Engineering* 85(2): 188–198
23. Dwivedi, P., Gaur, V., Sharma, A., Verma, N. (2004) Comparative study of removal of volatile organic compounds by cryogenic condensation and adsorption by activated carbon fiber. *Separation and Purification Technology* 39(1–2): 23–37.
24. Fuertes, A. B., Marban, G., Nevskaia, D. M. (2003) Adsorption of volatile organic compounds by means of activated carbon fibre-based monoliths. *Carbon* 41(1): 87–96.
25. Pradhan, B. K., Sandle, N. K. (1999) Effect of different oxidizing agent treatments on the surface properties of activated carbons. *Carbon* 37(8): 1323–1332.
26. Kaneko, K., Ishii, C., Ruike, M. (1992) Origin of superhigh surface area and microcrystalline graphitic structures of activated carbons. *Carbon* 30(7): 1075–1088.
27. Dubinin, M. M. *Chemistry and Physics of Carbon* (Edited by P. L. Walker, Jr.) (1996) Marcel Dekker Inc., New York.
28. Barton, S. S., Evans, M. J. B., Halliop, E., MacDonald, J. A. F. (1997) Acidic and basic sites on the surface of porous carbon. *Carbon* 35(9): 1361–1366.
29. Gaur, V. (2012) Reduction of chloramine from drinking water using catalytic carbon. *WC&P International.*
30. Schukin, L. I., Kornievich, M. V., Vartapetjan, R. S., Beznisko, S. I. (2002) Low-temperature plasma oxidation of activated carbons. *Carbon* 40(11): 2028–2030.
31. Boudou, J. P., Paredes, J. I., Cuesta, A., Martınez-Alonso, A., Tascon, J. M. D. (2003) Oxygen plasma modification of pitch-based isotropic carbon fibres. *Carbon* 41(1): 41–56.
32. Gaur, V., Shankar, P. A. (2008) Surface modification of activated carbon for the removal of water impurities. *WC&P International.*
33. Gaur, V., Swamy R., Knoll, J. (2017) Filtration media for removal of arsenic from potable water with iron-impregnated activated carbon enhanced with titanium oxide. US Patent. US0225968A1.
34. Gaur, V., Pontious, E. (2016) Surface-modified carbon and sorbents for improved efficiency in removal of gaseous contaminants, US Patent. Provisional application No. 62/416,899.

Index

Page numbers followed by f and t indicate figures and tables, respectively.

B

C

D

M

Milton Keynes UK
Ingram Content Group UK Ltd.
UKHW021624071024
449327UK00020BA/1177

9 780367 570934